Placing the Enlightenment

Placing the Enlightenment

Thinking Geographically about the Age of Reason

CHARLES W. J. WITHERS

The University of Chicago Press
Chicago and London

Charles W. J. Withers is professor of historical geography at the University of Edinburgh and a fellow of the British Academy. He is the author of *Geography, Science and National Identity: Scotland since 1520* (Cambridge, 2001) and, among other recent books, has with David Livingstone coedited *Geography and Enlightenment* (1999) and *Geography and Revolution* (2005), both published by University of Chicago Press.

The University of Chicago Press, Chicago 60637
The University of Chicago Press, Ltd., London

Printed in the United States of America

16 15 14 13 12 11 10 09 08 07   1 2 3 4 5

ISBN-13: 978-0-226-90405-4 (cloth)
ISBN-10: 0-226-90405-9 (cloth)

Library of Congress Cataloging-in-Publication Data

Withers, Charles W. J.
Placing the Enlightenment : thinking geographically about the age of reason / Charles W. J. Withers.
p. cm.
Includes bibliographical references and index.
ISBN-13: 978-0-226-90405-4 (cloth : alk. paper)
ISBN-10: 0-226-90405-9 (cloth : alk. paper)
1. Enlightenment. 2. Philosophy, Modern—18th century. 3. Geography. 4. Space. I. Title.
B802.W58 2007
910.9′033—dc22

2006032563

♾The paper used in this publication meets the minimum requirements of the American National Standard for Information Sciences—Permanence of Paper for Printed Library Materials, ANSI Z39.48-1992.

For Anne

Contents

Illustrations

Color Plates (*after page 170*)

Preface and Acknowledgments

This is a book about the Enlightenment understood geographically. Historians, literary scholars, and others have debated its "what," "when," and "why." Until now, relatively little attention has been paid to the "where" of the Enlightenment. This book aims to correct that by offering an account of the Enlightenment as a geographical phenomenon.

Admitting that the Enlightenment is a topic of enormous significance commanding the attention of many disciplines, scholars do not agree on its definition or its significance. Neither, as it happens, did eighteenth-century contemporaries. Even so, the Enlightenment is commonly thought of as a historical phenomenon, a matter of ideas, and is as commonly examined at the national scale—the French Enlightenment, the Enlightenment in America, and so on. Recent work from a variety of quarters has challenged such thinking and begun to consider the Enlightenment as geographical in several respects. Geographers have traced connections among the Enlightenment, the practices of geographical knowledge, and the nature of Enlightenment geography. Historians have sought to understand the Enlightenment "above national context" and have explored questions of Enlightenment sociability. Historians of science have recognized the geographical nature of the sciences in the Enlightenment. Literary researchers have pointed to the key role of travel and of translation in making the Enlightenment "move," in "mediating" the Enlightenment in different social communities.

This book is an attempt to bring these and other ideas together. It is, in one sense, a synthesis of existing work, one in which I have relied on the work of modern others in a variety of disciplines as well as on texts and other works produced in the Enlightenment. In another sense, it is an argument, a proposal that to think about the Enlightenment not simply as an intellectual movement and moment in time but as something that happened in place and over space is to enlarge our understanding of it and to be more faithful to the nature of its making. The focus in what follows is with material in the natural and human sciences, with relatively little being said about literature, religion, and politics as central features of the Enlightenment. But I contend that my arguments about the geographical nature of the Enlightenment extend to these aspects of the intellectual dimensions of the Enlightenment just as much as they do to the sciences.

I have incurred numerous and lasting debts in the course of this work. I trust that the many people whose scholarship I have drawn on will be comfortable with my use of it, recognizing of course that the interpretations I have placed on their findings are my own, and that they and others will find stimulus in my overall arguments. I have been encouraged and challenged by scholars in different disciplines, not only from those whose concerns inform my views about the importance of geographical thinking. In no order other than the conventionally alphabetical, I am sincerely grateful to Thomas Ahnert, Guyda Armstrong, Michael Bravo, Dan Clayton, Denis Cosgrove, Stephen Daniels, Felix Driver, Roger Emerson, Patricia Fara, Diarmid Finnegan, John Henry, Paschalis Kitromilides, David Livingstone, Robert Mayhew, Miles Ogborn, Mary Terrall, Paul Wood and, not least, the two anonymous readers of an early draft for their support and critical encouragement.

A number of people have given graciously of their time and expertise in helping secure the illustrations. I acknowledge with thanks Sheila Noble of the Edinburgh University Library Special Collections; Christopher Fleet of the National Library of Scotland Map Library; John Dallas and Iain Milne of the Royal College of Physicians, Edinburgh; Bob Karrow of the Newberry Library, Chicago; Elisabetta Dal Carlo of the Museo Querini Stampalia, Venice; the Marquis and Marquise of Breteuil; Philippe Petout of the Musée de Saint-Malo; Helen Trompeteler of the National Portrait Gallery, London; Philip Oldfield of the Thomas Fisher Rare Book Library at the University of Toronto; Sylvia Carr of the National Library of Australia; Laura Whitton of Tate Enterprises, London; Susan Harris of the Bodleian Library, Oxford; Emma Lauze of the Paul Mellon Centre for Studies in British Art, London; the staff of the British Library and of the Bibliothèque National de France; and the anonymous owner of the lovely portrait by William Hoare which here appears as plate 7a. I am particularly grateful to Ray Harris for drawing figures 6 and 9 and to Richard Dagorne of the Musée Girodet for his kindness with respect to the Girodet painting here used as the cover illustration.

The receipt of a British Academy Research Readership gave me the time to sit still, read widely, and think hard, and it is a pleasure to acknowledge this support. I spent much of the readership as a visiting fellow at the Institute of Advanced Studies in the Humanities at the University of Edinburgh, and I am grateful to the many fellows for their insights and to the directors during my time there, John Frow and Frances Dow, for affording me that opportunity.

Other academic institutions and libraries, in Cambridge, London, Oxford, Philadelphia, and Trondheim, have provided time and space for me to collect my thoughts and rehearse my ideas. The research on which this book is based was made possible primarily by funds from the British Academy as well as by funds from the University of Edinburgh, the Carnegie Trust for the Universities of Scotland, and the Strathmartine Trust. I am pleased to acknowledge this assistance. At the University of Chicago Press, Christie Henry has been a wonderful source of encouragement and a model editor. It has been a pleasure to work with her and her colleagues.

1

Introduction

The Enlightenment—Questions of Geography

This book is about understanding the Enlightenment—or to use one parallel term among many, the Age of Reason—geographically. It is, in several senses, an argument about and extended review of the Enlightenment's "where," about how we may "place" the Enlightenment, and about why it matters that we think about the Enlightenment in geographical ways. The Enlightenment was that period—conventionally the "long" eighteenth century in Europe, ca. 1685–ca. 1815—when the world was made modern. For its contemporaries it was then, and for modern scholars it is now, an intellectual movement distinguished by critical, analytic, and scientific concerns and by claims that the power of reason could improve the human condition. Rather than being a fixed set of beliefs, the Enlightenment—as a moment and a movement—was a way of thinking critically in and about the world. Ancient authority and tradition were open to challenge. Philosophical inquiry would yield useful practical ends; science social benefit.

So understood, the Enlightenment is usually approached in one or more of several ways. Much is known about its defining features and principal characters—its "what" and its "who." The Enlightenment's origins and motivations—its "how" and its "why"—fascinate given the continuing importance of ideas of reason, truth, and scientific progress. For the same reasons, the Enlightenment's legacy—its "so what"—is an enduring concern. In one significant respect, however, the Enlightenment has been neglected. The Enlightenment, a historical phenomenon, was also a geographical one. The Enlightenment—the ideas embraced by the term, the people who formulated them, and the things they did—was a moment and a movement in space as well as time.

To consider it thus is in keeping with work on the Enlightenment, for in fact, understandings of what the Enlightenment was and of what is today embraced by "Enlightenment studies" have varied considerably. Let me, for convenience's sake, distinguish three strands of inquiry, each identifiable by a single central question. Enlightenment contemporaries asked "What is it?" In the last fifty years or so, Enlightenment scholars have been asking "What was it?" Others, concerned with the Enlightenment's legacy, want to know "What's left of it?" Implicit in each of these, of course, is the question "When was it?" I do not pretend to do full justice to each here in this introduction, but it is helpful to have some idea of the answers to these questions before engaging with that strand of inquiry concerning the Enlightenment—"Where was it?"—that is central to my concerns.

The Enlightenment—Questions of Definition

"What is Enlightenment?" When the German philosopher and writer Moses Mendelssohn posed this question in December 1783, he can hardly have imagined that his query would continue to enjoy the significance it has. We still ask his question, of course, because it speaks to fundamental issues to do with political ideals, truth, and individual freedom. Mendelssohn's view then was that enlightenment had to do with how public affairs should be governed by reason, with how humanity had the capacity to resolve society's ills. Responding to his fellow German's query, Immanuel Kant famously considered enlightenment to consist of "mankind's exit from its self-incurred immaturity"—freeing the human consciousness from ignorance and error. Kant's motto for enlightenment was simple: *Sapere aude*—dare to know. For others at the time, enlightenment was about censorship, with how much a citizenry could or should be "enlightened," and with the tensions between public knowledge and religious faith and between new ways of thinking and established authority.[1]

It is possible to claim, then, that (the) Enlightenment began not as a definite "thing," or even as a chronological period, but as processes concerned with the central place of reason and of experience and experiment in understanding and improving human society. The emphasis on reason itself was not new: the ancient Greeks argued that. What distinguished the Enlightenment was the belief that in its application, reason mediated through direct encounter and not blind faith in ancient authority would bring practical social benefits. European intellectuals at the end of the eighteenth century understood that important changes were occurring—and had already happened—to do with reason, tolerance, and scientific progress. Contemporaries spoke of "this enlightened century," of "lumières" (lights), of "Aufklärung" (enlightenment), of "illuminismo" (illumination), and following Thomas Paine in 1794, of the "Age of Reason." But they did not then see those changes as defining any particular period or set of essential features in the ways now commonly signaled by widespread reference to *the* Enlightenment. "The word Enlightenment is being used by everybody now," one observer noted in 1789, "yet we have not found a defining concept which encompasses the entire movement, and which is at the same time appropriately precise."[2]

The idea of "the Enlightenment" as a clearly defined historical object is a modern invention. This is true at least in the sense that later interpreters have often tried to impose a unity upon it. In most modern interpretations, the Enlightenment is a distinctively eighteenth-century phenomenon concerned with new ways of thinking as means to intellectual and social reform. The Enlightenment was about criticizing social injustice, about exposing the roots of authoritarian ignorance and theological dogma, and about recovering intellectual nerve. Unquestioning faith in the Bible or automatic reliance on "the Ancients" (Greek and Roman writers) would not suffice. The goals—if it is appropriate to think in such a final and complete sense—were both a fuller understanding of the natural world and its meanings, and a "Science of Man," a practical comprehension of humanity's potential rooted in firsthand encounter, not secondhand authority. Where possible, such an understanding of the human world, of the human sciences, was to be arrived at by the same methods as were being applied to scientific study of the natural world. From the location of its leading thinkers and of the cultures of intellectual inquiry that distinguished it, the Enlightenment is usually seen as a "European (good) Thing"—in terms of its origins and "core" expression anyway—to have been largely urban and centered on the philosophical reasoning of great men. It was, for some, defined as the period between the lives of the German philosopher-scientists Gottfried Wilhelm von Leibniz (1646–1716) and Immanuel Kant (1724–1804). For others, the Enlightenment began with the life and work of two French writers, François-Marie Arouet, known as Voltaire (1694–1778), and Charles-Louis de Secondat, Baron de Montesquieu (1689–1755), and ended with Kant.[3]

Others have seen an altogether broader Enlightenment, with origins in the later seventeenth century or even earlier and in the philosophical enquiries of the Dutch and the English, notably in the philosophers John Locke and Sir Isaac Newton, rather than in French or German thinkers. Even this picture is further complicated. For although in the eighteenth century the enlightened of all nations revered England's society, government, and openness as the epitome of Enlightenment, many later historians have tended to see the Enlightenment as a unified, French, and primarily philosophical movement.[4]

Since the 1970s, interpretations of what the Enlightenment was have moved away from issues that preoccupied a few French philosophes to focus more on its complex origins, multiple nature, and uneven social consequences.[5] The antireligious sentiment of the Enlightenment has been scrutinized in different contexts. The enlightening potential of books—of the *Encyclopédie* of Denis Diderot and Jean d'Alembert, for example—has been stressed. Numerous Enlightenment topics—overlooked or ignored before—have been uncovered: the role of women, the place of science and medicine, sexual attitudes, race, and anthropology to name but a few. The Enlightenment's different national expressions have become a focus for study. The earlier view that there were many philosophers but only one Enlightenment has been overturned, the monolithic Enlightenment exposed as a myth—so much so that the notion of "enlightenments," not "the Enlightenment," of "pluralizing Enlightenment," is now more commonly accepted. Such a move was early apparent in the natural sciences. As long ago as 1980, Shapin pointed out how "against an older view that the 'new science' (and especially the 'Newtonianism') of the early and mid eighteenth century was the underpinning of 'the Enlightenment,' we now have a developing perspective which points out the existence of a number of species of natural knowledge, and a number of opposed 'Enlightenments.'"[6]

At the same time, a yet further strand of Enlightenment studies cannot be overlooked. This is one that takes seriously the contradictions and consequences of *the* Enlightenment and its continuing reverberations. Looking back on eighteenth-century claims concerning the power of reason, it is clear that enlightened thinking did not mean—and certainly has not resulted in—liberation from irrationality and ignorance. For Michel Foucault, the Enlightenment may have been about reason, but it was also about restraint. For the German philosophers Max Horkheimer and Theodor Adorno especially, the Enlightenment had then (and has now) a "dark side." The Enlightenment's power, they argued, was not that of tolerance and the potential for human freedom but the power of some people over other people. Knowledge was a means to challenge authority. But in the wrong hands, knowledge was power that could be used to control and to regiment—with dreadful consequences.[7]

Understood in those terms, and looking back on hollow claims to the rights of man and to imprecise notions of human nature and critical reason in the "Enlightenment project," there is an argument that it has failed. The social fruits projected to stem from the application of critical and modernizing reason to the political reform of ignorance and falsehood—freedom from political tyranny, an uncensored public, social, and scientific progress—have failed to ripen. The project of Enlightenment modernity was flawed from the outset, debilitated by its own contradictions. This may be so. But the claims made by its original proponents for the political project of enlightenment have continuing resonance nonetheless. This is one reason why the Enlightenment is relevant today. If the Enlightenment does have continued meaning and importance, distinction might be made between those for whom the "Enlightenment project," failure or not, is finished—it was *only* ever a historical thing—and those for whom the Enlightenment as an intellectual project and our understanding of it is still incomplete. After all, we may live in a world that has failed to enlighten its citizens equally, but commitment to improve people's lives through the application of reasoned reform is surely politically defensible. Continuing to subscribe to the philosophical and political concerns of the Enlightenment means continuing to understand the Enlightenment.

One thing is clear from even this summary review: the Enlightenment won't go away, whatever it was then or is now held to be. For some people, it was a key historical movement and moment of largely philosophical endeavor. For others, a distinct yet now much more diversified historical Enlightenment continues as an object of academic inquiry. For yet others, the Enlightenment, studied less as a formative historical moment and more for its continuing legacy, is an object of critical repudiation, even if, paradoxically, the vehemence of some contemporary critics toward the Enlightenment project only confirms its centrality to narratives of Western modernity.[8] The Enlightenment, scrutinized in all sorts of ways, may be politically dead—but it won't lie down intellectually.

It is precisely because the Enlightenment is a central feature in understanding the origins and the nature of the modern world that we need to think harder about it in a way that is consistently overlooked—namely, the Enlightenment as a geographical thing. What was a period of philosophical and historical inquiry was also a period of practical geographical investigation. People in the eighteenth century understood their world to be changing as a fact of geography, and as the result of processes of geographical inquiry—in the shape and dimension of continents, for example, in the types of human cultures making up mankind, in the reasons plants, animals, and humans were located as they were. Such things are true of the Pacific voyages of European navigators. They are true, too, of land-based expeditions to northern Europe and to South America designed to measure the shape of the earth. Geographical data provided much of the raw material from which philosophers and others fashioned theories in the Enlightenment about the human condition. And they are true at smaller civic scales, for at different times, in different places, and at different ways in the Enlightenment, people who called themselves geographers (and others who did not) published maps and wrote books to provide for geographically aware publics. Audiences—in different parts of Europe and not just in the larger towns, on the east coast of America, and in Africa and in Asia—read such works and voyagers' accounts with curiosity and wonder. In short, geography mattered in the Enlightenment.

In a different sense, the Enlightenment itself was geographical. Ideas of rationality and freedom, and practices of sociability did not "float free" from the ground. They were made by people in particular places, by being talked about, written down, read, and argued over, and they traveled between places to be worked with and thought about by other people. The Enlightenment took place *in* and *over* space—it had a geography, even geographies. It was also *about* space, about the earth and its geographical variety, and about how that variety—in plants and peoples, cultures and climates—could be put to order. So what about the Enlightenment's "where"? What, more exactly, is meant by "placing the Enlightenment," by understanding the Enlightenment geographically? What does *thinking geographically* about the Age of Reason involve?

Where Was the Enlightenment? Questions of Geography

At one time or another during the eighteenth century, the Enlightenment featured in almost every country in Europe. This was certainly so for England, Scotland, Germany, France, the Netherlands, Switzerland, Italy, Austria, Russia, and Bohemia. And while national Enlightenments have been noted and discussed for these countries, other studies have explored the Enlightenment in the Americas and in countries and regions on the periphery of the Enlightenment's English, French, and German heartland—in Greece, Hungary, Spain, and Portugal, for example. The Enlightenment varied across space, then, at the scale of the nation.[9]

Thinking about the Enlightenment as a national phenomenon is understandable, and important work has been done in this regard. As *the* approach to the Enlightenment considered geographically, however, it is problematic. For one thing, the idea of the nation was uncertain during the course of the eighteenth century. For another, national studies have often revealed more about differences in the making and reception of Enlightenment ideas *within* the nation—to do with social groups, religious affiliation, the nature of culture, or the impact of particular scientific practices—than they have about either similarities or differences *among* those nations. Further, studies that look only at Europe ignore the wider geographical context: the international connections Europe had with those countries "beyond Europe" in the eighteenth century, and the fact that places other than Europe may have had an Enlightenment or something resembling it at much the same time.[10]

Let me briefly illustrate these remarks with reference to the history—and the geography—of the Enlightenment in Scotland. The term "Scottish Enlightenment" was first coined in 1900. Yet what this label speaks to was initially delimited a century or so earlier by writers commemorating men whose work gave Scotland a prominent place on the map of eighteenth-century philosophical and scientific activity. Interpretations of the Scottish Enlightenment as an essentially urban and philosophical matter have been replaced by studies that reveal the Enlightenment in Scotland to have been about questions of practical utility, about agricultural improvement as well as aesthetics, gender and geometry, about chemistry and crop yields as well as philosophical speculation. To privilege philosophy or the sciences of Man is to tell only part of the story. Additionally, what was once taken as *the Scottish* Enlightenment differed in fact within and among the towns of Aberdeen, Edinburgh, Glasgow, and even smaller centers. Debates in what was termed "political economy" over the role of commerce in the workings of civilized society took place not just in Glasgow's more commercially oriented environment but between particular people in certain institutions in Glasgow and between that city's intellectuals and like-minded counterparts in Europe and in the Americas. In short, the Enlightenment in Scotland was "earthed" quite differently in different places. And what took place *in* Scotland as "Enlightenment," however and wherever understood, was also the result of connections *beyond* Scotland. Questions of the Enlightenment's geography matter not just as national pictures and local differences but also as connected accounts of international exchange.[11]

My point is a simple one. The Enlightenment was national *and* local *and* international. What is important to an understanding of the Enlightenment as a geographical matter is to show how these scales of analysis work and work together and how, if taken only singly, they may produce only partial "maps" of the Enlightenment's geographical constitution. Better still, we might explore the relationships between such scales. The Enlightenment took its shape not just from local or national circumstances "at home," but also as a result of connections between Europe and the "far away," with the Americas and, notably, with the cultures and geographies of the Pacific. As was long ago recognized, "the Enlightenment's self consciousness was to some extent a geographical consciousness based on the distinctiveness of the part of the world that came to be called Europe."[12] Much of this geographical consciousness derived from Europe's view of itself in relation to the exotic Otherness of the so-called New World. New information brought back to Europe—about human cultures or new plant and animal species—prompted new ways of thinking about the development of human societies, the role of God, and the nature of Nature. This interest in the "Science of Man" allowed new questions about the human condition and, in time, new disciplines to emerge. This is certainly so for systematic natural history, social anthropology, and comparative or conjectural history, given theorists' abiding interest during the Age of Reason in the different stages of human social development.[13] Encountering the world in such ways may even have allowed geography to take shape. The work of those scientific naturalists and illustrators accompanying global navigators like James Cook displayed, we are told, three features crucial to the formation of geography as a modern empirical science: realism in description, systematic classification in collection, and comparative method in explanation. Understood in such terms, "modern" geography itself may even be an Enlightenment product.[14]

These remarks are to note that where it has been considered at all, the Enlightenment understood geographically has embraced, unevenly, three main things: its national expression; partial recognition of the Enlightenment above and beyond national boundaries; and some attention to its local settings. If all this was both *simply* true and *all* there was to say about the Enlightenment understood geographically, I could stop now. Such an understanding is, I contend, geographical—but it is not geographical enough. Several things fundamental to knowing what the Enlightenment was and why it happened when it did will have been underplayed, even ignored, if we content ourselves with such a limited sense of the Enlightenment's "where." I do not deny that these elements are important. But I want to build on them to propose an altogether stronger notion of how the Enlightenment was placed and of how we may think of it geographically. To do so, I turn to ideas of place, space, scale, the "traveling" nature of knowledge, and to the languages and practices of geography. As will be apparent from what follows, these ideas help draw together that work in the Enlightenment and in Enlightenment studies, geography, and the history of science on which this book is based. They extend naturally from Shapin's and others' claims about the diversity of "Enlightenment" and from the view that considering the "variegated geography of the Enlightenment" can help recover a sense of the period as a whole.[15]

Enlightenment ideas, the people who proposed them, the books they wrote, and the audiences who read them were, of course, located somewhere. There is in this simple locational sense a geography that, even if only sketched out, might provide a working "atlas of the Enlightenment." "Dot maps" could plot Enlightenment institutions and other sites of significance. "Thread maps" with lines of different thickness could symbolize the strength of the connections between people as they exchanged ideas and traded the new scientific facts and their implications. Such a conception of the Enlightenment as a connected network of local points of activity reflects one view that, rather than perceive the Enlightenment as a coherent intellectual entity, it is better to see it as "a multiplicity of specific contexts, each one constituted by numerous (local and temporary) factors."[16] A geography of the Enlightenment as just a series of local places would indeed be one, perhaps partial, way of conceiving of its nature and understanding its history. Yet I want to make an even stronger claim than this. The geographical location of Enlightenment ideas, personnel, and artifacts and their movement over space—that is, questions to do with place and travel between places—are key elements in understanding how the Enlightenment was made and what, actually, it was.

This notion of the Enlightenment as geographically constituted being about the power of place, about different scales and local sites, and about the movement of knowledge is one in which knowledge itself has geographical expression. The idea of science and other knowledge being geographical is now an established feature of work in science studies and in the history of science, so much so that there is an evident "spatial" or "geographical turn" to such studies.[17] David Livingstone's recent review, *Putting Science in Its Place,* is even subtitled *Geographies of Scientific Knowledge.* One key feature of this and other literature is the dethronement of the universal and transcendental nature of science and an emphasis on its local making. Science, we are told, is always socially made, conditioned by its historical context *and* by its local geographical circumstances. The idea that science is indelibly marked by local contingencies and spatial settings is apparent in science's making in particular sites: in the botanical garden, in the zoological park, the library, the coffeehouse, the laboratory, the museum, or even the ship. Different places produce different sorts of scientific knowledge. Place—the "where" of science's making—is likewise central to the reception or consumption of science since what is understood by one audience in one place may not be so elsewhere. There is, then, a difference—one of geography and of interpretative significance—between the context of discovery and the context of justification. As Livingstone notes, "the geography of science . . . calls attention to the uneven distribution of scientific information." And this matters, since "what has been promoted as scientific objectivity, as the 'view from nowhere,' turns out to have always been a 'view from somewhere.' The recognition that rationality is not disembodied but positioned has significant implications for understanding science and scientists."[18]

The idea of place in these terms is crucial to the Enlightenment understood geographically. So too is the "traveling" nature of knowledge. Recognizing that knowledge is made in specific places is insufficient unless we are also able to chart the movement of knowledge between places. And we need to recognize, too, that as things move over space, so their meanings may change. Knowing that knowledge moves across space is not necessarily to presume that it moves "evenly" or equally. Migration always involves modification. Take Said's notion of traveling theory, for example. For Said, the entry of a new theory or idea "necessarily involves processes of representation and institutionalization different from those at the point of origin. This complicates any account of the transplantation, transference, circulation, and commerce of theories and ideas." And because this is so, the idea or theory in question "is to some extent transformed by its new uses, its new position in a new time and place." Thus, for Said, we can distinguish four stages in the travels of a theory or an idea: the point of origin—the initial circumstances in which the idea originates or enters discourse; the distance traversed—the passing of the idea through a variety of social and intellectual contexts; its sets of conditions—the degrees of acceptance or rejection faced in its new context of use; and its new situated position—the extent to which the theory or idea is accommodated or incorporated in other places.[19]

Consider, as an example, the key role in the Enlightenment played by illustrated travel accounts of the New World. Many, but far from all, derived from firsthand observations of what were for Europeans hitherto unknown peoples and unexplored environments. New geographical knowledge transformed what European philosophes and their audiences knew about human diversity, even about the shape of the world itself. But not everyone could travel. For those persons who could not encounter new geographies for themselves, taking seriously the claims of others about the geography of the unseen depended on trust in the written word and in the depicted world, and in the writer and illustrator. Knowing about the world in these ways depended on "geographically privileged persons"—not just those who saw what they wrote about but reliable persons who did not and whose social and geographical situation made them credible sources. Furthermore, in its making of "out there," geographical knowledge relied on tacit acceptance of others' knowledge systems—locals acting as guides or translators, for example. Geographical knowledge about other places also depended on the artifacts in which knowledge was invested—plants, instrumental calibrations, even human "specimens"—traveling securely. Even when new geographical information in the Enlightenment arrived "home"—that is, was produced as a textual account for European audiences—people there did not always believe what they were told or how they were told it.

Like Said's notions, Bruno Latour's ideas of "centers of calculation," "immutable mobiles," "combinability," and "cycles of accumulation" are potentially valuable here. Latour reconceptualizes eighteenth-century voyages of navigation—effectively, Enlightenment encounters with global geographies—as attempts to present and re-present knowledge of the far away. Thus, in metropolitan "centers of calculation," such as museums, mapmakers' rooms, and military institutions, knowledge of distant peoples and geographies was brought back to Europe, to be interrogated as to its meaning and used as the basis for new exploration. Accumulating such new knowledge about the far away required that distant objects and their meanings be stable and combinable. That is, the real features—coastlines, mountain ranges, botanical relationships, human cultures, for example—could not travel, but their abstract substitutes could: as maps, as geological specimens, as plant collections, as individual "natives." In this Latourian model, as Miller has shown of Joseph Banks and his role in coordinating natural knowledge in Hanoverian London, Enlightenment knowledge making through voyages of exploration "involved mobilizing and disciplining economic and political resources, administrative procedures, and museums and other centers of collection, as well as astronomers, botanists, zoologist and the like."[20] Producing and consuming any knowledge—and geography is a good illustration—always involves questions of travel and of translation: linguistically (will what is said be understood?), epistemologically (how do things mean what they mean?), and geographically (what are the differences between places, and why?). The Enlightenment understood geographically is no exception. There is no science without such issues of translation or displacement.[21]

Something of what I mean by the importance of place, of the connections between places, and of the forms that knowledge takes in—and in between—its places of making and reception is apparent in Steven Harris's discussion of the "geography of knowledge." Harris, who acknowledges his debt to Latour and to others, sees this term as distinct from "knowledge of geography." For Harris, there are "three related approaches to the geography of knowledge."

> In the first instance, it means a static geography of place: where did people "do science"? Where were they when they aimed telescopes and recorded observations, logged positions and sketched in charts, performed dissections or prepared medicaments, executed experiments and calculations, or wrote and published the accounts of their activities? In the second, it means a kinematic geography of movement: whence came the constituents of scientific practice and knowledge, the measuring or cutting instruments, the authoritative texts or latest correspondence, the exotic natural curiosity or well-wrought experimental apparatus, the returning botanist or navigator? In the third sense, geography of knowledge also means the dynamics of travel: why and by what means did all these movements take place? What was the *anima motrix* responsible for the multiple peregrinations of the elements of knowledge?[22]

In contrast to Harris, I here want to bring knowledge of geography—in the Enlightenment—into this understanding of the geography of knowledge. To do so means we can not only consider the geography of knowledge to be about questions of venue, movement, and knowledge's artifacts, but we can apply it to geography as a particular form of Enlightenment knowledge—with its maps, books, audiences, and so on.

This is also to make a general but nonetheless useful distinction between geography as discipline and geography as discourse. The first term encompasses what contemporaries held the subject to be. The second embraces those practices—observing, mapping, collecting, comparing, writing, sketching, classifying, reading, and so on—through which people came to know the world. In short, as for the Enlightenment itself, geography in the Enlightenment was not a uniform thing. Enlightenment geography books in France, America, and Britain, for example, used different political languages to support opposing political and intellectual agendas. There are discernible differences between English and Scottish textual traditions in Enlightenment geography. In France, the story of geography and of geographers in the Enlightenment was different still. Such variations matter. And because geography in these terms—and as a subject of formal instruction in universities and schools as well as in the public sphere—was around well before Cook and others encountered the Pacific, we cannot simply see the Enlightenment (or the Pacific) as the birthplace of "modern" geography.[23]

The language of the Enlightenment was suffused with the language of geography. Geography was a subject that enlightened publics took seriously. This is apparent in geography's books, in geographical practices such as mapping, and in polite and sociable *conversazione* in which geographical matters were aired and debated. Contemporaries used geographical language to understand and to place the Enlightenment's interest in useful knowledge. For Diderot and d'Alembert, their *Encyclopédie* (1751–65), and the challenge its individual articles collectively represented to authority, was "a kind of world map which is to show the principal countries, their position and their mutual dependence, the road that lies directly from one to the other." The map was a common classificatory device used to put Enlightenment knowledge to order. The language of the map was also used to describe those historical differences in human society encountered during the Enlightenment. This is the sense in which Edmund Burke was using it in 1791 when he wrote "but now the Great Map of Mankind is unroll'd at once; and there is no state or Gradation of barbarism, and no mode of refinement which we have not at the same instant under our view." Burke—and others were doing the same—was conceptualizing a chart of the world that was at once chronological and geographical. The peoples and places that lay beyond Europe were, to Europeans, before Europe. Those who have seen the Enlightenment as being only about understanding the world historically have too readily overlooked the fact that such an understanding depended on encountering the world geographically, of knowing who and what was where.[24]

Mapping was a material form of state governance and spatial ordering in the Age of Reason, not just a metaphorical procedure for the classification of knowledge. At different scales, mapping as a practical activity was something that landowners used to rationalize their estates, that government and military bodies carried out as a form of territorial surveillance; it everywhere depended, in local context, on incorporating local knowledge into different geographical forms. Yet in Enlightenment homes, in coffeehouses, and in public lectures, a different sort of geography altogether was being undertaken. Eighteenth-century geography books—gazetteers and grammars, dictionaries and descriptions—allowed the public to know about their world, but such works did not provide a manual for encountering it at firsthand. Maps were objects by which the Enlightenment's geographical publics would have understood their world, but few in such audiences would ever have drawn one.

Bringing such questions of geography and of geographical thinking center stage—not least, understanding the difference that place makes—frees us from the common view that Enlightenment ideas were somehow dislocated, not grounded at all in a geographical context. It allows us to think beyond the notion that the Enlightenment had geographical expression only in as much as its ideas varied simply either between or within European nations. Further, thinking geographically demands sensitivity to contemporary perceptions of the Enlightenment as geographically different. In fact, right from the start, the question "What is Enlightenment?" could not help but raise the question "Where is Enlightenment?" Mendelssohn, for example, distinguished between the terms "culture," the everyday practices of civility in society, and "Enlightenment," which he took to embrace the best social means for peace, freedom, and intellectual advance. Mendelssohn discerned even then what we can consider a geography of Enlightenment: "Nürnbergers have more culture, Berliners more enlightenment; the French more culture, the English more enlightenment; the Chinese much culture and little enlightenment. The Greeks had both culture and enlightenment." So, too, for Monsieur André in Voltaire's *L'homme aux quarante écus,* who drew, as Roger Chartier points out, an ironic map of Europe unequally open to the progress of the Enlightenment:

> He recently remarked to me that Reason travels by slow journeys from north to south, in company with her two intimate friends, Experience and Toleration. Agriculture and Commerce attend them. When Reason presented herself in Italy the congregation of the Index sternly repulsed her. . . . She has sometimes met with cruel foes in France; but she has now so many friends in that kingdom that she stands a good chance of at length becoming first minister there. When she presented herself in Bavaria and Austria, she found two or three great wig-blocks that stared at her with stupid and astonished eyes. Their greeting was: "Madam, we never heard of you; we do not know you." . . . I have been well received at Berlin, at Moscow, at Copenhagen, at Stockholm. It is long ago that I have been naturalized by Act of Parliament in England, through the labours of Locke, Gordon, Trenchard, Lord Shaftesbury, and a number of others of the same nation. . . . When she passed over the frontiers of Spain and Portugal, she blessed God on observing that the fires of the Inquisition were less frequently kindled. She rejoiced on seeing the Jesuits expelled; but was afraid that, while the country had been cleared of foxes, it was still left exposed to the ravages of wolves.[25]

Questions of geography matter, too, in terms of our making connections across different scales—in being able, that is, to capture different enlightenments: the daily hubbub in the small spaces of enlightening salons, the transactional and circulatory geographies of books, people, and ideas on the move, and the vast silent spaces encountered by oceanic navigators. Thinking about the Enlightenment geographically as global encounters locally articulated means looking beyond the presumed centrality of Europe as *the* center of Enlightenment. It allows us, for example, to conceptualize studies of the Enlightenment not just as histories of familiar literate intellectuals but of others, slaves even, whose existence was intellectually framed in Enlightenment theory but whose place in the Atlantic system of

Enlightenment has not been fully examined.[26] Europe in the Enlightenment had already been part of a globalized world for over two centuries. The Enlightenment's geographies inherited earlier legacies as they simultaneously determined new ones. James Cook, for one, recognized the Enlightenment's geographical consequences beyond Europe when he noted how contact with the Americas had brought the inhabitants disease, vice, and wants "which they never before knew."[27]

The idea of geography as active agency—as practices carried out by people in and over space—captures a key sense of what the Enlightenment, understood geographically, was about, namely, how contemporaries came to terms with the extent and contents of the terraqueous globe—the earth—as home. New geographical information challenged established views on natural theology. It helped solve questions about the earth's shape, size, and age. Voyages of exploration enlarged the contemporary geographical consciousness. Even the books of the Enlightenment had discernible geographies. Particularly after 1780, the *Encyclopédie* became a primary vehicle for the public legitimation of philosophical criticism in France, especially in the larger towns and in the provincial capitals.[28] On a different scale, Enlightenment mapping brought what we now think of as a recognizably "modern" shape to North America and, for Europeans, brought Australia into being. By contrast, Africa's external shape was known, but not its interior, and much of Asia was understood more from secondary accounts than direct experience.[29] New conceptions of human difference were everywhere rooted in the facts of geography. Just as geography was a subject of Enlightenment inquiry—in schools, in private drawing rooms, in academies of one sort or another—so the world itself in the Enlightenment was an object for geographical inquiry.

Such concerns with space, place, and the mobility of knowledge illustrate the concerns of numerous disciplines—and may even confirm "space" as the master metaphor of late twentieth-century epistemology. Certainly, my concerns here echo what one leading historian of science has termed the important question of "knowledge in transit," in which science can be understood in terms of communicative practice and in which, among many exciting challenges, we need to know more about patterns of circulation and use in the appropriate local settings in order to create a history—and a geography—"that keeps the virtues of the local but operates at a unit of analysis larger than a single country."[30] Rather than treat such notions as straightforward and unproblematic, I look at these geographical notions in regard to the Enlightenment as an interpretive framework, one that is necessarily empirical and investigative but one that is also suggestive of the

Enlightenment's practices and sites of meaning. As the historian Daniel Brewer has noted in speaking of "spaces of enlightenment," this is not just a matter of physical space and its representation in, say, maps, buildings, and "any number of discrete concrete eighteenth-century spaces—palaces and roadways, coffeehouse and ships, bedrooms, drawing rooms, laboratories, libraries, gardens and restaurants." It is to pay attention to "social space," to the ways that society and differences in social authority *produce* notions of space—such as salon culture and the sites of urban spectatorship, of private reading and public debate, of geographical inquiry and Enlightenment sociability. If, then, space is not neutral but a social product, so we can also think, argues Brewer, of "epistemological space"—forms of knowledge making that order—and encapsulate their worlds of meaning. Thus, for Brewer—and he cites an example noted earlier and explored at greater length in chapter 8—"The epistemological space of the *Encyclopédie* is universal in the sense that there exists nothing outside it; . . . the *Encyclopédie* is the hallmark of Enlightenment thinking in that it is a perfectly utopian space. Like the mappemonde or world map, . . . the *Encyclopédie* is an imaginary space that signifies the real but does not reduplicate it."[31] Like Brewer, my intention in thinking geographically is also to open up a space of reflection on variegated processes of spatialization that took place in the eighteenth century.

The ideas and terms outlined here certainly present a stronger sense of the Enlightenment as a matter of geography than current work. They draw from—and they draw together—the current concerns of numerous disciplines. My argument about the placed and mobile nature of the Enlightenment goes beyond the nation as the sole scale of inquiry, and above and beyond the merely local. Thinking about the Enlightenment geographically demands attention to questions of space—as real territory, as "imagined" space, as social space, and as epistemological space—and to the difference that space makes. Questions of place, to sites of production and of reception, become important. We have to consider the language of geography. And thinking geographically demands attention to the power of mapping, as a material and a metaphorical resource, and to the processes by which the world was encountered and represented. It distinguishes between the subject of geography and those practices used to make sense of the world. Contemporaries' views can be incorporated. There is a power to "thinking geographically" in using terms like "place," "space," and "scale" to inquire into the located and the traveling nature of Enlightenment knowledge. And there is a strong sense in which the world itself in the Enlightenment was revealed as an object of geographical scrutiny.

Thinking about the Enlightenment's "where" in these ways allows an enriched understanding of what and when it was. The Enlightenment considered geographically began in the late seventeenth century in practical inquiry rather more than in philosophical speculation. One expression of this was the appearance of "how to" guides instructing travelers in what and how to collect overseas and in the appropriate language of and for geographical description. Another was Edmund Halley's voyages of the 1690s in the *Paramore* to test magnetic variation and Newtonian theory, voyages later seen as the world's first government-sponsored ocean-going science. Yet another was the mapping of France by the Cassinis. Europe's first geographical society, the Accademia degli Argonauti, was begun in Venice in 1680, established by the map and globe maker Vincenzo Maria Coronelli. These things are not coincidental: government support, the institutionalization of natural knowledge, and the recognition of geography's power to provide new world data were all deeply implicated in the rise of "modern" science and in the birth of the Enlightenment.[32] In such things taken together, I see the start of the Enlightenment as a geographical matter.

The Enlightenment understood geographically closes not in any single date but with several events by which the world was realized in more or less its full extent. One was the trans-American exploration of Lewis and Clark. Another was the Louisiana Purchase of 1803, in which the acquisition from Spain and France of 830,000 square miles provided for the emerging United States an imperial colony of alien peoples between the Mississippi and the Rocky Mountains that significantly expanded the definition of "American."[33] At the same time, southern continents were for the first time being brought into detailed view. Between 1801 and 1804, the Lincolnshire-born hydrographer and explorer Matthew Flinders was circumnavigating Terra Australis in HMS *Investigator* and providing for New Holland—what would become Australia—the first definitive outline of its shape. In his two aptly named ships (*Géographe* and the *Naturaliste*), the French captain Nicolas Baudin sought to do the same in 1804. Between 1799 and 1804, the German geographer Alexander von Humboldt and his French companion, Aimé Bonpland, explored South America, bringing new specimens to Europe and ushering in new forms of systematic regional science altogether.[34] The shape of the world was a crucial question—literally so from the 1730s, given competing views about the earth's form—and it demanded geographical fieldwork for its resolution. With these and other events, the world was made modern in the Enlightenment through geographical inquiry. As a comparison of figures 1 and 2 and plates 1 and 2 shows, what was known about the world changed between 1724—and Lafitau's map showing "the little known

FIGURE 1

Map of America, from Joseph-François Lafitau, *Moeurs des sauvages amériquains, comparées aux moeurs des premiers temps* (Paris, 1724), vol. 1. Lafitau is here venturing an early anthropogeography of the Americas. By permission of Edinburgh University Library, Special Collections.

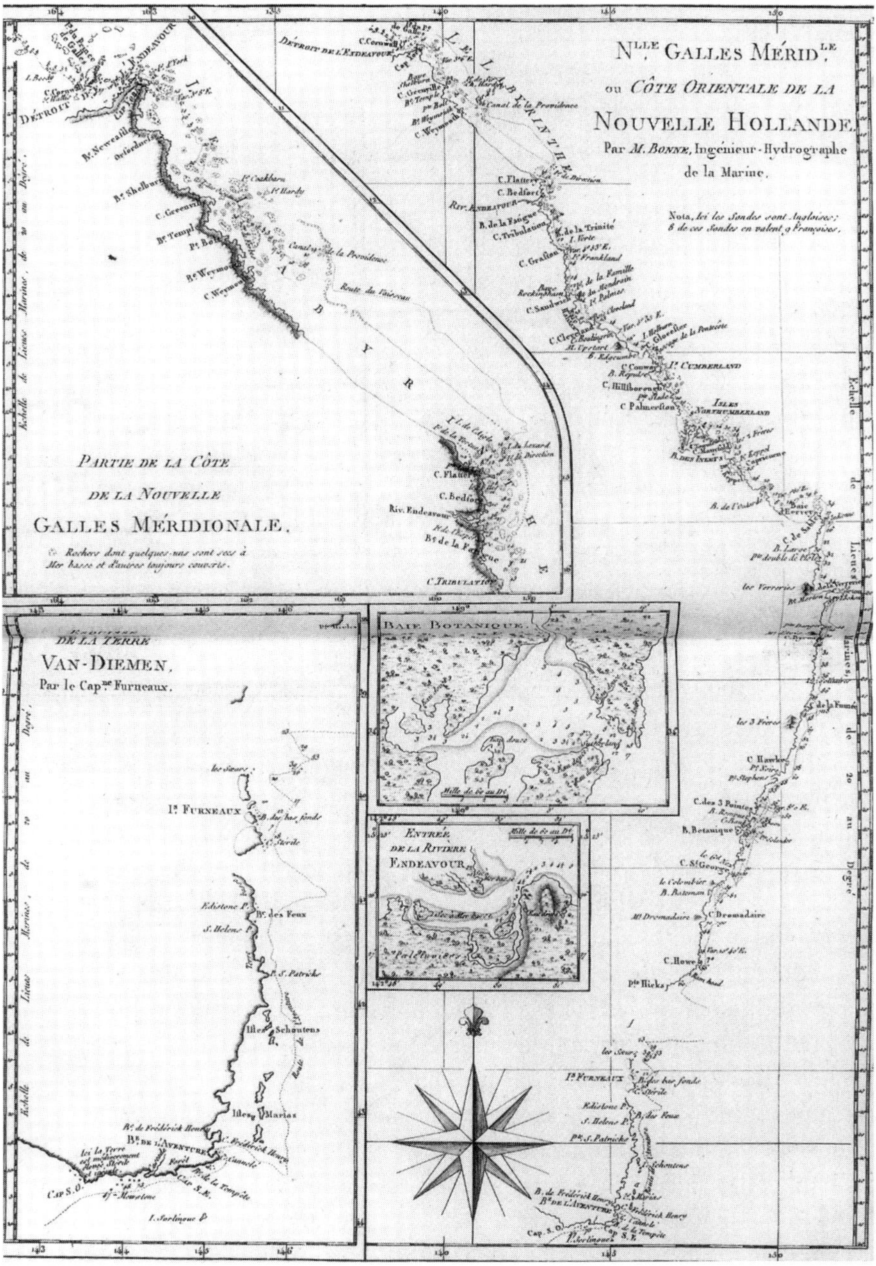

FIGURE 2

The mapping of New Holland, by Rigobert Bonne, French hydrographer royal, from Rigobert Bonne and Nicolas Desmarest, *Atlas encyclopédique* (Paris, 1787–88), vol. 1. Note the "gaps" in the maps of the Australian coastline as Bonne records the work undertaken by the voyages of European navigators. By permission of Edinburgh University Library, Special Collections.

lands to the south" (Australia)—and 1788, in the emergent certainty of that continent in the maps being then brought together by the French marine hydrographer Rigobert Bonne.

This book is divided into three parts, each with three chapters as well as a short summary that brings together the concerns of the constituent chapters. The three chapters of part 1, "Geographies of the Enlightenment," together explore the geographical scales—national, international, and local—through which it is possible to make sense of the Enlightenment. I pay attention to the difficulties of notions of *the* Enlightenment, to the problems of using terms like "core" and "margin" uncritically, and to the idea of the Enlightenment as a cosmopolitan Republic of Letters *and* as networks of connected social settings.

The focus of part 2, "Geographical Knowledge and the Enlightenment World," is on the ways in which the world was revealed geographically among the work of Boyle, Coronelli, Halley, and their contemporaries, and Flinders, Baudin, and their counterparts. The first chapter of this section is about exploring, traveling, and mapping, which leads to a discussion of the world's physical extent and contents and its human occupants, the subject of the other two chapters in part 2.

In part 3, "Geography in the Enlightenment," consideration is given to geography as a subject and practice of Enlightenment inquiry. Attention is paid to geography's books, to geography's place in enlightened institutions, and to ways geographical thinking was used in the Enlightenment. The book's structure—with three parts looking, respectively, at geography *of, and,* and *in* the Enlightenment—is thus designed to move from an understanding of the Enlightenment as a geographical question at different scales and in different places, to the impact of geographical ideas on Enlightenment thinking, to geography as an Enlightenment subject.

What follows is, of course, inevitably partial and personal. It is one attempt to think differently about one of the world's defining moments and movements. It is a summary synthesis of much other work: to do with the earth as the object of geographical scrutiny, of different places figuring in the Enlightenment, of the Age of Reason as geographical in its ideas, individuals, and institutions. Not everyone who has an interest in the Enlightenment will share my concern to place it and to site it through geographical events and processes or to see it, indeed, as geographical in any of these senses. The point might legitimately be made that there is a danger here: by extending the diversity of ideas that the label "the Enlightenment" encompasses, I risk reducing its power to explain anything. But understanding the Enlightenment geographically is not to make it historically any less real. Quite the opposite. The earth and its peoples as objects of intellectual inquiry were written about, understood, and depicted quite differently in 1804 than in 1690 (cf. figs. 1 and 2; plates 1 and 2). That is a fact of geography. The Enlightenment took place *somewhere.* Thinking about the Enlightenment as a locatable and not just datable thing—as dynamic, mobile, and cosmopolitan, not just static and national—is to enlarge our own consciousness as to what the Enlightenment is and was and how it should be thought of.

PART ONE

Geographies of the Enlightenment

2

The Enlightenment in National Context

"There were many philosophes but there was only one Enlightenment." Admittedly this is not the same as saying there were many nations but only one Enlightenment. Yet for Gay, the Enlightenment's "what" and its "where" did mean much the same thing. It was an essentially philosophical and French enterprise, embodied in those Voltairean radicals—"the party of humanity," that "little flock of *philosophes,*" as he variously terms them—bound together from the 1730s by the promotion of atheism, republicanism, and materialism. For others too, "the Enlightenment is ordinarily thought of as a French affair," "primarily a French phenomenon." The Enlightenment had a capital, even a spiritual home—and it was Paris. It is just such a view, after all, that Robert Darnton, one of the high priests of Enlightenment studies, uses to defend his call to "deflate" the Enlightenment in the face of adjectival excess and to restate a case for the Enlightenment as an intellectual movement that is philosophical, is about men, is "elitist, Voltairean, and incorrigibly Parisian."[1]

As matters of geographical and historical fact and in relation to recent trends of historiographical interpretation, such views merit scrutiny. Not everyone at the time thought of the Enlightenment as a singularly French thing. After all, Voltaire in 1733 looked first to the English—"the only people upon earth who have been able to prescribe limits to the power of Kings by resisting them; and who, by a series of struggles, have at last established that wise Government, where the Prince is all powerful to do good, . . . and where the People share government without confusion." In England a quarter of a century before, Lord Shaftesbury had anyway observed how "there is a mighty Light which spreads its self over the world especially in those two free Nations of England and Holland; on whom the Affairs of Europe now turn." Looking for the origins and "defining" features then of what, with prescience, Shaftesbury referred to as "the Establishment of an intire Philosophicall Liberty," most commentators now discern diversity, not single beginnings, plurality not essential unity: "Complex revisionisms mark our times."[2]

A key feature of this revisionism, even if it is not one shared by all those who work on the Enlightenment, has been the pluralizing of Enlightenment, the rejection of the definite article—"a dangerous linguistic tool for historians." To talk of *the* Enlightenment creates, we are told, the presumption of a single unitary process, displaying a uniform set of characteristics that define a series of phenomena or processes, responding to the same conditions, arising from the same causes, and even performed by the same or a continuous set of agents, in what is otherwise a diversity of temporal, spatial, and cultural contexts. For Pocock, the point is not that there *cannot have been* a single Enlightenment shared by the cultures of Europe and America; it is that we should not take it for granted that there *was.* The use of the definite article encourages the presumption that there was such a thing as *the* Enlightenment, and that how we should think about it historically is to trace its origins and its dissemination and modification in different cultures. The merit of omitting the definite article is that it is liberating, not constraining, and allows us to remember "that the term 'Enlightenment' is a tool, which we use to isolate a variety of phenomena which we suspect were similar, were interrelated, were the product of a shared history."[3]

The definite article is no less dangerous a tool in thinking about the Enlightenment as a matter of geography, shared or otherwise. It may even be more so, for where the Enlightenment has been thought about at all in such terms, it has been commonly discussed in national context—and "the nation" is likewise a dangerous tool, for historians and geographers both. Two variants are common in treatments of the Enlightenment as a national phenomenon. The first is signaled by "the Enlightenment in France," "the Enlightenment in England," "the Enlightenment in the Netherlands," and so on. The second is apparent in "the Italian Enlightenment," "the Scottish Enlightenment," "the Russian Enlightenment," and the like. A third but more minor variant also exists—in which a nation is considered "in the Age of Enlightenment." In such cases, however, the focus is more commonly on material changes during the Enlightenment—in agriculture, industry, communications networks, and so on—than it is with addressing the geographical expression of the Enlightenment.

In the first of the two variants, the Enlightenment is made to have both essential national features *and* variant national expression by virtue of its unproblematic movement, in the form of people, ideas, books, and other intellectual artifacts, from one nation to another. Over time and over space, the Enlightenment "travels," if you will, as an essential thing, and even is held to do so from an originating Enlightenment "core" or "hearth" to its "margins" or "periphery." In the second variant, more distinctive experiences are averred, but the particular conjunction of definite article and single national referent hints at a loss of any communality above the nation while obscuring differences within the nation. Yet revealing local variations below the national has been a notable feature of Enlightenment studies that have taken "the nation" as their initial frame of reference. So, too, nations differently drew on Enlightenment influences from beyond themselves. In studying the impact of the "Enlightenment in America," for example, not one but four Enlightenments have been disclosed. The moderate Enlightenment (which might also be called the rational Enlightenment), which preached order and religious compromise, was peculiarly English. The skeptical Enlightenment, British in part, was distinctively French: its method wit, its grand master Voltaire. The Revolutionary Enlightenment had its origins in Rousseau, and its culmination in Thomas Paine and William Godwin. And the didactic Enlightenment—opposed to its revolutionary counterpart—had Scotland as its chief center. And in "the Italian Enlightenment," internal geographical identities and differences between, for example, the south and the north, town and country, and within the Papal States were so distinct that few contemporaries thought about Italy—or the Enlightenment—as a whole.[4]

These issues to do with *the* Enlightenment, Enlightenments, the utility of "the national" as a scale of geographical inquiry; with the production, movement, and reception of Enlightenment; and with the local sites and spaces of Enlightenment form the subject of part 1. My concern in this first chapter of part 1 is with the national scale as one way—a prevalent yet problematic way—of thinking geographically about the Enlightenment. Two main sections give structure to my argument. In the first, I consider questions to do with national context and with *the* Enlightenment, using Porter and Teich's 1981 edited book as a guide of sorts. The second considers the idea of an Enlightenment "core" and "periphery" with reference to Portugal and Spain, and to the Greek-speaking regions—geographies omitted from consideration in 1981 but the subject of attention since. The conclusion, which reviews these issues in order to connect with the other chapters in part 1, offers several answers to the question "Why not the national?" in order to pose a further question: "If not the national, then what?"

National Enlightenments?

The nation has been a common means of understanding the Enlightenment. Even as Enlightenment studies have embraced new thematic concerns—music, for example, economic thought, social theory, even the body—their further elucidation is often couched in terms of national difference.[5]

Why the national figures so centrally is not clear. Indeed, it is obvious that different ends are being sought in using the nation to frame the Enlightenment. For nations and historians recently released from others' hegemony, as in the Ukraine's relationship with Russia, study of the Enlightenment in national context—hardly possible even to consider a few years ago—and the recovering of a national Enlightenment are being used to serve a resurgent modern national identity. In Russia itself, national self-awareness was a significant factor in that country's Enlightenment, but it is not the case that Russianness now looks back to the Enlightenment for historical approbation. Likewise, the Enlightenment's contemporary significance is less valid as an explanation for Enlightenment studies in Hungary and Romania, not least because in those countries meanings of the "nation" dissolve in the face of enduring regional differences between, in Romania anyway, Wallachia, Moldovia, and Transylvania. In Hungary too, regional differences between the northeast, including parts of Transylvania, the northwest centering on Pressburg (Bratislava), and Transdanubia and the coast make it difficult to speak of a single Enlightenment in a single Hungary.[6]

These matters—of the significance of Enlightenment and of differences between and within nations—are problems of wider significance. Whichever variant they present, Enlightenment studies have a tendency to assume the nation as the appropriate unit of measurement. Moreover, modern conceptions of the European nation are often "read backward," as it were, from the now to the then, even as Enlightenment figures then become invested with modern meaning to serve political purposes now. Yet neither "the nation" understood as a geographical entity nor the ideas embraced by that term or by its surrogates, "the national" and "national consciousness," were straightforward matters in the Enlightenment. The map of Europe was redrawn more than once between the 1713 Treaty of Utrecht, which settled the War of the Spanish Succession, and the 1815 Congress of Vienna, which ended the Napoleonic Wars. It is a truism that Enlightenment Europe intellectually positioned itself in this period in relation to southern continents, imagined and real, and to the Americas. But *national* self-awareness was not a constant or a consistent thing. Neither Europe nor its constituent nations were ever securely fixed labels. They were worked at and worked out through the ebb and flow of what we might think of as "civilizational geopolitics," through the gradual accrual of meaning to the idea of the nation through language, culture, historical sentiment, and claims to territory and space.

Terms like "civilization" and "progress" had external and relative geographical significance, being used to contrast Europe—at the presumed pinnacle of human development—with, say, the Americas or the Pacific islands. They had internal geopolitical meaning for the changing entity that was Europe. As Prussia expanded south and eastward under the acquisitive policies of Frederick William I between 1713 and 1740, Russia grew westward, first under Peter the Great and, more extensively, under Catherine the Great from 1762 to 1796. Squeezed in the middle, Poland lost further land to Austria and effectively ceased to exist by 1795 before being restored as the Grand Duchy of Warsaw in 1815. Italy in the eighteenth century was a territorial patchwork of church lands and lands under Austrian and Bourbon dominion. What people took to be France grew and shrank with the fortunes of war and the capacities of its own geographers. After the Revolution, a new political landscape within France—of *départments* as administrative units—reflected both the imposition of a Cartesian grid over national space and the centralization of the French geographical and political imagination upon Paris. More than others, the French used the power of geography to sweep away the old: equating liberty with freedom in space, political authority with the mastery and rationalization of space. Charles Monnet allegorized this power of the new in 1791, showing Reason, aided by "the genius of Geography," outlining the new administrative divisions of France (fig. 3).[7] Even in Scotland, whose political identity altered in union with England after 1707, geographical work of one sort or another during the Enlightenment gave shape to the nation—literally so in terms of competing mapped outlines for the country's true extent, but also in terms of topographical painting, natural history, and antiquarian survey.[8] On a larger scale, shifts in the nature of what Europe was held to be reveal a gradual "Westernizing" of Russia and the acceptance of the Ural Mountains as the continent's eastern border. Yet Catherine the Great's 1762 claim that "Russia is a European power" was more or less synchronous with the emergence within the European political imagination of a West–East divide. This perceived division paralleled western European views toward the Ottoman Empire and, like those views, has underlain European political sensitivities since.[9]

Enlightenment historiography, thus distorted by hindsight over what a nation is, has also been characterized by too easy an elision between great French thinkers (or Scottish, or German) and *the* Enlightenment in France (or Scotland, or Germany). Voltaire, Rousseau, and Montesquieu do not stand for France any more than Hume, Smith, and Reid do for Scotland, but they have often been made to do so. If we are to speak of broader notions of Enlightenment and include not just philosophers but chemists, natural historians, different publics—and geographers—we must be cautious about the seeming equivalence of the personal and the national.[10]

FIGURE 3

Allegorizing Enlightenment. Charles Monnet's *Pièce allegorique* of 1791 shows a new France inscribed by Reason with the aid of Geography (note the prone figures, symbols with their maps of an ancien régime geography). By permission of the Bibliothèque Nationale de France (Coll. Hennin 11075).

The 1981 book *The Enlightenment in National Context* was an important moment in thinking about the Age of Reason nationally and geographically. The editors recognized that "comparatively little has been written about its [the Enlightenment] geographical, social and political *location* as a cultural movement" and that many studies treated the Enlightenment simply as systems of socially disembodied ideas. The book's purpose was "to grasp—while accepting a certain common identity in the Enlightenment—what precisely the Enlightenment meant in thirteen national contexts, ranging from England to Bohemia, from Russia to America, from Italy to Sweden."[11] Twelve countries were studied: America, Austria, Bohemia, England, France, Italy, the Netherlands, Russia, Scotland, Sweden, Switzerland, and, lastly, Germany, the subject of two essays.

In Italy, the Enlightenment was shown to consist of a series of small groups—"learned marquises, liberal priests, angry anti-clericals, cultured bibliophiles"—but claims that this was the case "in every other country" were denied by others' accounts.[12] In both Austria and Sweden, the mark of Enlightenment was educational reform, state led or monarch led. Reforms in Austria took as their targets an improvement in the intellectual levels of the rural priesthood. In Sweden, science teaching in the universities was introduced to counter the dominance of classical learning. In neither country was anticlericalism much apparent: there was hardly a sign of a liberal priest, "not even the disappearing tailcoat of a learned marquis."[13] England, an Enlightenment exception for earlier commentators by virtue of its very absence, moved from nowhere to first, in some respects at least. "The short answer . . . to the thematic question, 'what are the unique features of the English Enlightenment?' would be: None, because practically all its currents subsequently irrigated Enlightenments elsewhere. England's special role was that in many areas—freethinking, empiricism, utilitarianism—she came first." Holland was "the Enlightenment's first ray," but English Enlightenment thought—practical, pietistic (unlike the militantly secular French philosophes)—provided "the parentage of so many of the continental children of light."[14] In Scotland, practical concerns likewise figured. The universities and clubs and societies provided a focus for a culture of critical sociability in which "the principal intellectual achievement" was the application of a language of civic morality as an instrument for the discussion of the moral, political, and economic organization of commercial civilization at large.[15]

What of Enlightenment matters on the other side of the English Channel? Hampson's essay in *Enlightenment in National Context* views the Enlightenment in France as a much less essential and philosophical movement than was experienced elsewhere: "we are dealing with a widely disseminated attitude of mind rather than a specifically literary or philosophical movement."[16] Even so, the Enlightenment there was distinguished by skeptical and anticlerical philosophes who saw themselves in a state of perpetual opposition: to the Church, to ancient learning, to the power of absolutist monarchs, to the status quo. Yet more evidently there than elsewhere, such men and the few women involved became divided: between, on the one hand, Rousseauian and other radicals and, on the other, men of a practical bent, keen like their Scottish, English, and German counterparts to equate enlightenment more with immediately utilitarian ends and less with future utopias.

In the Netherlands, by contrast, "that spirit of Voltarian pagan rationalism which set the tone for the Enlightenment in France" was absent from all but the most *savant* circles. Dutch moralizers did not look to France. They echoed Swiss writings in the same vein, even mirrored the language of English reform and, in the universities, the language of German idealism. Unlike the English view of the Dutch, some French refused to admit that their northern neighbors—dour merchants, speakers of a guttural language, unrefined and unimaginative—could be enlightened at all. Yet the Enlightenment in the Netherlands had three features: an emphasis on toleration and free inquiry, the promotion of these ideas in books and journals, and the slow filtering of such ideas into Dutch vernacular culture. Further, geographical difference was social difference. Because the language, idiom, and sensibility of the lumières did not find easy reception in the United Provinces beneath the patrician elite, "the Enlightenment in the Dutch Republic, then, was not at all the same phenomenon as the Dutch Enlightenment."[17]

In the twenty or so associated cantons that made up Switzerland, enlightened thinking had the greatest impact in urban Protestant areas and in the academies. It was "the child of Liberal Protestantism, open to foreign influences and above all to the Cartesian and then the empirical, Newtonian revolutions."[18] In Protestant Germany, the Church fostered enlightened ideals. In Catholic Germany, it did not. Cities everywhere were enlightened centers for academies, the book trade, reading clubs, and reform, and the universities were important torchbearers of Aufklärung. In Bohemia, Enlightenment prompted a revival in nationalism—one expression of which was a now lost geographical survey—and the foundation of learned societies, educational reform, and the beginnings of "modern" Czech historical scholarship.[19] Russia may have inched its way toward inclusion in Europe by the later eighteenth century, but whether its Enlightenment was similar to that in Europe was debated by other Europeans, some of whom—like a few of the French vis-à-vis the Dutch—refused to equate the Russians with enlightenment. The Enlightenment in Russia, a "local phenomenon in its own right" and a consequence of external "civilizing influence" with its initial expression under Peter the Great before 1725, effectively began and ended with the reign of Catherine the Great. At first an enthusiast, she had by 1785 dropped Montesquieu and most of the other pilots for the course steered nearer the beginning of her rule.[20]

For America, J. R. Pole distinguished between two competing views of the Enlightenment. The first was of an essential unified Enlightenment "imported" to America: "The Old World imagined, invented, and formulated the Enlightenment, the New World—certainly the Anglo-American part of it—realized and fulfilled it." The second was of a more complex Enlightenment, to some degree a foreign object, to some degree not. This was an Enlightenment given local shape—for example, in emphases on Protestant Christianity, in debates on rights and liberty, on the applicability of others' enlightened ideas to the making of nationhood—by the fact that America was, literally and metaphorically, a different soil from Europe's.[21] And within America, others revealed regional differences to be important. The skeptical Enlightenment was stronger in the slave-owning "Stoical South" than ever it was elsewhere. In New England, by contrast, the new nation was forged out of the tension between the moderate and Revolutionary versions of Enlightenment.[22]

Although an established work within the Enlightenment canon, *Enlightenment in National Context* does not pretend to be the last word in thinking geographically about the Enlightenment. Quite otherwise, in fact, for the volume's prolegomenal nature is stressed in conclusion. The approach to national Enlightenments is selective: Latin America was not included, nor was Portugal, Spain, or the Greek-speaking regions. The essays in *Enlightenment in National Context* do not represent the first time "national" Enlightenments had been thought about. Other earlier geographical studies are extant, for countries and regions covered by the collection and for others not—Corsica, for example, Wales, even Japan.[23] Yet the essays in *Enlightenment in National Context* signal a willingness to embrace matters of geographical difference beyond ideas that referred unquestioningly to *the* Enlightenment when considering simply the "differentiated rhythm" with which ideas moved across Europe (with the exception of England) from Paris, "the home of the Enlightenment."[24] Despite this, the book is an opportunity missed. There is little discussion about what the Enlightenment's "common identity" was. If there is a unifying feature, it is a failure to think about what geographical difference was and what it meant. For as the idea of *the* Enlightenment spreading outward from a single originating (French) hearth has foundered on the rocks of geographical difference, so the idea of national Enlightenments has broken up in the face of differences within nations.

Work since the mid-1970s has added to and refined these claims.[25] It is now possible to chart more complex national geographies of Enlightenment and even to discern different trends in their interpretation. Take Italy, for example, whose "national" Enlightenment has been refined and challenged. Enlightenment in the south centered on Naples and diffused from that center—in debates about Newtonianism, geological inquiry, and on the need for rural reform—at the same time as the city improved itself. In the Papal States from the 1740s, significant features of Enlightenment included the scientific reform of agriculture and topographical surveys designed to know the contents of the papal lands. In northern cities like Bologna, the urbane scientific and courtly climate encouraged the participation of women in ways unheard of in the south.[26] In assessing such work, an altogether more detailed picture of Enlightenment in Italy emerges—but not of Italy as a nation.

Consider too Scotland, whose Enlightenment has been the subject of considerable revisionism. Enlightenment there was never just characterized by the language of civic morality. Nor should we think of that as the nation's principal intellectual achievement in that period. It was distinguished by a practical and scientific turn of mind—the working-out of the "Science of Man"—and often had very local expression: in some towns and universities more than others, in experimental laboratories as part of domestic spaces, and in individuals' daily circulatory geographies.[27] So dynamic was this conjunction of practical inquiry with polite sociability that one commentator would have us look there and not to France, to Edinburgh and Glasgow and not to Paris, to sober religious moderation rather than to philosophical skepticism, for the birth of the Enlightenment and of modernity.[28]

In America, Enlightenment ideas and ideals were found in the "Stoical South," on the western frontier, and in the backwoods of New England—among fewer people admittedly—but they were discussed more conspicuously among the publics of enlightened urban centers such as Philadelphia.[29] In noting that the rays of the *siècle des lumières* were sighted as far from Paris as the Mississippi Valley and that a didactic print culture was being passed from hand to hand and door to door in New England's "village Enlightenment," are we talking about the same thing that was circulating in the wynds of Edinburgh? Did early Americans talk of the same thing? Was it *the* Enlightenment? The *American* Enlightenment? Or did particular sites—New Orleans' drawing rooms, Puritan meeting halls in Boston and Marblehead, and, on the other side of the Atlantic, Parisian salons and Genevan pulpits—always lend the Enlightenment local purchase, no matter that the same books were read, the same notions of intellectual freedom aired?

One other way to approach these questions is to consider a further sense in which the Enlightenment has been considered to have had national geographical expression, namely, the idea of its having a core and periphery. This too is an often-assumed feature of Enlightenment studies. Sufficient is now known, of course, to dismiss the idea of a single *national* originating point or characteristic—in English conservatism or pietism, Dutch tolerance or French skepticism. And enough evidence has been gathered for us to be at least skeptical about the Enlightenment simply moving between nations and about national Enlightenments being everywhere and at once the same. I return to these issues later in the volume. Here I consider Enlightenment in Portugal, Spain, and in the Greek-speaking regions of Europe, in part because their geographies have been consistently overlooked but principally because they illustrate in different ways the difficulties associated with thinking of the Enlightenment in terms of centers and edges as well as in national context.

Enlightenment Margins?

It is possible to speak of the Greek Enlightenment, but only just. Until recently this field was a neglected area of scholarship; the term meant little until late in the eighteenth century. Greece did not formally exist until 1821, although the Greek-speaking regions within the Balkans, part of a multinational society under Ottoman rule, extended beyond the boundary of the nation itself. In Greece, the nation and its Enlightenment helped establish one another. One reflection of this was the appeal made by Greek intellectuals to both ancient and modern Greek as the languages of philosophical and scientific enlightenment and of national identity. Another was the movement into the Greek-speaking regions of new scientific ideas from western Europe. When looked at more closely, however, and for the sciences especially, the idea of a donating western European Enlightenment "core" and a passively receptive Balkan–Greek "periphery" does not easily hold. The transmission of scientific knowledge took place through a number of channels. Paris was the principal source of the new sciences. Yet for many Greeks, especially those in university, Italian and German institutions and cities, not Greek ones, were the receiving sites for the Enlightenment's new ideas. Vienna not Athens was the leading publishing center for Greek books and the leading reception point for influences from the German Enlightenment.

The Enlightenment in the Greek-speaking world was a project of social and intellectual criticism defined not just by national needs but by making work in local and particular contexts the expressions of a cosmopolitan humanism that had its origins elsewhere. Interestingly—a point I return to in chapter 8—books of geography produced by Greeks, modeled on European trends, and published in Vienna were key expressions of these trends.[30]

In short, what became the Enlightenment in what would become Greece transcended national limits. A tributary of the European Enlightenment perhaps, the neo-Hellenic Enlightenment cannot be absolutely restricted to national or geographical borders or to philosophy alone. The Enlightenment, Diaphotismos, in and beyond Greece presents a complex picture: of expatriate Greeks in cities such as Venice, Paris, and Vienna; of indigenous intellectuals cultivating Western science for its own sake—some to highlight its advances and some to protect Aristotelianism as a native product; and of different sites in and beyond the nation's boundaries, such as the Greek communities in southern Russia and in Hungary. Further, argue some Greek scholars, thinking in terms of the simple transfer of ideas is inadequate. A better term is "appropriation," signaling as it does "the measures devised within the appropriating culture in order to shape the new ideas within the local traditions which form the framework of local constraints." Seen in these ways, the principal role of Greek Enlightenment scientists and other intellectuals was less to produce new work along the lines of their Western counterparts than it was to "appropriate" such work for educational purposes within Greece, to make others' claims serve a different national purpose.[31] In the eastern Mediterranean then and in the Balkan regions, the concept of an Enlightenment core and periphery breaks down not just because of the difficulties associated with the "nation" as a frame of reference but because of the particular ways in which the Enlightenment worked over and above national borders and in local context.

In Portugal, Enlightenment was characterized less by its nationals working without and more by foreign intellectuals acting within. In the first half of the eighteenth century in particular, Enlightenment was constituted through the *estrangeirados,* loose networks of resident foreigners and Portuguese who had contacts with European intellectual circles, either through their own education and travel abroad or through correspondence. What held the *estrangeirados* together was the desire to introduce into Portugal the rationality of an Enlightenment they took to be happening elsewhere. One early leading figure, for example, the Franciscan friar Manuel do Cenáculo, became aware of Enlightenment matters only when in Rome at a Franciscan conference in 1750. Enlightenment was facilitated too by a few private academies. The later Enlightenment in Portugal is chiefly associated with the work of Sebastião José de Carvalho e Melo, the Marquis of Pombal. Effectively the country's civic ruler during the reign of Joseph I, Pombal secularized science and technology in Portugal between 1750 and 1777. This involved banishing the Jesuits after 1759, revitalizing universities such as Coimbra, proposing a General Congregation of Sciences, overseeing Lisbon's reconstruction after the 1755 earthquake, and laying the foundations of a culture of literate criticism and scientific inquiry (including the Royal Academy of Sciences in 1779). But in Pombal's mind and for his plans on the ground, notions of Enlightenment and its associations with progress, rationality, and modernity in later eighteenth-century Portugal had very different meanings than was the case elsewhere. Increased state control not individual liberty, tolerance of change but with institutional oversight, "Order and Progress" not "Order and Freedom": for such reasons, the story of Pombal's enlightened despotism in later eighteenth-century Portugal is of an antidote to Enlightenment.[32] Much the same is true for Spain. The authority of the Catholic Church was challenged by the advocacy of the "new philosophy" among some Spanish intellectuals from the end of the seventeenth century, notably in Seville, and there and elsewhere in private rather than in the universities, but it was not much weakened. The intellectual life of eighteenth-century Spain was backward looking, agricultural reforms extensive in intent but limited in effect, modern philosophy everywhere a perceived threat to Catholicism.[33]

Can we speak, then, of an Iberian and neo-Hellenist margin—even a romance Enlightenment—on the edge of the Enlightenment's northwest European French, Dutch, and British "center"? If we think only in terms of *the* Enlightenment having definable points of origin and of it moving with some sort of organic unity to have later expression elsewhere, the answer might be a qualified yes. More properly, the answer must be no. Just as it took different forms elsewhere, Enlightenment was a different thing in Portugal, Spain, and the Greek-speaking regions. It was made as a set of ideas through processes of creative exchange, from the appropriation in local context of given ideas, not simply as the consequence of relationships of dependency with other "leading" areas. Furthermore, Portugal and Spain were colonial powers, both core to an even more distant periphery, and in that context, Enlightenment was something both exported and made locally.

In the countries then forming the Spanish American empire, the expulsion of the Jesuits from 1767 permitted a freer intellectual climate than previously. From the 1780s, local economic societies spread the message of material progress through the application of the new science. Campaigns against scholastic natural philosophy in which some clergy took a prominent role were fought in universities and in private academies, in Mexico and most evidently in the viceroyalty of Peru. There, moves to embrace modern science and oppose Aristotelianism were first proposed, in Lima in 1767, by the priest Toibio Rodríguez de Mendoza, who later established a modern curriculum in a former Jesuit college to teach the scientific thought of Descartes and Newton among others. Natural philosophers such as Hipólito Ruiz and José Pavon undertook geographical and natural history surveys of Chile and Peru, doing in and to South America what men like Peter Simon Pallas, Thomas Pennant, Carl Linnaeus, and Déodat de Dolomieu were doing for Siberia, Britain and Iceland, Lapland, and France, respectively. Voltaire, Rousseau, and Condorcet were all read—but their works were also at moments proscribed. Sites of reform, proposals to modify the curriculum—to use Spanish instead of Latin as the language of instruction, for example—and programs for useful national knowledge through natural inquiry everywhere met with opposition. Rodríguez fell foul of the Spanish Inquisition. The Economic Society of Lima, of which he was a leading member, was closed by the government in 1794. Progressive journals later became outlets of repression, Enlightenment ideals a justification for wars of independence. Contrary to some prevailing notions, Brazil likewise had an Enlightenment—a locally modified Portuguese model—built around academies to develop useful knowledge, beginning with one in Rio de Janeiro in 1772 (although it too was disbanded by suspicious state authorities in 1794). In Brazil as in Spanish America and in the southern regions of the emerging United States, the French influence was strong—in terms of books owned and read, and their authors' ideas discussed. But the Enlightenment in Brazil was also strong in its own terms: "the inferior of only a few and the equal of several regions of Spanish America."[34] South Americans in general did not so much receive the Enlightenment as "reproduce it from the sources upon which its exponents in Europe depended." The themes that were significant to the public in Spanish America had less to do with building new religious and political languages than with constructing alternative, critical epistemologies: "The Spanish American Enlightenment was a dual process of creating such discursive space and consolidating a public sphere."[35]

Terms such as "core" and "periphery," "center" and "margin," have only weak descriptive value. Used only in relation to nations and then simply and in comparative context, they are limited and limiting. Using them has the effect of establishing a simplistic view of a superior Europe—of a superior urban northwest Europe even—and an inferior distant Other. Such a polarized geography is not sustained either by the evidence of variation within and beyond national contexts or by the fact that Enlightenment took place well beyond Europe. It is true, of course, that if we think in relation to certain influential philosophical and scientific works, then Enlightenment ideas were received later in parts of Scandinavia, the Greek-speaking regions, in Russia, or in Spanish and Portuguese America than they were in Paris, London, Vienna, or Berlin. In such terms, geography becomes a matter not of space but of time. Hours even days and weeks, not miles, are the meaningful units of its measurement, the Enlightenment's geographies calibrated by the time ideas took to circulate, not just in the fact and site of their reception. Considered thus, it is tempting to equate "early" with "essential," "later" with "derived," *the* Enlightenment as something only originating from outside. Some contemporaries did think like this: the Portuguese intellectual who noted that a distance of ten leagues from Lisbon took you back one hundred years, or the French artist Duclos who wrote that "those who live a hundred miles from the capital [Paris] are a century away from it in their modes of thinking."[36] We could, on the basis of the Enlightenment's time geography, replace "center" and "margin" with notions like "leading" and "lagging." We might then proceed to map Europe and the wider world in relation to its various enlightened "iso-chrons," lines in space connecting that moment in time at which, say, the works of Newton or of Voltaire were first bought, read, or translated in a place or news of Lisbon reached other towns and cities.

Such a view fails to address adequately the nature of the local sites in which the Enlightenment was made and received, the social differences and connections associated with the crossing of geographical space, or the ways in which what was held to be Enlightenment resulted from the "appropriation" of others' views. To map the Enlightenment thus in respect of a single and unchanging entity is to give credence to a "flat earth" conception of Enlightenment space, as if the earth was unmarked by its geography, a featureless globe permitting ease of movement for knowledge and people in any and every direction. Such "chrono-geographies" of Enlightenment between an early center and a later edge were apparent everywhere—within nations, within cities, within social institutions, within and between groups of philosophes and others communicating one with another. Whatever the scale of analysis, describing the timing of the Enlightenment's geographical expression is not the same as explaining its how and why: intellectual engagement cannot be equated with geographical and temporal distance.

There is a further reason why we ought not to use the idea of Enlightenment margins and edges uncritically, certainly in national context. In several key respects, the Enlightenment's core was actually at and more usually beyond Europe's margins. That is, ideas in Europe about the geographical extent of the world, about human difference, about what it meant to be "enlightened," were framed by geographical encounters with the "New" World, with the Pacific especially. This claim—that the Enlightenment crucially depended on geographical knowledge—is explored at greater length in part 2. Here, it is sufficient to note that Enlightenment Europe's view of itself as core was significantly shaped by its margins. Frequent use, for example, was made of China as a means of satirizing Europe's institutions by affecting to view them through the eyes of foreigners—a literary device that helped evade the censor—Montesquieu's *Lettres persanes* (1721) and Goldsmith's *The Citizen of the World* (1762) being notable examples.[37] So common was this attention to places on the "edges" of the Enlightenment's world—precisely because such places were the limits to Enlightenment knowledge about the world—that we might reverse, or even equate, the terms applied to the geographical locations: the margin becomes the core. Considered in such ways, the Enlightenment's "originators" are not Europe's nations and prominent individuals—Locke, Newton, Shaftesbury, Voltaire, Rousseau, or Montesquieu, nor even James Cook, Joseph Banks, George Vancouver, or the Forsters, Johann Reinhold and Georg—but the peoples and places they encountered and charted. The Enlightenment's geography becomes less a map of national boundaries fixed in space than one of myriad lines of movement across space, lines connecting sites of knowledge's making and reception.

Thus the nation and Europe both come to be displaced. The revisionism in Enlightenment studies since the mid-1970s, as evident in the move away from the Enlightenment-as-philosophy and toward Enlightenment-as-the-social-history-of-ideas, has anyway been geographical of sorts. After all, to see the Enlightenment as social history—whether of Dutch burghers, New England congregations, Glaswegian chemists, Viennese publishers, academicians in Seville and Lima, or journal readers in a London coffeehouse—is to discern its variegated geographies at least in local outline. Yet this is only a weak form of geographical revisionism. Its principal feature has been that the nation still figures as the natural unit of assessment by which the Enlightenment can be considered in comparative context. For the reasons explored here, this geographical conception of the Enlightenment is inadequate. The nation was not a fixed space in the Enlightenment. Nor indeed, as we shall see in part 2, was the extent of the world itself. The nation was not the start point for Enlightenment. Neither should it be our end point in thinking about the Enlightenment geographically. The nation does not represent a fixed scale of geographical analysis. For the more it is scrutinized, the more its integrity as a unit of measurement fractures in the face of differences—between town and country, within towns, in social spaces—in what was held to be Enlightenment. Then and now, geography makes nations not as immutable objects but by constituting through active processes—mapping, describing, picturing, and so on—processes that have expression above and below the national scale.[38]

This is to say, then, that the idea of *the* Enlightenment, certainly of it as something singularly French, urban, and philosophical, no longer holds. There was not "only one Enlightenment"—there were many. We ought now to dismiss the definite-article conception once and for all, and consider it of interest only as a historiographical classification, not as an accurate description of something more complicated in its substance and in its geographical making and expression. So too, and as an inevitable welcome consequence, we should reject that related conception of *the* Enlightenment as simply moving outward and downward from originating national hearths or cores to peripheral regions of Europe and from thence to colonial or marginal spaces.

My proposal for the rejection of these ways of thinking is, in effect, a call for the more accurate qualification of the Enlightenment in geographical terms. It is to take up in more general ways—and it is also to refine—the two variant forms of Enlightenment at work, and that point I noted earlier, by Schama, in discussing the Netherlands: "The Enlightenment in the Dutch Republic, then, was not at all the same phenomenon as the Dutch Enlightenment." In the first sense, the Enlightenment was an intellectual moment and movement that transcended the simply national. In the second, how we can better understand the Enlightenment is to consider not an essential unity but its diverse plurality, to consider the particular manifestation and "appropriation" of the ideas of freedom, tolerance, and the recovery of intellectual nerve. The national context is not the only one available to us. Others have called for the more accurate geographical qualification of science in much the same ways in challenging the "sturdy indefensibility" of the nation as an analytical category.[39] Rather than think of the nation as a "container" for the Enlightenment, we should think of space in different ways and as something actively made by those processes constituting the Enlightenment. Rather than begin and end only with the inevitability of the nation in thinking about the Enlightenment geographically, it is I suggest more fruitful to consider both the movement of its ideas above and beyond national contexts and to explore their regional, local, and social manifestations within the nation. Let me turn to the first of these claims.

3

Above and beyond the Nation
Cosmopolitan Networks

Dismissing *the* Enlightenment in national context as *the* way to understand the Enlightenment geographically is consistent both with that spirit of critical revision now apparent in Enlightenment studies and with that disquiet expressed at the supposedly universal nature of science. In considering a possible "end to national science" and the uncertainties attaching to "nationhood" in the history of science, for example, Pyenson, through his engagement with the "regional manifestations of transnational practice," emphasizes the importance of local conditions over national ones. And in calling for attention to "the Enlightenment above national context," even as the national scale for its interpretation was becoming established, Robertson has significantly enlarged the nature of inquiry about the Enlightenment not by insisting on national particularity but by looking at connections and differences transcending national boundaries.[1]

Thinking about a more "decomposed" Enlightenment (to use Reill's term) as a matter of transnational connections *and* as local sites and social spaces—the subject of the following chapter—may appear contradictory, but the issues are closely connected. In stressing the international dimensions of political economy in the Enlightenment, Robertson focused on free trade thinkers in Scotland—in Edinburgh, Aberdeen, and Glasgow—and in Naples, on individual men, works, and certain institutions, as well as on the connections and comparisons between them. Robertson also aimed to extend the work of Franco Venturi, whose view of the Enlightenment in Italy and across Europe was of an international intellectual movement made manifest in the specific and patriotic concerns of different local adherents. What interested Venturi were the ways in which ideas of political economy articulated elsewhere crossed linguistic, cultural, and confessional boundaries to be adopted and adapted in Italy to meet local needs.[2] For Robertson and Venturi both, political economy was at the heart of the Enlightenment's local making and its international currency. Robertson makes his case for a comparative and connected Enlightenment with reference to Scotland and Naples. Pyenson illustrates his argument with reference to the southern United States, specifically to Louisiana, to New Orleans, and to men such as Antonio de Ulloa y de la Torre Giral, natural historian, geodesist, naval officer, first Spanish governor of Louisiana, and arguably "the most distinguished savant in all of North America during the eighteenth century."[3]

What connected Neapolitan political economists, Scottish philosophers such as David Hume and Adam Smith, and a Spanish statesman-scientist working on the Mississippi frontier was their cosmopolitan outlook. This cosmopolitanism was apparent in shared interests that moved above and beyond national context. It was apparent too in the local and patriotic uses to which Enlightenment knowledge could be put at levels "beneath" or "below" the nation. As these examples illustrate, the men and women of the Enlightenment in different locations and in different social spaces did not necessarily privilege one discourse above another—political economy over philosophy, jurisprudence over natural history. They communicated with one another about all sorts of things and, anyway, did not so readily see neat distinctions between such topics. Communication expressed social status and gave shape to the Enlightenment's public sphere. Communication over distance also effectively "closed" geography, or at least greatly reduced the facts of geographical separation. Because this is so, and because we can discern networks of correspondence between and among botanists in different countries, among medical men, women in Parisian salons, as well as from natural philosophers on the move, we can recognize the Enlightenment as something actively constituted in place and over space.

My central concern here, then, is in thinking about the Enlightenment working geographically over space, as something dynamic, mobile, and varied. I want to reveal what I loosely term the Enlightenment's "traffic"—the mobility of its personnel, ideas, and artifacts—and to illustrate how the Enlightenment was constituted by such traffic. Many Enlightenment thinkers themselves traveled of course. The long-distance travels undertaken by oceanic navigators, explorers, and mapmakers significantly altered the geographical consciousness in and of the Enlightenment, and these are issues discussed in chapter 5. Here my focus, in the first half of the chapter, is on two closely related and seemingly more mundane artifacts: the letter and the book. In the final section of the chapter, I briefly consider the circulation, collection, and display of other artifacts—natural history specimens, instruments, and even other humans themselves—within this Enlightenment republic.

Many influential Enlightenment works were published as letters: Voltaire's *Lettres philosophiques sur les anglais,* for example, in 1734, or Montesquieu's 1721 *Lettres persanes.* The move from correspondence in private to print in the public domain, and thus from one epistemic and social space to another, was common in the Enlightenment. The Enlightenment was in general terms a cosmopolitan Republic of Letters whose boundaries, sustained by the production, circulation, and reception of correspondence and books, extended well beyond those of national territories.[4] Seeing the Enlightenment in terms of the Republic of Letters, book geographies, and in respect of other artifacts extends and complicates the notions of Harris and Latour and others concerning the dynamic nature of the geography of knowledge. Yet neither the letter nor the book or other printed media is a simple thing. The move from manuscript into print was far from straightforward. On the whole, men of science used the letter differently than did the habitués of Parisian salons. Scholars of eighteenth-century epistolarity and historians of the book have shown that attention to such differences in the forms of knowledge making in the Enlightenment demands consideration not only to the production of ideas, the context of making and discovery. It also requires attention to their transmission over space and to their reception by readers and audiences in yet different places—to the context of justification and disclosure. Correspondence was not just a social practice and a means of communication. For some, it was an epistemic requirement over what could legitimately be held to be truth. Correspondence was what many Enlightenment men and women of letters and science craved: between their observations in one place concerning facts in nature and those of distant correspondents, or between experimental and instrumental results differently arrived at in other places. Letters were of little use as sources of enlightened knowledge if one could not regard as reliable the word of one's correspondent. Secure knowledge, an epistemic question, depended greatly on the social standing of one's correspondents, on how what was known was known (by eye witnessing as opposed to being told it, for example), and on the strength of the links that sustained one's networks of correspondence.[5]

Similarly, books do not necessarily contain new or useful knowledge until they are received, read, and acted upon by others. Reception may involve translation in one or both of two senses: linguistically, from one language to another, and geographically, from one place to another. Either way, reception and translation depend likewise on correspondence: between what was said in one language or context being read in the same way in another elsewhere. Because of their local context, however, readers often had a local agenda in mind in interpreting others' works. Intended meanings do not always travel well.[6] There are numerous examples in the Enlightenment of the translation and reception of books and ideas changing their meaning. Likewise, cabinets of specimens displayed distant geographies but did so only by separating them from their originating sites of meaning. Letters, books, and other artifacts thus demand our attention as part of the Enlightenment's "traffic" not because they were immutable forms of knowledge making that transcended different geographies but precisely because they did not.

The Enlightenment as a Republic of Letters

The idea of a res publica litteraria dates to antiquity but has particular currency from the Renaissance. Enlightenment thinkers knew that they stood in this longer-run intellectual tradition and respected it even as they disputed its authors' findings. Yet the ideal and the idea of the Republic of Letters and the practices that sustained it found renewed expression in the Enlightenment in several ways. One was the new social status of the critical intellectual. A second was the inherently collaborative and public nature of engaging with others in the challenges posed by new knowledge and in the claims made concerning its social benefits. Another was the language of this collective engagement. Latin, hitherto the common medium of European scholarship, was replaced as the language of intellectual exchange by French. Books and letters circulated among people of certain social ranks. "The Republic of Letters was, then, a conceptual space defined in terms of cosmopolitanism and universality, although in reality its membership was limited to the educated elite and was, therefore, almost exclusively male, university-educated, and European." It was also "an actual space" defined by its languages, practices of exchange, and networks of communication.[7]

Seen thus, the idea of the Republic of Letters has value in explaining the Enlightenment as a dynamic entity because it allows us to see better now the connections then between social space and epistemic space and at scales above and below the nation. The term certainly works against the idea of the Enlightenment having definitive national characteristics because the cosmopolitan networks that sustained it were international and collaborative, not the preserve of one nation more than another. This does not mean that "the national" should slip from view altogether as a scale of inquiry. There were differences between the ideal and the reality of the Republic of Letters and between that republic and the emergence of the nation-state in the eighteenth century. The Republic of Letters was not a "free country" as it were, affording equal access to all. Further, as Lorraine Daston recognizes, although considerations of nationality were not absent from Enlightenment science, "it would be anachronistic to parse the Enlightenment intellectual scene in terms of national divisions, for other divisions of confession and training were far more telling at the time." What merits our attention is the nature of the relationship in the Enlightenment between expressions of nationhood and the Republic of Letters. "How did this transnational confederation of the learned make its peace with the nationalism that was often the key to its financial support?" "If the citizens of the Republic of Letters refused their ultimate allegiance to their respective nations, to whom or what did they swear fealty?"[8]

For Daston, answers lie in the intent of individuals to sustain international endeavor even when their nations were in conflict—correspondence between Sir Joseph Banks and his French counterparts and shared interests in mapping during the Napoleonic Wars being two such instances.[9] Answers are to be found in the many academies and institutions (discussed in the following chapter), in the emergence of new journals, and in the rise of the periodical press within the Enlightenment's public sphere. Beginning in the mid-1680s with Pierre Bayle's trendsetting *Nouvelles de la République des Lettres* (News from the Republic of Letters), periodical publication and dissemination in other closely related genres such as the multivolume encyclopedia characterized the Enlightenment as no other period. The number of French-language titles in Europe rose from fewer than 30 in 1710 to 167 by the 1780s. In Germany, 718 new titles appeared in the 1780s alone. These new periodicals reflected a diversification of subject matter and shared interests in making new knowledge public. Letters made up a large part of the contents of many of the journals.[10] Above all then, answers lie "in the voluminous correspondence of its far-flung members."[11] The Republic of Letters was about cosmopolitan networks of individuals corresponding and of institutions collaborating. It was not about nations affiliating.

What distinguished the Enlightenment man of letters? Voltaire saw him more as an encyclopedist than as the expert or specialist of modern parlance, as a man of learning and, importantly, of science. He was also a man free to offer public criticism. In that, of course, there lay a difficulty, for the demands of private patronage could compromise free expression of critical opinion. Voltaire was dismissive of those who had to write to live, since such a status brought dependence on publishers and audiences. Better, he argued, to be among the "many men of letters who do not publish anything. They are probably the happiest of all. They are spared the humiliations that the profession of author sometimes bring with it. . . . They live in greater concord with each other, they enjoy sociability more, they are the judges while the others suffer judgement."[12]

Not all writers fitted this role. Judging from the position of the man of letters in Voltaire's France, we find that the ideal figure he portrayed was far from common. Taking as a guide the listings in *La France littéraire*, we can identify three groups of men of letters. For the first, writing activities were backed up by income from another post, a title, or a benefice. Clerics, for example, accounted for about one in five literary men in France in the first half of the eighteenth century. In the second, writing activity was linked to professional endeavor of one sort or another—lawyers, physicians, professors, and so on. In the third, the least Voltairean by type, were writers under the protection of an influential patron. But what was true of France was not so of Germany—where the population of men of letters was twice as large by the 1780s as it was in France—or of Italy. There, the literary community was smaller in size and proportionately more clerical in membership until the last quarter of the eighteenth century.[13]

Such social and geographical differences within the Enlightenment Republic of Letters were paralleled by variation in the nature of letters and in the practices of letter writing. Analysis of letter writing as a social practice must for any period distinguish between the texts, the participants, the activities, and the artifacts involved. In the Enlightenment, official genres of correspondence were quite different in their texts and purpose—but not necessarily in their participants when men of letters held government positions—from epistolary formulae between learned men of independent means who knew one another's work and interests but who had never met. In this latter case, certain conventions—"Esteemed Sir," "Your Respected Servant," and the like—needed to be employed to elicit the support of one's distant correspondents. The Voltairean image of the Enlightenment man of letters as a man of science did not always sit comfortably with such men's other responsibilities or with an individual's self-image as, say, "Godly Naturalist" or "Moral Philosopher" (consider Diderot's self-cultivation of his image as a "Man of Letters": for example, plate 3).[14] Letter writing between women was by and large different still, in consequence of the different forms of sociability among women, who were often excluded by their gender (or expected to be so) from public discourse.[15] What traveled was as much a reputation, a perception of social standing or of scientific prestige as any clear conception of "reason" or "truth."

These facts of social and geographical variation do not weaken the idea (and ideal) of the Republic of Letters as a way of thinking about the Enlightenment as a dynamic transnational phenomenon. Indeed, the opposite is the case. They remind us of the need to be attentive to social, epistemic, and geographical detail, to recognize both "the social grounding" of different genres of letters,[16] and the located and specific nature—the "geographical grounding"—of those "networks of networks" making up the Enlightenment's cosmopolitan public sphere. And they raise new possibilities methodologically, given that different epistolary practices help reveal the Enlightenment's constitutive geographies by highlighting the where and the who as well as the how and the why sustaining the cosmopolitanism of its communities.[17]

For the physician and botanist Albrecht von Haller, for example, letter writing was a means to secure his own social position as a professor in Göttingen between 1736 and 1753. While there, Haller established and benefited from a web of correspondence across northern Europe especially, with correspondents in Hanover and in Berne particularly important, but with southern Europe figuring hardly at all. Haller's work with Swiss botanists in Zurich and in Basle on the Swiss flora, the *Enumeratio Methodica Stirpium Helvetiae Indigenarum* (1742), one of many such books on native flora based on local fieldwork and agreed taxonomic principles, depended on correspondence. Haller drew upon others' works in this respect—on the Russian Johann Georg Gmelin's *Flora Siberica,* for example—as part of his plans to establish a botanical garden in Göttingen. Although Haller's botanical network was strong, his medical one was weak, with only one man of science south of the Alps. His letters to the Padua anatomist Giovanni Battista Morgagni were of a different sort: currying personal favor, keeping the Italian abreast of the state of anatomical research in German institutions. Only toward the end of his career did letters cease to be crucial to the nature of Haller's science (although they remained so to its organization through his far-flung associates), being replaced then by articles in learned journals.[18]

Where Haller's network "knotted together" botanists and physicians in northern Europe, André Thouin's botanical correspondence positioned him at the Parisian center of a global botanical network whose corresponding members ranged from princes to peasants. Thouin was head gardener at the Jardin du Roi between 1764 and 1793, the year in which the King's Garden was transformed by revolutionary decree into the Musée d'Histoire Naturelle. Thouin's correspondence network, more evidently than Haller's, was one of patronage and exchange. Letters in both directions often included seeds: from overseas and provincial gardeners to Thouin, as he made the Jardin du Roi a center for national and colonial botanical knowledge, and from Thouin outward as he placed actual and potential correspondents under the ties of reciprocal obligation. Cultivating plants required cultivating the right sort of people. Nor were these only men of status: Thouin's correspondent in Toulouse paid "païsant botanophiles," effectively local field collectors, to collect plants and provide information on their uses. Such facts of local travel were repeated on the grander scales of oceanic navigation—by men like Jean-François Galaup de Lapérouse, who took one of Thouin's protégés with him on his Pacific voyages to help record and preserve the plants found. Correspondence between the collected plant, dried and pressed by the time it reached Paris, and information recorded in the field about its habitat in the wild—what we would now see as its ecology—had to be as exact as the letters that made such botanical knowledge possible and its travel successful or not.[19]

Such examples can of course be multiplied many times over. Cosmopolitan networks of correspondence and the republic they sustained extended the social and epistemic worlds of the Enlightenment beyond the botanical garden and the salon and over and above national boundaries. The Enlightenment's "traffic" in letters, its "epistolary commerce"[20]—genres of letter writing, letters' movement over space, the ways they facilitated action at a distance, the things they reveal about the strength of links within and between its communities—provides us with interconnected maps of the Enlightenment's public and private spaces.

For Alessandro Volta, the leading Italian figure in the history of electricity during the late Enlightenment and inventor of the battery, three such maps may be drawn. The first, the "expert's map," was the map Volta relied on as an electrician: "it was the map of the region of the Republic of Letters to which he felt he belonged as a natural philosopher specializing in that field." Several countries and capitals figured more prominently on this map than did others, for this was a chart of scientific preeminence, and that changed during his lifetime. Volta's second map, the "enlightened lay person's map," although broadly the same as his first, charted the well-run public administration and the commercial and cultural institutions necessary to sustain critical and useful inquiry. Volta's third map was "the civil servant's map—the map of allegiance and power." Unlike his other two, the features on this map changed dramatically after 1796, when Lombardy, Volta's home region, passed from Austrian to French rule. "Both before and after 1796 the main features, borders and capitals marked on this third map did not coincide with those of the maps Volta relied on as a natural philosopher and a citizen of Enlightenment Europe."[21]

Volta's "expert's map," with its capitals in London, midland England, and Paris, was a map of movement from Italy and reception across northern Europe in the late Enlightenment. For the Swedish astronomer Anders Celsius, by contrast, his travels to Italy in the early Enlightenment sought to learn from scholars in Rome in order to benefit Uppsala. But international travel revealed only local similarity, for in both places what was needed were fewer idle priests and more useful science.[22] And as men like Volta and Celsius moved south to north across Europe, other myriad lines of movement may be traced to illustrate the Enlightenment's cosmopolitanism.

Book Geographies: Translating and Receiving Enlightenment Knowledge

Book history is an established if relatively recent field of scholarly inquiry, even if the term and its variant, the history of the book, is misleading. What is covered by it is not just the history of books but the "social and cultural history of communication by print . . . how ideas were transmitted through print and how exposure to the printed word affected the thought and behavior of mankind during the last five hundred years."[23] A number of influential, now almost canonical texts have shown how printing changed the nature of culture and have helped pose questions regarding the history of books and printing as a social history—and a social geography.[24] By whom were books produced? For whom? Were books equally available or the preserve of a few? How did books and print culture move through society? Who were the audiences? Where were they? How were books read—silently in private, aloud to others? Robert Darnton's model of the "communications circuit" has achieved prominence as one way of addressing these issues. Darnton proposed this model as a way of tracing the cyclical life history of a book or other printed work: from author to publisher (if the bookseller does not assume that role) to printer to shipper to bookseller to reader and, since authors respond to their audiences, back to the author.[25] Darnton's work on the *Encyclopédie* has traced the urban and social geography of that work's ownership and readership throughout France—an example to which I return in chapter 8. In doing so, of course, he mapped then an urban-provincial geography of Enlightenment in stark contrast to his more recent "narrow" or "deflationary" view that the Enlightenment was an elite and philosophical movement centered upon early eighteenth-century Paris.[26] Issues of publishing history—the production part of the communications circuit—have been accompanied by work on the reception of print in different contexts and new questions concerning reviewing cultures and reading practices in different countries and social spaces.[27] Thus book history, and Darnton's formative role within it, has become a key element in explaining the Enlightenment's literary and social contours.[28]

What, then, of book geography and its variant, the geography of the book? Is it useful in thinking about the Enlightenment geographically? The term, we should remember, is not a new one. In their *L'apparition du livre*, arguably the bible of book history, Febvre and Martin considered "the geography of the book" to include printers' journeys, the locational geographies of publishing, and the diffusion of the printed word in the Slav countries, the New World, and the Far East.[29] Initial use of the term thus incorporated questions of location and of movement compatible with viewing the Enlightenment in terms of sites of knowledge making and the mobility of people, ideas, and artifacts. My use of the term here is designed to extend its application to cover the reception of books and printed media, and to emphasize the movement of texts between the sites and social spaces in which books were read, reviewed, and acted upon in the Enlightenment. The difference between book history and book geography is, I suggest, more one of approach than of substance. Where the first looks at temporal dimensions, the second addresses the spatial, to include the displacement of texts, reading, and reviewing practices in different physical and social spaces and the questions of meaning and epistemic significance that arise from such matters of geography. Humboldt's Mexican work as an Enlightenment geographer, for example, was read differently—and in consequence Humboldt as a whole was seen quite differently—by British, French, Spanish, and German reviewers.[30] My use of the term "book geographies" thus echoes Rupke's remarks on "geographies of reviewing," Secord's attention to the "geographies of reading," which charts Chambers's *Vestiges* in its different social and geographical spaces with reference to sources that highlight its private reading and not alone its public reviewing, and Secord's wider interests in the study of those communicative practices by which knowledge moves.[31] Thus understood, "book geographies" can illuminate the Enlightenment's making, movement, and reception as productive activities in different social and epistemic spaces—points returned to in chapter 8 in looking at books of geography.

Consider the central role of translation in spreading the Enlightenment. Translation as a practice reflected the transformation of the book and printing industries across Europe. Enlightenment Europe's great centers for translation were Paris, London, and, after 1760, Leipzig. Smaller centers included Amsterdam, Zurich, Hamburg, Lisbon, Naples, Edinburgh, Copenhagen, Dublin, Saint Petersburg, Stockholm, and Berlin. In Göttingen, by far the most Anglophile of German universities, professors and students there came into contact with British colleagues and English books more regularly than was the case in other German towns. It was largely through translations that Voltaire found a readership in Budapest, Adam Smith an audience in Lisbon as well as in Naples, and so on.[32] The history of translation in the Enlightenment is mainly a history of the relationship between French and English, the first the Republic of Letters' lingua franca, the second only rising to prominence later in the eighteenth century. Almost every important Enlightenment work not originally written in French was translated into it. When political economists in Naples read the works of Scots, they did so in French translation. "More than any other European culture, the German Enlightenment transformed its literary standards under the guidance of translated texts."[33]

Questions of the geography of translation thus go to the heart of the Enlightenment understood as sets and processes of cosmopolitan networks but not because translation easily equates to the reception of Enlightenment ideas. Translation does not directly ensure correspondence between one language and another or between the intentions of the writer and the needs of his audience. The David Hume, Adam Smith, Thomas Reid, Adam Ferguson, and John Millar who "left" Scotland were not, for instance, the same as those who "arrived" and were worked with in German cities. The Enlightenment prompted by these men's books did not arrive at the same time. Hume's *Treatise on Human Nature* (1740) was not translated into German until 1790–91, although his *Enquiry Concerning Human Understanding* (first published in 1748) appeared in German in 1755. Adam Smith's *Wealth of Nations* had almost no impact in its first German translation. Smith achieved belated success only after his book's second translation in 1794–76, but the cartographic contours of "Smithianismus," his reputation as a political economist, varied between different universities and anyway lagged behind those of his fellow Scot and political economist James Steuart for at least two decades.[34]

Thinkers in Germany were drawn to Scottish Enlightenment authors for their work on civil society, moral philosophy, and aesthetic theory. But translators there acted as "mediators" for the spread into Germany of a Scottish Enlightenment understood in these and other ways. In Adam Ferguson's case, the quality and timing of the translation of his university textbook, the *Institutes of Moral Philosophy,* overshadowed the poor translation, by a different translator, of his more important and original book, *An Essay on the History of Civil Society* (1767).[35] Unlike Hume and Smith, the works of the Scottish Enlightenment historian William Robertson with their themes of European national histories framed through comparative social analysis were on library borrowing-lists in Göttingen within months of their publication in London. By the time favorable reviews had appeared in German-language periodicals, German editions had already left the printing house. Whereas in Göttingen and elsewhere in Germany, Robertson's *History of Scotland* and his *History of the Reign of Charles V* were particularly praised, in Paris it was his *History of America,* in French editions, that secured his reputation.[36]

Depending where one was on the map of Enlightenment Europe, one encountered a different Voltaire. In Croatia and in Serbia, separated by religion and by language, Voltaire was "known" neither in person nor in translation, nor even by the presence of Voltaire's works in French since only a few copies of his works were held in libraries there. In those parts of eastern Europe, what was on the move first was Voltaire's reputation, his books only second, the author not at all. Perceptions of him and his writings were shaped by the intermediary role played by Italian translations of his work, by the presence of a few radically minded sympathizers, and chiefly by his opponents, Roman Catholic and Eastern Orthodox priests alike, and by the city authorities in towns like Dubrovnik. In Serbia, Voltaire's books were held in the libraries of a few, but these Serbs received Voltaire's writings and ideas through a much more diverse set of geographical intermediaries—via German translations, from booksellers in Vienna and Prussia, as smuggled copies—than was the case for his readers in Croatia and Slavonia.[37]

In Enlightenment Ireland, by contrast, few took seriously the view of their central European counterparts that "the reading of books such as Voltaire's injects hidden poison into the young people, which little by little extinguishes all the responsibilities of religion."[38] There, several leading churchmen were among subscribers to a London edition of Voltaire's epic poem *La henriade*, and the Irish took a keen interest in Voltaire's theatrical writings as well as his historical and philosophical works. Discussion about him in periodicals such as the *Dublin Magazine* (extant 1762–65) and, at greater length, in the *Hibernian Magazine* (1771–83) kept the Irish public informed about Voltaire's life and writings. In private individuals' libraries, the works of Voltaire were the most owned of the French Enlightenment writers, by Anglican and Catholic alike, although, as in England, it was Voltaire the historian and Voltaire the epic poet more than Voltaire the deist and religious critic who interested his Irish readers.[39] To judge from the ill-informed material on that country that was circulating in France, Ireland, a land of squabbling clerics and once-pagan Celts, did not interest the French in general or Voltaire in particular much at all.[40]

Such thoughts about the body of Enlightenment thought being geographically mobile and unevenly so may be extended to thoughts in the Enlightenment about the body. Encountering medical knowledge in the Enlightenment—whether in theory or in practice, as a professional or a "quack" doctor or as the patient, via peasants' beliefs, specialist text, or popular "cure-all" pamphlet—was in many ways a matter of geography.[41] Medical students moved to universities—Scots to Paris or to Leiden, for example, before the opening of Edinburgh's medical faculty in 1726; American students to London and Edinburgh; eastern Europeans to Padua, Bologna, or to one of the many medical faculties in France. On completion of their training, physicians and surgeons moved back out into society, often returning to countries and towns that would have little to offer in the way of formal medical education until the later Enlightenment. Doctors and their patients moved one to another when they could afford to do so. Medical schools and faculties and their teaching programs were far from similar. In Leiden, under Herman Boerhaave, one would have been exposed to views that depended on seeing the human body as a "hydraulic" object, for Boerhaave was much influenced by Newtonian concepts of force and pressure. In Göttingen, by contrast, under von Haller, or in Edinburgh under William Cullen, or in Montpellier with Théophile de Bordeu, a more vitalist medicine was subscribed to in which the body's health was coordinated and governed by the related well-being of each of its organs. In Halle in the early Enlightenment, the German Georg Ernst Stahl's advocacy of "animism" was based on the presumption of a God-given soul as the mover and regulator of living beings. Each of these theorists and systems had its opponents. In these and other ways, eighteenth-century scientific medicine was far from monolithic. Popular medical beliefs and customary treatments were more varied still. Then as now, where a person lived and worked, and what social stratum he or she occupied were major determinants of that person's health experience and what he or she could reasonably expect to have available as medical provision.[42]

One of the things that united the Enlightenment's professional medical thinkers and practitioners across Europe and in the early United States was their interest in the "public's health"—the importance of medical knowledge to the well-being of individuals and of nations alike. We should allow, of course, for differences between expressed intentions and actual results—in urban-rural terms, in social terms, in, for example, the acceptance and timing of inoculation against smallpox, in what medical luminaries said and wrote and what patients experienced. Yet everywhere the connection between medicine and human emancipation, between Enlightenment and bodily improvement was strong. In the works of a Swiss and a Scot, my arguments here about the translation and reception of the Enlightenment, here as a medical-humanitarian enterprise, can be illustrated through its book geographies.

Samuel-Auguste-André-David Tissot, who studied at Geneva and Montpellier and maintained an extensive medical correspondence throughout Europe, is best known for his works on improving public health, which were aimed at popular audiences. His *L'avis au peuple sur la santé*, published in Lausanne in 1761, is the most notable. Like William Buchan, author of the best-selling and much reprinted *Domestic Medicine* (1769), Tissot's emphasis was public self-education through health, social improvement via a careful physical and moral regimen. To that end, both men exemplify "the Enlightenment's goal of popularizing information to enhance public knowledge and 'know-how.'"[43] But as their books and medical-cum-moral messages moved geographically, so they had to be adapted in order to be adopted. Take Spain, for example. In Spain, Tissot's work was first published in translation in 1773 (and was available there in a further seven editions by 1795), Buchan's in six editions between 1785 and 1798. Spanish translators supplemented their editions of Tissot with "local" remedies and preparations in vulgar Spanish as well as with notes on how to deal with scorpion stings and snakebites, given their prevalence in Spain. To the Spanish version of Buchan's *Domestic Medicine* the translator added sections designed among other things to "adapt the medical precepts to the particular climate and way of life of this Kingdom" (fig. 4). Geographical facts required textual modification. Whereas in Spain, Tissot's and Buchan's books were modified to fit the local geography—and the political climate of the Inquisition—in Hungary they reinforced already established concerns about medical education as a basis for social reform. In Britain and in America, Buchan was read differently still, for there his message of enlightened populism was thought to threaten the medical establishment.[44]

The movement of Enlightenment ideas was thus a far from smooth passage from letter to book, from writer to private reader to public audience, from printed word to talked-about review, from advisory text in one context to actual practice in all. What is revealed is not just the importance of translation as processes of mediation between Enlightenment thinkers, their books, and their audience, but also the "untransferability" of Enlightenment ideas. Ideas in one language or context were not always readily understood in another. Meanings were transformed as they moved. Where you were as well as who you were mattered in terms of when and how you got to know of others' work. The Enlightenment's cosmopolitan networks facilitated the spread of Voltaire's ideas, of course, just as they did those of Newton, Hume, Smith, Tissot, and others. But these are networks of "misreception" too, of the uneven "appropriation" of Enlightenment ideas in different geographical contexts and social spaces.[45]

Considering the Enlightenment's "traffic" to be socially mediated and geographically grounded means asking questions concerning which social *and* which geographical spaces are important in relation to what the Enlightenment was and how it moved. Where did networks begin and end?—in private libraries, physician's studies, public journals, conversations in a salon, as letters evading the censor's gaze, in gardens and public parks

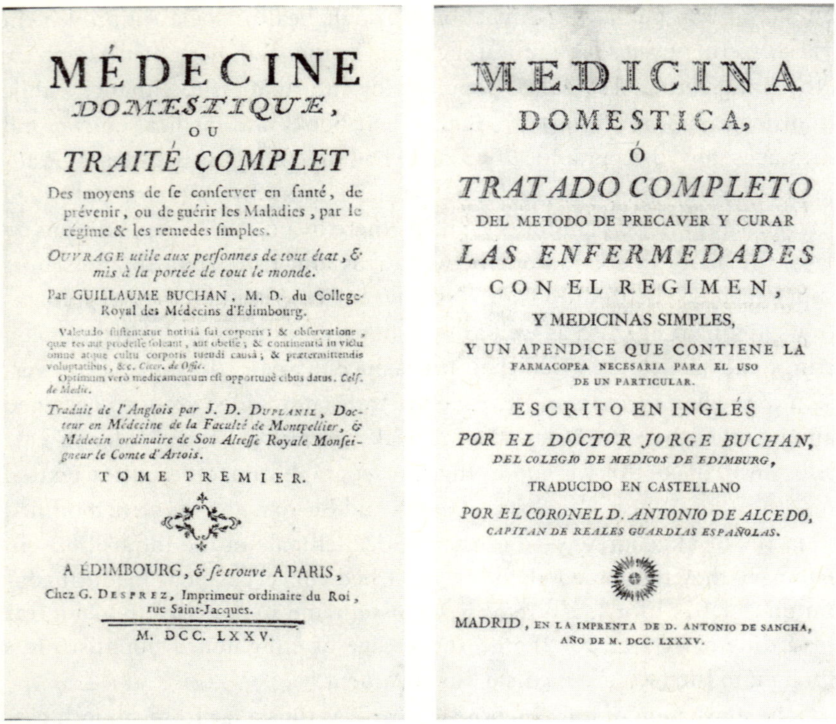

MÉDECINE
DOMESTIQUE,
OU
TRAITÉ COMPLET
Des moyens de se conserver en santé, de prévenir, ou de guérir les Maladies, par le régime & les remedes simples.
Ouvrage utile aux personnes de tout état, & mis à la portée de tout le monde.
Par GUILLAUME BUCHAN, M. D. du College Royal des Médecins d'Edimbourg.
Valetudo sustentatur notitiâ sui corporis; & observatione, quæ res aut prodesse soleant, aut obesse; & continentiâ in victu omne atque cultu corporis tuendi causâ; & prætermittendis voluptatibus, &c. *Cicer. de Offic.*
Optimum verò medicamentum est opportunè cibus datus. *Cels. de Medic.*
Traduit de l'Anglois par J. D. Duplanil, Docteur en Médecine de la Faculté de Montpellier, & Médecin ordinaire de Son Altesse Royale Monseigneur le Comte d'Artois.
TOME PREMIER.
A ÉDIMBOURG, & *se trouve* A PARIS,
Chez G. Desprez, Imprimeur ordinaire du Roi, rue Saint-Jacques.
M. DCC. LXXV.

MEDICINA
DOMESTICA,
Ó
TRATADO COMPLETO
DEL METODO DE PRECAVER Y CURAR
LAS ENFERMEDADES
CON EL REGIMEN,
Y MEDICINAS SIMPLES,
Y UN APENDICE QUE CONTIENE LA FARMACOPEA NECESARIA PARA EL USO DE UN PARTICULAR.
ESCRITO EN INGLÉS
POR EL DOCTOR JORGE BUCHAN,
DEL COLEGIO DE MEDICOS DE EDIMBURG,
TRADUCIDO EN CASTELLANO
POR EL CORONEL D. ANTONIO DE ALCEDO,
CAPITAN DE REALES GUARDIAS ESPAÑOLAS.
MADRID, EN LA IMPRENTA DE D. ANTONIO DE SANCHA, AÑO DE M. DCC. LXXXV.

FIGURE 4

Translating the Enlightenment. Frontispiece plates from the French and Spanish editions of William Buchan's *Domestic Medicine* (1769). Unlike the Spanish edition, neither this French edition nor an earlier Dutch edition (not shown) changed the text to reflect the facts of different geographies. By permission of the Trustees of the National Library of Scotland.

even as people gathered to hear spoken the news they could not read?[46] Notions of *the* Enlightenment as an essential Parisian and philosophical thing become further eroded in light of these networks and what they disclose about the making and mobility of Enlightenment ideas. Greater geographical and classificatory accuracy becomes necessary. For instead of the Enlightenment being unproblematically made and as unproblematically received, we can see the Enlightenment republic in greater thematic detail in terms of political economy, botany, moral philosophy, and Newtonian physics and consider it geographically qualified: political economy in Scotland, Edinburgh, Naples, Lisbon; botany in Uppsala, Göttingen, Zurich; and so on. The Enlightenment's cosmopolitanism moved across national boundaries and between languages. In the Greek or neo-Hellenic Enlightenment, characterized as we have seen by Greek scholars using the works of others in local context, Newtonian science was encountered not via Newton's works but through Italian editions of 1751 and 1752 of works by the Dutch Newton scholar Peter van Musschenbroek.[47] At the same time, Italian earth science, in translation, was traveling north across the Alps and being modified in the process, even as it remained the preserve of a few.[48] Publishers, audiences, and the circulation of knowledge, not just "great" authors and their ideas, become important in thinking about the Enlightenment as dynamic and mobile. As Charles Elliot, the Edinburgh-based bookseller who controlled an international network of authors and agents from his Parliament Square bookshop noted of himself, "I am the principal Man Midwife (in the literary sense) here, to Man Midwives, Physicians, Apothecaries, &c &c &c."[49] And as books and letters embodying the Enlightenment traveled, so did other things.

Artifacts and Instruments: Collecting and Displaying the Enlightenment

In his Edinburgh University rooms only a short walk from Elliot's bookshop, the Rev. Dr. John Walker, professor of natural history between 1782 and 1803, gathered his teaching collections from across the globe. To his students, Walker's lectures emphasized the utility of natural knowledge. To his network of acquaintances, his cabinets held the world in ordered miniature. Shells came from the conchologist Emmanuel Mendes da Costa—whose own reputation was then being secured through travel, collecting, and correspondence—Siberian animals via Simon Peter Pallas, mineralogical specimens from Spain courtesy of the British ambassador in Madrid, Brazilian plants from one of his students.[50] As for Thouin in Paris, Haller in Göttingen, and many others, Walker's Enlightenment was a constant process of trafficking in the world's productions—of declaiming, requesting, receiving, and ordering—and doing so for given audiences. Collecting, dispatching, and displaying—whether of seeds, shells, or other specimens—bound natural philosophers into networks on the basis of social status and mutual exchange.

For Sir Joseph Banks, new knowledge of the world's diversity was largely accomplished not by being mobile but by staying still to gather in the results of others' endeavors and by helping yet different others witness new things for themselves. In opening up his house in London's Soho Square as public display space, in managing the Royal Botanic Garden at Kew—from whence he dispatched "plant hunters" to Britain's colonies—and in coordinating Britain's overseas interests from his Whitehall offices, Banks was a sort of individual "center of calculation," for the botanical networks of Enlightenment Britain anyway.[51] To Banks, botanical knowledge carried national significance, was an emblem of Enlightenment and empire, and had economic benefit as well. With it, the king and his ministers could make "acceptable presents to crowned heads." Banks coordinated—and even wore—the trappings of natural knowledge, but to some of his peers and for some modern scholars he was no scientist. This "meer toad eater to the King," as one leading contemporary put it was, for one modern commentator, at the center of a web of "territorial and agricultural espionage," ringmaster to blundering and plundering agents of empire whose primitive empiricism on the margins did not translate into good science at the center.[52]

Power over space is thus always predicated upon power over people. First and foremost, the Enlightenment's cosmopolitan networks were social things that made knowledge, in whatever form, move in and over geographical space. Correspondence as a written social practice did not always result in epistemic correspondence between writer and reader. Mobility allowed different meanings to accrue. This happened even when the artifacts moving in the Enlightenment's traffic were themselves human. The Ulaietan islander Mai, known as Omai, who was brought back to England by one of Cook's captains in 1774, began his travels as a political refugee. On the voyage he became an able seaman whose shipmates knew him as Jack. Upon arrival in England, he was received as a human artifact invested with significance as a symbol of distant places and an embodiment of opposing Enlightenment theories concerning human nature. Traveling changed him. To his audiences, he was an unknown. Was he cannibal or gentleman, brute primitive or noble savage? Was he capable of refinement or was he innately civilized, a symbol of "natural" good manners? Like the Tahitians Tupai before him and Aoutourou, who had returned with Bougainville, Omai carried within himself notions of social rank and significance and knew that such things mattered in England. Yet how he was "understood" locally depended on the social spaces and places through which he moved during his reception—on botanizing trips with Banks, meeting King George III, at society dinners, as a pictured symbol of human nobility and individual enlightenment which made him at once domesticated yet exotic (see plate 6).[53]

My point, that seeing the Enlightenment's cosmopolitanism as networks demands attentiveness to the complex connections between knowledge's production, movement, and reception and not their separation and that these connections have geographical constitution and expression, holds for the nonhuman instrument as well as for the human agent. Consider the case of Volta's "invention." At first no one knew what to call what we now know as the battery, its inventor least of all. In announcing the invention in London, but in French, in April 1799, Volta proposed two names—the "organe électrique artificiel" and the "appareil électro-moteur." By 1800, others in London were speaking of the "electric pile," in Paris and Geneva of the "l'appareil électrique ou galvanique," and in Germany of "Volta's column." Different interpretations were put on the device, and what one could do with it, in London, Lombardy, Paris, Jena, and Copenhagen, mainly in consequence of the fact that the Voltaic battery—as it was generally known after 1801—was easy to replicate. Volta's "expert's map," partly a matter of Volta's own mobility and reception across Europe, was marked out by erratic transmission lines of news about his invention: from Como and London in March and April 1799, word reached Vienna before it reached Paris, Halle before Glasgow, Copenhagen before Geneva.

Given this different appropriation of the device over time and space, how should we understand "the geography of the new continent opened up by the invention of the battery"? For Pancaldi, the answer lies in several things. Circles of people, experts and lay, and located in different places, shared Enlightenment notions concerning natural philosophy and "useful knowledge." Through the press, through correspondence, and through personal travel, "members of those circles constituted loose but effective networks, . . . often crossing Europe's national borders, interested in assessing innovation in natural philosophy and new instruments like the battery." Within such networks, instruments like the battery were subject to "repeated and varied assessments, in the course of which frequent permutations of . . . Enlightenment notions and their ranking took place, leading to widely different interpretations of what was nonetheless regarded (and named) as the same basic instrument." Yet different geographies of an instrumental Enlightenment underlay Volta's "maps" of mobility and personal reception: a European network of shared interests, news of the invention and the contingencies of local testing, the movement of his electrical instrument and its different placed interpretations. What was in London seen as a chemical machine with industrial applications was in Paris taken as a device of mathematical physics and in Copenhagen as an icon of Romantic natural philosophy.[54]

For Benjamin Franklin, who many times crossed the Atlantic, mobility as a statesman-scientist within the Enlightenment's cosmopolitan networks afforded an opportunity for a yet different form of instrumental experimentation. In 1769, officials in Boston had drawn to Franklin's attention the fact that the Falmouth–New York packets, ships used for mail rather more than passengers and goods, took generally a fortnight longer in their passages than did merchant ships from London to Rhode Island. Written years later to a Parisian chemist friend, and after his own shipboard measurements of the Gulf Stream using a thermometer, part of Franklin's 1785 letter illustrates my more general claims about seeing the Enlightenment geographically as a matter of networks.

> It appearing strange to me that there should be such a difference between two places, scarce a day's run asunder, especially when the merchant ships are generally deeper laden, and more weakly managed than the packets, and had from London the whole length of the river and channel to run before they left the land of England, while the packets had only to go from Falmouth, I could not but think the fact misunderstood or misrepresented. There happened then to be in London, a Nantucket sea-captain of my acquaintance, to whom I communicated the affair. He told me he believed the fact might be true; but the difference was owing to this, that the Rhode-Island captains were acquainted with the gulf stream, which those of the English packets were not. We are well acquainted with that stream, says he, because in our pursuit of whales, which keep near the tides of it, but are not to be met with in it, we run down along the sides, and frequently cross it to change our side: and in crossing have sometimes met and spoke with those packets, who were in the middle of it, and stemming it. We have informed them that they were stemming a current, that was against them to the value of three miles an hour; and advised them to cross it and get out of it; but they were too wise to be counselled by simple American fishermen. When the winds are but light, he added, they are carried back by the current more than they are forwarded by the wind: and if the wind be good, the subtraction of 70 miles a day from their course is of some importance. I then observed that it was a pity no notice was taken of this current upon the charts, and requested him to mark it out for me, which he readily complied with, adding directions for avoiding it in sailing from Europe to North-America. I procured it to be engraved by order from the general post-office, on the old chart of the Atlantic, at Mount and Page's, Tower-hill; and copies were sent down to Falmouth for the captains of the packets, who slighted it however; but it is since printed in France, of which edition I hereto annex a copy.[55]

My point is a simple one. What may matter more than claiming that Enlightenment knowledge moved above and beyond the nation is to show how, where, and in what form it moved—through personal contact and correspondence, in books and in articles in learned journals, in conversations in taverns, and through the mobility of instruments whose results could be used to make local claims travel. For Franklin, the question "how and where was oceanographic and navigational knowledge made in the Enlightenment?" would allow different answers: at sea, as ships crossed the ocean and one another's path, even if English captains refused to acknowledge the experiential knowledge of their American counterparts whose navigational practices were based on natural observation—whale watching; on a map, produced in a London tavern as a result of the verbal testimony of a Nantucket sailor (and allowing for the fact that the map in question was later slighted by its intended audience); in Paris as a letter received; and in Philadelphia as an article read by fellow members opening their *Transactions*.

Terms such as "Republic of Letters" and "network" as a social, intellectual, and geographical conception thus help describe the Enlightenment's traffic. Terms like "translation," "mediation," "reception," and "appropriation" may help better explain it. As others have indicated, the Enlightenment in this sense emerges as a period of intellectual mobility when "a general transformation of different kinds of knowledge into different forms of knowledge took place, of the world of knowledge into society and sites of power, and vice versa."[56] In developing my claims that the Enlightenment geographically understood did not "float free" and that what was understood by its knowledge was made, moved, and received differently, it is appropriate to turn to the variety of locales, institutions, and spatial settings in which the Enlightenment was grounded.

4

Doing Enlightenment
Local Sites and Social Spaces

Even as their revolution raged, some citizens of the new French Republic found time—and space—for late Enlightenment. For on the morning of 10 November 1793, or 20 Brumaire year II in the Revolutionary calendar, Parisians flocked to Notre Dame to witness the Festival of Reason. Actresses from the Paris Opéra paraded as the goddesses of Reason and of Liberty. In the choir of the church the assembled crowd erected an artificial mountain on whose top stood a small temple inscribed "To Liberty." Elsewhere in Paris that day, others were constructing different monuments to the power of reason. Nearby, in what would become the Bureau de Longitudes, surveyors and mathematicians were poring over statistics collected as part of a 1791 plan for a new cadastral map of France (see fig. 3).[1]

For Parisian actresses and their watching crowds as for revolutionary engineers in the bureaucratic quiet of their calculating spaces, the Enlightenment was not a disembodied suite of philosophical dispositions, something that "floated free" above the streets and working places of everyday life. The Enlightenment was something that could be celebrated, performed, inscribed, something that could be put to work—in the street, in a church choir, and in mapping projects that aimed to shape a new type of nation altogether. For men and women and in different ways, Enlightenment was a lived and situated experience, made and encountered in different local places and social spaces.

The importance of local sites and social spaces—of local sites *as* social spaces—has been recognized in other contexts. Science has been "put in its place" by reference to questions of region, circulation, and notably, site—experimental laboratories, cabinets of natural history, museums, botanical gardens, expeditionary tents in the field, hospital wards, even public houses, to name only a few.[2] Modernity too had its local geographies. In eighteenth-century London, for instance, what was held to be modernity was differently constituted between and within five sites in the capital. It meant different things for the prostitutes and reformers of the Magdalen Hospital, for agitated politicians in Westminster's streets, strutting macaronis on display in Vauxhall Gardens, and anxious clerks in the Excise Office and the Universal Register Office.[3] Even the regulation and synchronization of modern universal "standard time" are accounts of certain relative spaces.[4] In combination, this and other such work has established the "premisses of premises," namely, that local settings matter in the conduct and explanation of social life.[5]

But in what sense is "local" being used? In these and other studies, "local" is used not simply in relation to locality, the geographical placing of given institutions and personnel—and, in that sense, as a matter of location and of scale. It is also being used to embrace questions of social and intellectual constitution. Local places are social spaces by virtue of the practices and relationships undertaken there. So interpreted, understanding the Enlightenment in terms of its local settings means looking at matters of practice made *in* place, accounting for social and intellectual configurations that made work what people in different places and sites held Enlightenment to be. This is not to see the Enlightenment's local sites and social practices as separate from its transnational geographies, somehow disconnected from the cosmopolitan dynamics of letters, people, and ideas on the move. Neither does thinking about the Enlightenment in terms of local sites, practices, and social spaces mean that we should lose further sight of its national dimensions. Many local institutions in the Enlightenment were concerned with practical and patriotic outcomes, with improving the nation in question in terms of particular pursuits—agriculture, chemistry, language, belles lettres, fine art, the people's "happiness." To follow from the previous chapter, thinking about the Enlightenment locally is to see the Enlightenment's international Republic of Letters, its networks, as "grounded," made in place in contexts in which textual and social practices came together with, say, particular patrons, because of shared intellectual interests or from a focus on one form of knowledge more than another. In short, what others have seen as the Enlightenment's "idioms," its "communities of discourse," can be seen as "idioms" and "communities of practice," as locatable sites of endeavor in and from which people corresponded, experimented, and so on.[6]

The Enlightenment Locally: Sites of Practice

My focus is not just on local sites and practices but on the connections between the "local" and the "distributed" as Harris has it, to reveal something of the relationships between ideas and practice by examining various Enlightenment sites and social spaces, where they were and how they operated. Such a concern with the Enlightenment's cosmopolitan making at local scales is a distinguishing feature in the reconsideration, the reconceptualization even, of the Enlightenment in at least one geographical context—the Netherlands.

What Schama saw in the Dutch Enlightenment as "an unpromising subject," others have since considered more fully. For Jacob and Mijnhardt, "the Enlightenment began in England and the Dutch Republic because in both places by the 1690s the presses were freer than anywhere else."[7] The relationship between this early Enlightenment and the Dutch Republic's later intellectual decline has been consistently explained in terms of the Enlightenment's cosmopolitan and national geographies of ideas. "Decline and Enlightenment together produced a new national consciousness, directed to the past, which fortified the intellectual traditions that had already decisively influenced the reception of early Enlightenment ideas." Because in the Dutch Enlightenment, the recovery of enlightened civic virtue was virtually impossible without the support of enlightened religion, intellectuals there were less receptive to other versions of enlightened discourse. Thus, "the ideas of the radical French Enlightenment that were gaining currency after 1750 incited fierce opposition in the Netherlands." Further, "the ensuing intellectual isolation of the republic within the European Enlightenment was only strengthened by the loss of its position as one of the centers of the European republic of letters. After 1750, the Dutch Republic served less often as a refuge for intellectuals driven from their own countries, not because of changes within the intellectual climate of the Republic itself but as the result of greater tolerance elsewhere."[8]

In this interpretation, the Dutch Enlightenment is to be explained by its own innate features—its relatively early occurrence, its tolerant press, its reputation as a haven for foreign intellectuals, and its receptivity to Newtonianism and to other new ideas. *And* it is to be understood in relation to the Enlightenment beyond itself—the radical French, the later decline in foreign students in Dutch universities, the Netherlands' "reduced position as a European clearing house for French books and journals."[9] In such an interpretative view, notes Lissa Roberts, "the Dutch were significant precisely because and so long as they linked English and Continental centers of enlightenment by educating a generation of Newtonian do-gooders and publishing works that others dared not touch. To be Dutch according to this scenario meant not so much to engage in a complex set of culturally distinct practices, but to serve as purveyors of foreign-born knowledge, be it for the sake of truth or profit. Throughout, the emphasis is on ideas."[10]

Rather than treat the Enlightenment as a corpus of self-contained ideas circulating without reference to their local sites of social making, it is possible to consider it as situated and multiple practices, concerned in dif-

FIGURE 5
An Enlightenment anatomy theater. Note how members of the audience converse as they are instructed by the dead. Courtesy of the National Library of Medicine, Bethesda, Maryland.

ferent places and in different ways with different conceptions of practical reason, natural philosophy, and so on. This is Lissa Roberts's view of the Dutch Enlightenment in looking at four sites—anatomy theaters, the Leiden physics theater, private lecture courses, and scientific societies. In each, different audiences and particular expressions of what "enlightenment" meant are identified, and certain intellectual practices of demonstration and social practices of display are disclosed. In the anatomy theater, Enlightenment was a lived experience using the dead for entertainment and for moral instruction about the "body politic" as well as for medical advancement (fig. 5). Like Dutch anatomy theaters, the university physics theater was a site for demonstration and display, enlightenment there a process of advancing science through professional practice and audience participation. For those who did not get to the theater, dead or alive, traveling science lecturers such as John Theophilius Desaguliers made enlightenment public through experimental demonstrations—a mixture of Newtonian mechanics, lecturing, and personal showmanship. What connected all such sites and practices was an emphasis on utility, what Enlightenment could do.[11]

Such an approach has several implications, for the Dutch Enlightenment and more widely. The Dutch Enlightenment is revealed not simply as a matter of comparative intellectual history, nor even of cultural or social history. It can be shown to be a historical geography—of institutional and social sites, audiences in place and on the move, meanings of Enlightenment made in place. Particular themes in the Enlightenment—different practices *of* Enlightenment—can be more clearly discerned: public anatomy teaching, chemical improvement via experimenting and publishing, lecturing in astronomy, performing Newtonian principles via demonstration. Questions about what the Enlightenment was become questions about how it worked, where, and for whom. For Roberts, siting the practices of science and medicine offers insight into the local doing of Enlightenment in the Netherlands and how it was understood differently by practitioners and audiences because of its different settings. Institutional and locational differences in practice mattered. In learned scientific societies, for example, whose members emphasized the practical benefits of enlightened knowledge to counter Dutch economic decline, new ideas were debated, experiments conducted, and competing propositions about national utility aired.[12] Where you were in Holland and who you were mattered in regard to what you took Enlightenment to be.

Understanding the Enlightenment to have such local expression and constitution—in a Leiden physics theater, a learned body in The Hague, or in a small-town agrarian society—is consistent with the claims of those who have referred to "multiple Enlightenments" among different social groups and to the multiplicity of local social contexts. For Vartanian, "what is required is a taxonomy, along with a comparative anatomy, of Enlightenments."[13] Golinski proposed "the analysis of the Enlightenment as a system of cultural meanings," an analysis that might "read back from local manifestations of the Enlightenment to the wider cultural formations which made them possible."[14] Others have recognized how a "new geographical consciousness of locality and distance has been seen to have emerged in the eighteenth century as an integral part of the experience of enlightenment itself."[15]

Thinking locally about the Enlightenment is not to separate that scale of geographical understanding from the national and the inter- or transnational but to consider how its practical articulation in diverse sites and social spaces connected—or did not—with the Enlightenment's making and movement elsewhere. Questions of taxonomy and comparative anatomy in regard to the Enlightenment and its local meanings become, in effect, ones of geography. Where exactly did people do Enlightenment? In what sort of institutional location or other site and social space? What does the map of institutions promoting science and other forms of useful knowledge in the Enlightenment look like? If our emphases are toward "the local" and toward multiplicity, to what extent does it make sense to think of *the* Enlightenment as a shared public enterprise?

Ideas concerning the public production and consumption of knowledge in the Enlightenment owe much to Habermas's *Structural Transformation of the Public Sphere* (1962). Habermas explained the rise of democratic polity within eighteenth-century western European urban society with reference to the rise of a literate and bourgeois politicized public sphere. Its members bought and read books and newspapers, socialized one with another, established institutions and journals of polite learning, and, above all, promoted reasoned argument and tolerated criticism in the public domain. The men and women of the Enlightenment Republic of Letters wrote letters, published pamphlets, periodicals, and books, and engaged in polite conversation. They especially came together through public discourse in particular spaces and places—coffeehouses, circulating libraries, anatomy and other theaters, concert halls, reading clubs, public gardens, and scientific societies. Habermas's conceptualization of the public sphere in these ways has been the subject of considerable attention, not all of it favorable. His relative neglect of women, of differences in social rank, and his partial treatment of science as a form of public culture have each been noted in this respect.[16] Even so, the utility of his model is widely recognized. The public sphere of the eighteenth century set one of the preconditions for the emergence of modern democratic society in the West. "The heuristic of the public sphere undergirded by civil society enriches our vision of the Enlightenment, expanding it from the family of a dozen or so philosophes and a series of canonical texts."[17]

With this enrichment comes the challenge of describing and explaining enlightened cultural practices, philosophical and textual, nonphilosophical and nontextual. Part of this challenge includes knowing the local discursive dimensions of these practices as well as their wider purchase. To illustrate these issues, I consider two sets of sites and practices: experimentation and instruction in scientific academies and societies; and conversation and demonstration in coffeehouses and salons. This is not to see these sites as socially separate, nor is it to see the practices in them as peculiar to them. They are, simply, examples of what it means to think locally about the Enlightenment.

Improving Spaces: Learned Academies and Scientific Societies

Between the foundation in 1660 of the Royal Society of London and of the Académie Royale des Sciences in Paris in 1665 and the early 1790s, about seventy official academies and societies of science were in existence. This Age of the Academies, as one contemporary put it, permits several observations to be made concerning a modern taxonomy for these local sites of and for Enlightenment.

Official academies and societies had corporate status in a legal charter promoted by one civil authority or another. They operated under written rules. For societies, following the London model, the state offered members no remuneration and did not, in principle anyway, interfere with that body's business. In academies, the Paris model, the institution was normally state-sponsored, its members paid, the agenda established in part by government. Official academies and scientific societies had their private counterparts, different by virtue of their nonofficial status rather than by any lesser engagement with science. Both were paralleled by a variety of "Renaissance academies": some were carryovers from an earlier institutionalization of natural and cultural inquiry, others the particular product of eighteenth-century patrician patronage. Scientific societies of one sort or another were only part of a wider universe of learned societies in the Enlightenment—bodies for agriculture, the antiquities, literature and belles lettres, medicine, painting and the fine arts, philosophy, and so on. Some scientific bodies did only that, others no science at all. Many bodies had a mixed portfolio. Yet each saw its activities as promoting useful knowledge through corporate activity. In one guise or another, the learned society or official academy was the defining feature of the institutionalized public sphere in Enlightenment Europe, a key site in which the Enlightenment was locally fashioned and from which claims issued to the practical benefits of useful knowledge.[18]

The scientific societies established between 1660 and about 1800 were international in geography but not so alike in character (fig. 6). The seventy or so bodies in existence by about 1789 were found across Europe, from Lisbon and the Academia Real des Ciências de Lisboa (founded in 1779), to Saint Petersburg's Academia Scientiarum Imperialis (1725), and from Naples and its Reale Accademia delle Science e Belle-Lettere (1778) to Trondheim's Kongelige Norske Videnskabers Selskab (1760). A few bodies were established further afield. Most institutions in North America were concerned with promoting useful knowledge. Others in Asia reflected the colonial interests of European nations. This is so of the Bataviaasch Gnootscap van Kunsten et Wetenschappen in Batavia (modern-day Jakarta), for example, begun in 1778 through the support of the governor general of the East Indies and the Dutch East India Company and which focused on science and the economy, and the Brazilian Academia Cientifica do Rio de Janeiro, which ran from 1772 to 1775 with support from the local viceroy.

What we may term the outline map of scientific societies (fig. 6) illustrates one set of local sites sustaining the Enlightenment as a cosmopolitan network.[19] Contemporaries certainly saw it so. For one academician in Berlin, "literary and learned Europe makes up . . . only one single society, united

FIGURE 6

Enlightenment academies and societies. This "outline" location map of institutions is adapted and expanded from James McClellan, *Science Reorganized* (New York, 1985), and David Goodman and Colin Russell, *Rise of Scientific Europe* (London, 1991).

by a common goal, which is the progress of the sciences and letters." As a Dijon academician put it in 1763, such bodies were "colonies in the Republic of Letters."[20] But not all such colonies were everywhere the same. For McClellan, academies and societies were functionally similar "but characteristic of two different cultural spheres: the society form is typical of maritime, Protestant, relatively more democratic Europe; the academy form is typical of Continental, Catholic, and relatively more authoritarian regimes."[21] France was dominated by provincial academies, the majority of which were multidisciplinary foundations established in the early eighteenth century. Many acted to promote local economic interests or specialized in one science or another—Dijon in anatomy and chemistry, Montpellier in mathematics—but such emphases and proclamations of utility did not spare them or their Parisian counterparts from being dissolved by revolutionary decree in 1793.[22] In Germany, two models existed: the royal court academies, such as in Berlin and in Mannheim, and the more research-active university academies, beginning with and based in Göttingen, the country's leading Enlightenment school. The beginning of institutionalized enlightenment in Russia, with the foundation by Peter the Great in 1725 of the Imperial Academy in Saint Petersburg, was a matter of intellectual import. Its foundation mirrored German academies. Its academicians at foundation were from elsewhere in Europe, its first students mostly foreigners. This academy's establishment was one example of "opportunistic appropriation" from a variety of "Wests," a fact supporting the view that the Russian Enlightenment was not a nationwide engagement with philosophical preoccupations but was more a matter of social change among a metropolitan few.[23]

If the Republic of Science had an institutional headquarters in the Enlightenment, it was the Académie Royale des Sciences in Paris (fig. 7). The "anatomy" of this institution, the subject of considerable study, is not my concern here. Elsewhere, in chapters 5 and 10, I consider its role as a center for the geographies of the Enlightenment beyond Paris and look at the work of several of its leading geographers. Here it is sufficient to note that, like London's Royal Society from the 1680s, this local site sought to establish universal codes of practice for how secure knowledge was to be made. Particularly after its internal reform in 1699, the Académie established itself as the leading arbiter of scientific discourse in the Enlightenment. Words were replaced by deeds. Unexamined claims of ancient authority were abandoned in favor of direct experimentation. Individualism, especially where accompanied by personal ostentation, was replaced by collectively agreed codes of intellectual conduct. Certain judicative norms about what was sound knowledge and what was conjecture were established: in the language of reporting, the repeatability of experiments, personal observation over hearsay, the moral standing of one's informant, in the social bearing of the reliable "scientist." In thus connecting utility in science to the needs of the French state, and in turning *honnêtes hommes* into nascent professionals, the Académie more than any other body shaped its working practices and the spaces for their conduct in terms at once moral, social, intellectual, and political.[24]

What lends these observations additional significance is the fact that it was not possible to seed Parisian practices in foreign soils. Parisian academic Enlightenment did not readily move. Trying to transplant French ideas and intellectual ideals into Berlin in his role as president of the Berlin Academy of Sciences, the French mathematician Pierre-Louis Moreau de Maupertuis (see plate 5) failed in the face of Prussian politics. Where in Paris tolerance, reason, and appeals to the value of useful knowledge fuelled Enlightenment

FIGURE 7

The Paris Academy of Sciences, from Charles-Marie de La Condamine, *Journal du voyage* (Paris, 1751). The Paris body is here depicted as the mother of Enlightenment invention; note the child holding and inquisitively turning a model of the globe. By permission of Edinburgh University Library, Special Collections.

challenges to traditional authority, in the Berlin of Frederick II the same principles bolstered Prussian autocracy. What Maupertuis intended as "the freest academy in Europe" was never so: French ideals were lost in translation, and in dependence upon Prussian officials, scientific criteria lost out to royal will. At the same time, in his insistence on a view of metaphysics based on the principle of least action and not on matters of motion and equilibrium, Maupertuis distanced himself from his French counterpart Jean d'Alembert. Where, in Berlin, God regulated physical motions by endowing matter with forces, in Paris individual bodies moved independently: forces were simply not admissible in the same way as they were in Berlin. "What made the principle of least action foundational in Berlin, namely its reference to final causes, made it anathema in Paris, where metaphysics interfered with analysis rather than enhancing it."[25] Looked at locally, two contrasting views of God, Nature, and human agency are revealed in these twin capitals of enlightenment.

Anatomizing scientific societies and their practitioners' affiliations in such ways moves us away from a view of the Enlightenment as an original and unchanging specific system of values—and not just toward the "doing" of Enlightenment locally as a set of situated practices. It also requires that we recognize how "living" the Enlightenment in any one place was a form of local cultural production and consumption in which social worlds overlapped, and different discourses and practices, languages and values, contexts and representations interacted and influenced one another. It is just such an interpretation of the Enlightenment as multiple and contested, of course, that has underlain its reinterpretation as a social history of ideas. But it is also a social geography, for when we locate these ideas and experiences we better see how people put them to use in different contexts. When we look at the circumstances within given settings—in academic societies, universities, in occupational groupings—we see how social worlds interacted *in place.* This may, in turn, necessitate a particular emphasis on Enlightenment research not in terms of differences between places—Berlin and Paris, Padua and Saint Petersburg—but in terms of the different social worlds within one place and what the "extralocal" relationships were between public sociability and private inquiry.[26]

Understanding the importance of place in these ways should not be confined to the "metropolitan" spaces of capital cities or just to scientific ones. If we follow Roberts's lead further still and look at other sorts of institutional and social spaces and at their defining practices, a different sense of Enlightenment emerges. Enlightenment Britain's many voluntary clubs and societies each constituted local associational worlds in which mutual civic interests were played out, with advances in science often taking a back seat to moral reform, reason second place to popular sociability.[27] Enlightenment audiences were constituted in concert halls, reading rooms, and *cabinets de lecture,* just as much as they were in academies and salons. Far from "diluting" an essential Enlightenment, a fuller picture of the "lived" Enlightenment emerges if we look at "provincial" towns, rural society, and local "family practices."

In midland England, James Watt and his associates in the Lunar Society of Birmingham long looked to France for enlightenment, chemical and otherwise. The French looked to Birmingham and to the Soho Manufactory (the world's largest factory in the 1760s) as a model of the practical application of science in the state's interest. But models of institutional and academic specialization in France did not sit well with the lay model of scientific endeavor promoted in the West Midlands. In the face of revolutionary politics in France and changes in the cultural climate in midland England, what brought a "family of *philosophes*" together across geographical boundaries broke up in consequence of local circumstance. What had been celebrated by actresses and audiences during the Revolution—and by Watt, Boulton, Priestley, and others—was by 1795 not traveling at all: "Philosophical news I have none of any consequence. These cursed French have murdered Philosophy & continue to torment all of Europe."[28]

Just as Paris was not France, so Birmingham was not midland England. In Derby, natural philosophy and industrialization were likewise motivating partners in local conceptions of Enlightenment. Public lecture classes in Newtonianism and in practical mathematics helped further establish a culture of science motivated both by concerns for a local utility—in the silk mills, and in the technology of water supply—and by talks from itinerant lecturers. Many of those like James Ferguson, Desaguliers, Benjamin Martin, and others who made a living from Enlightenment by being on the move saw in small-town life better opportunities to disseminate Enlightenment via public performance than in the competitive markets of London.[29] Scientific culture in the Enlightenment throughout provincial Britain was based on similar elements—a culture of utility locally expressed, particular associational interests, dissent both religious and political—and took place in sites such as subscription libraries and academies in front of local audiences supplemented at moments by itinerant lecturers. Although the mix differed by place, enlightenment and utility often came in machine form, progressive knowledge through mechanical demonstration—power through the circulation of machine and knowledge alike.[30] So too it came, if in different ways, for the "village Enlightenment" in rural America. There the formation of a market for cultural commodities in the early republic involved the hand-to-hand transmission of books and pamphlets in order to promote self-learning and to democratize knowledge. Enlightenment was promoted not by engagement with the works of a philosophical few that were written, bought, and discussed in metropolitan centers of calculation, but through reading in the home and village school and in conversations around the water pump. In the face of such differences and complexities in site and in practice, the idea of multiple local Enlightenments is more compelling than any single and unitary model.[31]

In the many small and often short-lived economic societies of mainland Europe, enlightenment was synonymous with improvement, improvement with a vision of future geographies. In later eighteenth-century Tuscany, enlightenment was about science in the form of agriculture and practical chemistry, its object to advance the region's political economy. Academicians in the Accademia di Agricoltura in Pistoia, in the Accademia dei Fisiocritici in Siena, and in Florence's Accademia dei Georgofili worked "on the most solid foundations of experiment" to improve soils and cropping practices, extend cultivable margins, and instill a progressive spirit among landowners and land workers alike.[32] In Spain and its overseas colonies, the institutionalization of Enlightenment on foreign models came later and included sixty-eight Economic Societies of Friends of the Country between about 1775 and 1791. The Basque Society of Friends of the Country, which met first in February 1765, took root in a local culture of deep anticlericalism. Its advocates aimed at the practical realization of enlightened discourse: "this Economic Society is to develop, perfect, and advance the agriculture, rustic economy, the sciences and arts and everything that may be immediately directed to the maintenance, relief and ease of the human species." Its opponents considered it "dedicated only to profane diversion," a "school for idleness," or worse, of "libertinism."[33] Bodies like this one and the Madrid Economic Society, established in 1775, are not usually accorded a place on maps of Enlightenment science. Yet the view that "of the major European capitals only those of Spain and Austria were without scientific societies" ought to be qualified,[34] for in Spain at least these were important locales for formulating the Enlightenment's practical realization.

Further illustration of the institutionalized and local association between patriotism and profit, economics and Enlightenment, is to be found throughout Germany in the many patriotic and "physical-economic" societies established in the later eighteenth century. In their emphasis on *landeskunde,* "regional survey," these bodies compiled registers of resources and sought to promote useful knowledge via botany, mineralogy, fiscal studies even, as a basis to local economic advance. There was a distinctive geography to the institutionalization of economic patriotism in the German-speaking Enlightenment. Many of the physical-economic and patriotic societies of Enlightenment Germany (as with the provincial academies in France) were small-town affairs. The relatively large number of scientific and medical societies in the Protestant areas of northern Germany and Switzerland and in cities like Hamburg was in contrast to the negligible activity in southern Germany or in the Hapsburg territories. Yet wherever they were established, the goal of Enlightenment was commonly apparent in these bodies' collective preoccupation with *Oekonomie* as the geography-of-what-could-be, with cameralism or cameral science, the study and teaching of administrative practices geared to the state's future well-being.[35]

The fact that scientific and other institutions in Barcelona, Madrid, and Florence, in Holland, in Germany, and elsewhere had such emphases reveals a persistently utilitarian character to the Enlightenment. This, in turn, supports a view of the Enlightenment as something shared yet always locally expressed. In looking at scientific societies and learned academies and at their efforts to make Enlightenment work—through prize competitions, published transactions, experimentation, advocacy of future best practice for one place based on past principles established elsewhere—different pictures of the Enlightenment's local geographical expression come into focus.

Different but not contradictory. In one view, working locally helps further reveal the Enlightenment as a collective intellectual, social, and practical enterprise without geographical boundaries. Physical location was relatively unimportant: being in Göttingen or Glasgow, London or Leiden, made little difference. For in looking at how local institutional spaces worked to "ground" the Enlightenment, what we see are the worlds of Enlightenment men and women being constituted by contacts with like-minded others elsewhere, not just through social relations with immediate neighbors or in concerns for the future geography of their nation. The Enlightenment's cosmopolitanism transcended local contingency even as it "came to earth" in certain places and not in others.

In another view, particular settings and locales *are* crucially important precisely because upon their closer examination, the Enlightenment is shown to be not one but many. Thinking locally makes it possible to better place the Enlightenment geographically *and* to better qualify it thematically. Take chemistry as an example. The nature of the "chemical Enlightenment" is brought into sharper focus in knowing which societies investigated chemistry, which universities taught it, where experiments were performed—as "open" to the public, in teaching laboratories to a student audience, or even at home—where booksellers and chemical authors were when their books were written and disseminated, who read them, which instrument makers provided for specialists' needs. In Britain, shared claims to the public utility of new chemical knowledge obscured local differences in theoretical affiliation and experimental practice, and in the personal, institutionalized, and market relations sustaining the discourse of utility. Enlightenment in Joseph Black's and William Cullen's Edinburgh was different, chemically speaking, from Joseph Priestley's provincial England as it was from Lavoisier's Paris. Continental Europe was different still. Unlike in Scotland, the wholesale acceptance in later Enlightenment Germany of Lavoisier's "chemical revolution" from France helped create a wholly new German chemical fraternity from previously disparate groups of experimenters—so much so that by the late Enlightenment the emergence of a national discipline-oriented community weakened that very republic of letters through which Germany's chemical practitioners had earlier kept in touch.[36]

Similar stories might be told of natural history, astronomy, meteorology, agriculture, moral philosophy, political economy, and so on. The idea of multiple local Enlightenments is not diminished by our knowing that members of certain societies exchanged ideas with like-minded individuals elsewhere, that institutions with which they were more locally associated had counterparts elsewhere, or that what we now call specialist knowledge was then not always clearly defined. Scientific bodies and economic societies were not the only sites and social spaces in which Enlightenment was made and from which it traveled and was received. That being so, a further view of the Enlightenment as local may be revealed, namely, that practiced within popular sites and social spaces. If we want to understand better what the Enlightenment was locally and for whom it was so, it is appropriate that we look not just at how enlightenment knowledge moved "beyond the laboratory walls," so to speak, but at its forms within the "laboratory" itself.[37] By looking inward at particular social spaces and at different practices, as Roberts has proposed for the Netherlands, it may be possible to reveal how Enlightenment ideas were proposed and disseminated without our having to suppose that any one microcosm in one place was what was taken to be Enlightenment elsewhere.

Talking Places: Coffeehouses, Pubs, and Salons

Coffeehouses do not fit easily into any taxonomy of the Enlightenment locally understood, but they are a crucial part of it. Common throughout

Europe, London had thousands of such "penny universities" as they were later termed. Because they were so numerous, so there were many faces to the London coffeehouse: artists gathered at Old Slaughter's, Tories at the Smyrna, Jonathan's was popular with stockbrokers. In the Marine Coffee House in Castle Court, lectures on Newtonianism could be heard, accompanied and enlivened by various demonstration devices, some shop bought, some homemade. If you wanted to gossip, hear the news, be seen, engage in "flash talk," or be enlightened through other's declarations about natural philosophy, then you went to a coffeehouse: "You that delight in wit and mirth/And long to hear such news . . . Go hear it at a coffee house/It cannot be but true."[38]

Because in London especially these spaces for intellectual and social exchange were so many and so close, some had circles of people and circuits of movement associated with them. In the city's Royal Exchange, the names of the Merchants' Walks—areas where traders promenaded while dealing—on the floor of the Exchange reflected the names of nearby coffeehouses as well as particular trading connections across the world—Jamaica's and Barbados's "walks" to the west, the Italian "walk" to the east, and so on. A short walk outside the trading places of the Exchange took merchants to other places and social spaces, there to become students of natural philosophy. In giving classes in places such as the Marine, men like the mathematical lecturer James Hodgson not only promoted experimental philosophy—and, of course, himself—but helped establish new sites and new publics for experimental science in early Enlightenment London. There and in other places around Exchange Alley, such as the Swan, Jonathan's, Garraway's, and the Sword Blade coffeehouses, Newtonian principles were practically demonstrated, debated, and disseminated. At home too, through the use of Newtonian texts for children, the principles of natural philosophy could be talked about and shown to work: Enlightenment as practical education for the young. Among adult merchants and students, such advances in the new science had potential practical consequences in relation to astronomy and navigation and thus to trade. It is therefore not appropriate to make a sharp distinction between Enlightenment as a philosophical concern and its expression in economics and other discourses of material advance. All were commodities to be traded, whether on the walks of the Royal Exchange or in the talks of Hodgson and others. In early eighteenth-century London, coffeehouses "served as the perfect location to turn scientific principles and experiments into commodities. . . . The public performances of the mathematicians at the Marine and of the Hand and Pen represented the transformation of the social space of natural philosophy."[39]

So too for the political dissenters and radical chemists, Joseph Priestley among them, who made up the Coffee House Philosophical Society in London's Paul's Yard between 1780 and 1787. There, conversations centered around experimental philosophy—but not mathematics—and chiefly on the practical applications of chemical science. Here we can again see, but for a smaller space than in the Paris academy, how local space was a physical, social, and epistemic thing. The Coffee House Philosophical Society was where the Royal Society "didn't count," both "a supplement and an alternative to the Royal Society, a place where topics in experimental philosophy and its industrial applications could be discussed without entering into the politics of the Royal Society or, indeed, national politics, which the Royal Society also eschewed." The Coffee House Society was a closed space: its members met in a private room, controlled who could visit as a guest, and did not publish proceedings. Yet it acted as a center for foreign correspondence among like-minded men, a fact that facilitated the early introduction into Britain of European developments in the chemical sciences and afforded political and religious tolerance as well as "a philosophically motivated commitment to the Enlightenment goal of improvement."[40]

Those social and intellectual concerns for practical chemistry, radical politics, and Enlightenment-as-improvement that came together in the upstairs room of a coffeehouse in Paul's Yard were different from those that constituted the Spitalfields Mathematical Society, which met in pubs throughout east London from 1717. Initially a small group of literate craftsmen, the Spitalfields Society later had fellows of the Royal Society among its members and drew from a very broad range of occupations—apothecaries and seedsmen, brewers and teachers, actuaries and weavers. Their interests were mathematics and education, their chosen means public lectures and experimental instruction. Their Enlightenment, a mathematical one, was vocational.[41]

In this it shared much with aspects of the medical Enlightenment taking place in London. As in Holland's public anatomy theaters, medical lecturing in London was vocational and high quality, Enlightenment a matter of listening and witnessing not simply of reading. Unlike in Edinburgh, where high-quality medical classes to university students were often taught using model corpses, London's anatomists had no shortage of dead to inform the living. Elsewhere, viewing bodies—painted, sculpted, and parading—meant displaying one's own. Different publics were to be found in the spaces of London's art world. The emergence in the 1790s of new public art galleries in and around the Saint James district—social spaces promoting what contemporaries took to be a truly "public art"—not only altered London's urban topography but helped fashion a new type of political landscape. Going to the Shakespeare Gallery to view the paintings, not just to be seen, was an altogether more democratic experience than a trip to the Royal Academy. In the Shakespeare, the talk was of the liberty of the press, the political constitution of the people. In the Royal Academy, rational debate about the nature of English art excluded women and created a different and much less politicized audience. In London's Royal Society, just as in Paris's academic spaces, advancing experimental philosophy and the culture of practical utility was no less a political achievement. But Enlightenment there took different forms than it did for those cultivating history painting in Pall Mall, promoting mathematical education in a Spitalfields tavern, or witnessing an anatomical dissection in William Hunter's back room in Great Windmill Street.[42]

To map in full such local spaces of practice across Enlightenment London—or anywhere else—would be to plot overlapping circuits of movement and overlapping local worlds. To be at all complete it would involve tracing the many lines of communication that held like and unlike settings together, knowing the strength of links between them and showing how and why particular forms of discourse were privileged in some sites and not in others. Not all spaces of like type had quite the same function. In Paris, cafés were much less open spaces for Enlightenment than they were in London. Hardly any sold newspapers as did London's coffeehouses. French "public noise," in cafés at least, was overheard by spies and police informants. Talk—against the king, of religious tolerance, over unpopular taxes—was written down and reported on. That "map of café talk" that has been recorded from twenty-nine cafés in Paris between 1726 and 1729 represents but a fraction of that capital's sites and networks of popular conversation.[43] Yet even if partial, it highlights in topic, social status, and location a different geography of Enlightenment talk than was the case in London or in Paris's leading salons.

Whether predominantly literary, philosophical, or political, salons were primarily but not exclusively the domain of intelligent and well-to-do women who orchestrated conversation as a form of mannered inquiry. Talk there was not modulated gossip but shared, informed, and topical debate. In France especially, salons were key spaces for talking Enlightenment, sites of convivial and civilized critique, and intermediary spaces between the philosophes and the public. They offered polite spaces in which public and abstract ideas were, so to speak, domesticated. In being controlled by women and admitting women as equal participants, the salon offered a gendered space different from the masculine preserve that was the learned academy or society.

For Émilie, Marquise du Châtelet, distinguished mathematician and physicist and for sixteen years Voltaire's lover, being a society hostess at Cirey allowed her to welcome distinguished visitors and organize *conversazione* with scientific figures; it also provided an opportunity to find her way as a woman in the masculine world of Enlightenment science (see plate 4). For du Châtelet, the Académie Royale des Sciences was a known world but a closed space. Familiar with its work and personally acquainted with its leading mathematicians such as Maupertuis and Alexis-Claude Clairault, she was excluded by her gender from participating in Enlightenment debate there. Writing her Newtonian treatise *Institutions de physique* (1740) in Cirey, alone and in secret, domestic space was scientific space, a conversational arena, and a space of private intimacy as well as public entertainment. Du Châtelet was at one and the same time a public figure, a private scientific writer, an audience member, and a caring mother (her Newtonian translations and work were for her children). She was also a figure of disdain for those who thought that women should not do Enlightenment at all. Later in the century, but for chemistry not Newtonian natural philosophy, Marie Paulze, wife of Antoine Lavoisier, the prominent figure in the chemical Enlightenment, assisted her husband as translator and laboratory assistant. She continued as a practicing chemist long after his death by guillotine. The position of these women in France was most notably echoed in Italy, in the figure of Laura Bassi, who lectured in physics in Bologna from 1732 until 1778.[44] In the Netherlands, too, the elite women who came together from 1785 in the Natuurkundig Genootschap der Dames (Women's Society for Natural Knowledge) in the small town of Middelburg further illustrate the early domestication and institutionalization of science among women—and reinforce Roberts's view of the importance of local place and site.[45]

These remarks are not made to emphasize exceptional women or to acknowledge the gendering of space in the Enlightenment. Both are important elements in pictures of the Enlightenment revealed locally. Rather, they are to note that not only does it matter where we look for Enlightenment but it matters how "local" we get in thinking about the Enlightenment in terms of its sites, practices, and social spaces. In the case of du Châtelet and Paulze Lavoisier, it is important to our understanding of the spaces in which Enlightenment ideas and practices were made and from which they moved that we look at the "extra-local," the study and the domestic laboratory. Only by being done there and not elsewhere was their work possible.

This is of greater significance than just pointing out the uneven distribution of Enlightenment ideas and practices. Different sites and social spaces have different cognitive claims associated with the practices undertaken there. In the space that was the learned academy, establishing credibility in what was said, in how it was said, and by whom meant certain disciplining practices had to be put in place there that were not commonplace elsewhere. In the less exclusive space that was the café, different sorts of knowledge circulated and did so without the warrant that was demanded of debate in the academy. In the home, science could be domesticated. The home afforded a space for polite inquiry and sociable instruction, geography included. Recognizing then the distinction between spaces of production and of consumption and the need to distinguish the practices by which it was made, we see that Enlightenment knowledge was always the product of specific spaces.

What emerges from consideration of the Enlightenment in local context is a picture of unequal difference—in what Enlightenment was held to be, how it was undertaken, in its social spaces, its sites, and in the consequences of its practice. It is a more nuanced picture than that suggested by the distinction between "High" and "Low" Enlightenment, an essentially social stratification between the textual productions and philosophical concerns of an elite few, and the more diffuse but essentially "popular" Enlightenment evident in the circulation of books in provincial France or rural New England, mathematical demonstrations in a London pub, or the gossip in Parisian cafés.[46] Thinking locally demands that we afford the Enlightenment yet greater qualification in which matters of practice and questions of social meaning have significance because of their production and variation in place. This is not to say that we should further refine our taxonomies in terms of, say, a "metropolitan Enlightenment," the "provincial Enlightenment," a "village" or even an "institutional" Enlightenment. It is to insist on a greater specificity as a result of differences in theme, practice, and geography. Knowing what the chemical Enlightenment was demands an attention to differences among, say, midland England, Edinburgh, and Göttingen, between university chemical teaching and public lectures, between conversing in a coffeehouse and corresponding with distant associates, between public witnessing in a laboratory and writing and reading in the quiet of one's study. The same is true for the medical Enlightenment, for political economy, for natural philosophy, and so on. Knowing what the Enlightenment was demands that we pay attention to its local settings *and* to connections between its settings in order to reveal what was done there. The displacement of scientific knowledge between academies and salons, for some women anyway, afforded them an opportunity to be enlightened and to enlighten others. For some men, meeting to discuss enlightenment ideas entailed privacy and seclusion, safeguarding there the practices by which they set themselves apart from others. For other men and women, enlightenment was a convivial and lived experience, practiced with fellow students, discussed in a drawing room, shared in public spaces. "Grounding" the Enlightenment socially and epistemologically cannot avoid placing it geographically.

Several key issues have been addressed in part 1 in order to consider what is understood by the geographies of the Enlightenment. The uncritical attention to the nation has been noted. The Enlightenment's cosmopolitan features have been discussed. The importance of geographical scale has been stressed. The significance of movement and dynamism, the unequal "translation," "mediation," and "appropriation" of the Enlightenment as a Republic of Letters has been highlighted. The constitution of the Enlightenment has been revealed through its different practices—corresponding, talking, experimenting, translating to identify only a few. Particular sites and social spaces of production, reception, and engagement have been examined: academies, coffeehouses, inns and taverns, laboratories, salons, university teaching rooms, spaces of private withdrawal and of public sociability. The different claims of different people in different places over what the Enlightenment was and the ends it served have been illustrated.

The effect of such scrutiny has been to suggest that the notion of *the* Enlightenment as a unitary phenomenon—a French, urban, philosophical, largely masculine, and textual phenomenon—be replaced by the idea of Enlightenment not as one thing but as many, a multiplicity of ideas, practices, and meanings located in place and moving over space. Thinking geographically about Enlightenment in these ways is to recognize the significance of geographical scale, physical location, epistemic movement, and social space. It is to think in terms of a series of qualifiers, of geographical modifiers even. It is, of course, still possible to think and work in terms of the Enlightenment in national context. But if we do, we should be at the very least more attentive than has been apparent to an important initial distinction: namely, between *the* Enlightenment in national context, which presumes in ways unwarranted by the evidence an essential Enlightenment moving out from an originating European "core" to wholly receptive "margins," and *the national* Enlightenment, in whose context certain defining features are held to signify a nation's geographical particularity. Even if we are thus more attentive, we must also recognize that the nation was a far from standard unit of geographical "measurement" and that differences in scale, location, social status, and timing of what was taken to be Enlightenment offer "internal" contradictions within "national" space. At the same time, the cosmopolitan movement of ideas, of practices, of people provides evidence of "external" connections over and above national space.

Because this is all so, the term "geographies of the Enlightenment" should not be thought of as signifying static maps of national and political boundaries fixed in space, with the Enlightenment more or less "bright," more or less in shadow for different places and times. It is important, of course, that we site where and when the Enlightenment was being discussed and received in places and to know why those places and not others. But rather than see a single "thing" moving wavelike only across Europe, we should visualize a series of networks, myriad lines of movement across space. The lines sustaining these Enlightenment networks were made up of different concerns—practical chemistry, natural philosophy, mathematics as a vocational discourse, conceptions of metaphysics, anatomy and medical advance, political economy, and so on. In being of unequal strength and varying duration, they at moments traced out "maps" of expert knowledge among, say, leading medical writers or chemical experimenters in distant places. And they traced out "layperson's maps" as different people read or heard about the results of others' deliberations.

Whatever people held Enlightenment to be, its dynamic networks and line maps—of correspondence, plant exchange, café talk, of the movement of private individuals to witness an anatomical dissection and thus to form for a moment a "medical public"—came to earth, as it were, in certain places. They were made in and received in particular institutional and social spaces and places, in settings that demanded that certain moral codes be adhered to, certain social and intellectual practices followed. The Enlightenment as a conceptual space somehow "without" geography—a Republic of Letters, as transnational cosmopolitanism "above" national context—was an actual space defined by its practices, languages, and networks, by what people did in particular places, by how Enlightenment ideas about reason, criticism, and tolerance were lived and worked with.

PART TWO

Geographical Knowledge and the Enlightenment World

5

Exploring, Traveling, Mapping

Part 1 established some of the ways in which the Enlightenment can be thought of geographically—thought of, that is, in terms of its national expression, its international and cosmopolitan making, and its reception in local sites and social spaces. Attention was drawn both to the limited utility of *the* Enlightenment and to the Enlightenment as a national phenomenon, and to the differences that questions of place and scale and movement make to a richer understanding of the Enlightenment's where.

Part 2 is concerned with the world as an object of geographical inquiry in the Enlightenment and with processes of geographical discovery rather than with questions of scale. As the first of three chapters making up this part, this chapter explores the ways in which the world was geographically encountered. Where did European navigators go, and how did indigenous populations regard them? How were facts about natural diversity and other peoples made sense of? What, indeed, did it mean to be an Enlightenment explorer or mapmaker? These and other questions underlie the ways in which knowledge of the terraqueous globe was made and debated. Yet although the earth became in the eighteenth century the object of systematic scrutiny as never before, it will not do to associate such scrutiny and the resultant knowledge with loose, catchall terms like "globalization" or to see the making of geographical knowledge as an unproblematic enterprise. Oceanic navigators and mapmakers did chart the geography of new continents, and the earth itself was revealed as an object of geography—its contents, human and nonhuman, mapped and measured, classified as well as located. But it is also true that Enlightenment exploration touched only certain points on the earth's surface: in truth, continental margins more than centers. Certain parts of the globe figured crucially, others hardly at all. Voyaging did not translate into geographical knowledge in straightforward ways. What we now see as knowledge about the earth was often arrived at as a result of commerce, not organized science. Encountering the world empirically in an age of European empires meant that trade and learning went hand in hand, but they did not do so equally. The agents of European capitalism extracted others' natural resources and incorporated indigenous production systems into imperial commodity networks. The same is true of non-European knowledge systems. Because this is so, terms like "negotiation," "translation," and "mediation" are more useful than "globalization," with its connotations of everywhere and at once the same.

In discussing the ways in which the world was geographically revealed during the Enlightenment, I use three sets of related terms: encountering and imagining, mapping and inscribing, envisioning and publicizing. What connects them are matters to do with the ways in which knowledge was made—as maps, as narratives of travel—with the ways in which geographical knowledge traveled and with the fact that audiences back home received new geographical information about the "out there" in particular forms—as books, specimens on display, or paintings—more often than they experienced them firsthand. That this was so raises questions about the traveling nature of geographical knowledge in and of the Enlightenment and so reiterates the need to consider the sites and the forms of that knowledge's making and reception.

Encountering and Imagining

Enlightenment exploration and the individuals and institutions that promoted it sought to bring the globe under the sovereignty of science. Doing this meant getting out into the world, traveling safely, and returning to tell others—through books, correspondence, maps, and word of mouth. There is nothing new in that: Christopher Columbus's letter of February 1493 documenting his American discoveries was widely circulated and went through twenty editions by 1500. Curiosity about other places and the tales told of them are innately geographical matters, and Enlightenment exploration was rooted in these longer-running traditions of geographical discovery.

Enlightenment geographical exploration was also distinctly novel. For one thing, Enlightenment voyages of exploration were concerned to test earlier claims. For another, they were much bigger enterprises. Further, they emphasized method and description based on exact observation and reporting in ways not apparent before the late seventeenth century.[1] Their distinctiveness lay also in the size and cosmopolitanism of the Enlightenment's expeditionary endeavors between about 1760 and 1800, in the institutional and governmental support and in the intricacy of patronage networks, in the emphasis on the languages of instrumental precision and calculation, and in attempts at the theoretical explanation and systematic ordering of what was discovered. Traveling the globe in order to understand it did not necessarily mean leaving the laboratory and the library behind. Explorers' ships were themselves mobile "knowledge spaces"—floating instruments of science and European geographical expansion (see plate 7). The ships of

Jean François de Lapérouse's expedition in 1785–88—arguably, the Enlightenment's best-equipped expedition and akin to a seagoing university—were even named after instruments: *L'Astrolabe* (the astrolabe) and *La Boussole* (the magnetic compass). The ship was the Enlightenment's principal geographical instrument—the Enlightenment's "vessel of modernity."[2] Even so, doing Enlightenment science was difficult at sea. Johann Reinhold Forster and his son Georg, German naturalists on James Cook's *Resolution* voyage from 1772, complained of the "Oaths and Execrations, curses and Dam's" of drunken sailors and for weeks lived and studied in a cabin "beset by cattle and stench on both sides."[3]

Europe's principal geographical and scientific testing ground in the Enlightenment was the Pacific.[4] Pacific islands provided celestial observation sites in 1761 and 1769 for European astronomers anxious to view the transit of Venus in order to calculate the mean distance of the earth from the sun. The Pacific was also a laboratory for the testing of long-held ideas about the terrestrial globe. One concerned the supposed existence of a sea passage—the Northwest Passage—between the northern Atlantic and the northeastern Pacific. Western explorers had been attempting to locate a commercial sea route to the Far East by sailing north and west around the Americas since the efforts of John Cabot in 1497. Names in this part of the world testify to this genealogy of geographical encounter: for example, Hudson's Bay, named after Henry Hudson, who between 1607 and 1611 explored the waters that now bear his name; and Baffin Island, named after William Baffin, navigator on Bylot's voyage of 1615. From the 1740s, the British were concerned enough about French incursions into the area to pass an act (in 1744) promising £20,000 for the first merchant ship to pass from Hudson's Bay to the Pacific. The record of exploratory activity in the region from the 1770s includes James Cook, whose *Resolution* and *Discovery* crossed the Arctic Circle before abandoning the search in the face of ten-foot-high pack ice. Phipps's British polar expedition in 1773 perhaps best epitomizes both the international agenda behind polar geography and the "mediation" of books and theories about the polar regions. The expedition built on concerns expressed by Louis Antoine Bougainville (who in turn was influenced by reading Maupertuis) and by the Swiss geographer Samuel Engel, the second (1772) edition of whose 1765 *Mémoires et observations géographiques et critiques* influenced British naval and scientific authorities in arguing for "a vast and empty ocean" to the north by virtue of erroneous theories concerning the origins of ice. Aided by Joseph Banks, who instructed the navigators on what to collect, and armed with a vision that they would not "fail of being attended with many very interesting particulars in Geography and other branches of Science," Phipps's expedition indeed failed—although Engel refused to accept that the route to the North Pole was not open sea—and the voyage has been overshadowed by later encounters with the polar regions (fig. 8).[5] In respect of the polar world, Enlightenment voyages were less often voyages of discovery than they were "voyages of delusion."[6]

A further geographical context in which Enlightenment voyages must be understood was the idea of a great but unknown southern continent, which classical accounts had argued should exist to balance the European and Asian landmass. Unlike the Northwest Passage, which remained uncertain, Cook and others dismissed the myth of Terra Australis Incognita—at least, it was claimed that if such a landmass did exist, it was inaccessible and icebound. In the southern oceans, however, Europeans encountered other human cultures living virtuously and close to nature, free—or so it seemed—from the vices of "civilized" society. These peoples offered challenges to current thinking—on theories of human social development, on what it meant to be civilized, even on what it meant to be human. Yet they did not do so immediately. William Dampier's *A New Voyage round the World* (1697) contained important new observations on ethnography and natural history in the South Pacific, notably in New Holland (Australia). The book provided Jonathan Swift with material for his novel *Gulliver's Travels,* stimulated later voyages, and John Locke had Dampier among others in mind when in 1704 he noted how, through the activities of navigators, "the Empire of Europe is now extended to the utmost Bounds of the Earth." Dampier's "scientific" work has recently been reappraised and fits well with other philosophical observers of the natural world in the early Enlightenment. Yet there is still a sense that the rhetoric of early Enlightenment exploration in Dampier's and others' works has a hesitancy borne of geographical novelty. In remarking upon trade winds, for example, Dampier wrote, "thus have I finished what my own Experience, or relations from my Friends have furnished me with on this useful Subject, . . . which I humbly offer, not as a compleat and perfect Account, but as a rude and imperfect Beginning or Specimen of what may better be done by abler hands thereafter."[7]

This tone of "geographical wonder," even of astonishment and of the admission of partial capacity, is not so commonly shared by later exploration accounts. Dampier's accounts are very different, for example, from those of Bougainville, or Cook, or the Forsters. There is every reason why they should be. By the later Enlightenment, oceanic voyages—and the contacts made with Pacific natives in particular—provided Europeans with data that allowed speculation on the nature of the human condition. "Wonder" was replaced by a concern for "order" and a belief in "system." Like Diderot after

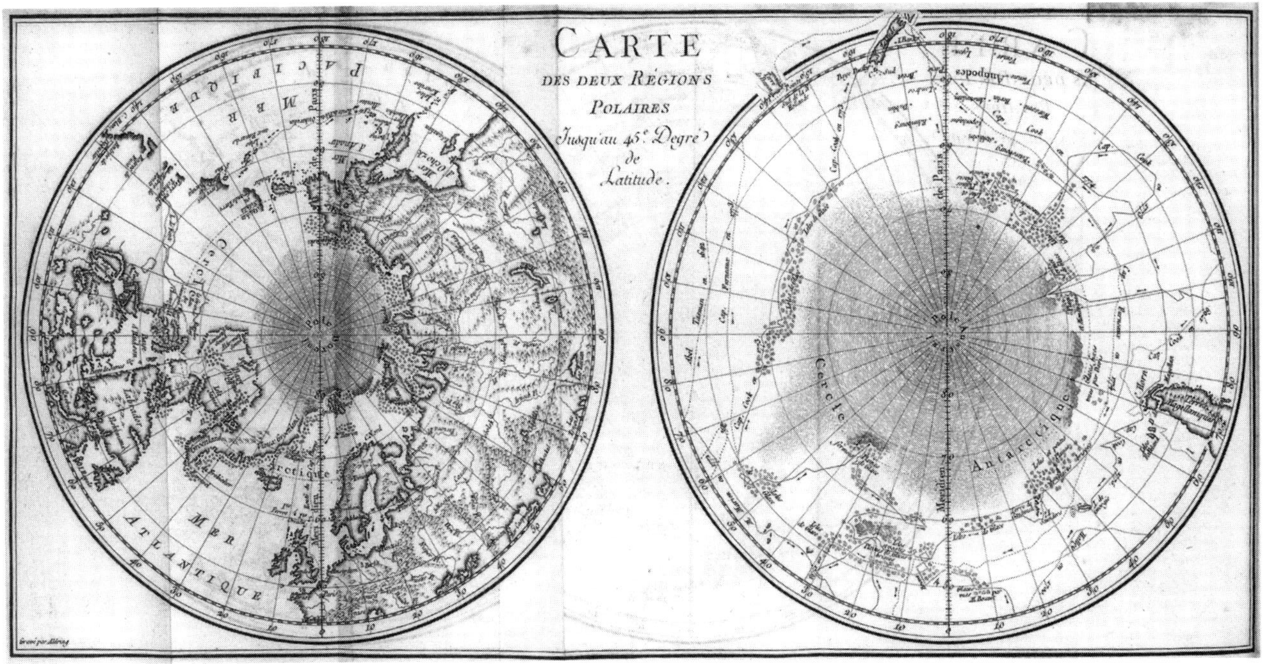

FIGURE 8

Map of the polar regions, from Georges Louis LeClerc, comte de Buffon, *Histoire naturelle, générale et particulière* (Paris, 1778), vol. 4. Note the "broken" lines for the coastline of northwest America and the "tracery" of lines plotting the voyages of Furneaux and Cook. By permission of Edinburgh University Library, Special Collections.

him, of course, Bougainville was struck by Tahiti's natural beauty and the beauty of its people, particularly the women. On the edges of the known world, "savages" living in a state of paradisal grace were taken to represent an earlier and happier stage of humankind's development, a state that Europe (if it had ever had it) had long lost. The Pacific, and Europe's encounter with it in the later Enlightenment especially, became the site of scientific empiricism on unprecedented scales—and of new ways of thinking and writing about the globe.[8]

The geographical exploration of the Pacific from the 1760s significantly enlarged the contemporary "planetary consciousness" in ways other than the rhetoric of geographical description.[9] Empire followed—was made by—the innately geographical processes of navigation, inscription, and theoretical prescription.[10] As philosophers speculated about human diversity, botanists and naturalists attempted to discern order in this new world's bounty. Artists endeavored to capture what, for them, was its essential strangeness. Even sailors took advantage of the facts of cultural difference. Staggered by the openness of islanders' sexual mores and with iron nails being the currency for sex, crewmen drew them from their own ships: the timbers of the *Dolphin* under the command of Britain's Samuel Wallis—in 1767 the first European to encounter Tahiti—actually began to separate. Elsewhere, in the metropolitan centers of Enlightenment Europe, poetry and books of travel described the faraway, satirized European society and caricatured distant "others" in their natural habitats. In London in 1785 and in 1788, the pantomimes *Omai* (named after the Ulaietan native brought back by Captain Tobias Furneaux) and *The Death of Captain Cook* provided audiences with stylized versions of those bigger geographical "performances" that, thousands of miles away, were transforming both worlds forever.[11]

More than others perhaps, mapmakers and geographers had to rethink their works and words in order to accommodate the enlarging world. Looking back on European maritime exploration in his *Modern Geography* of 1802, for example, the geographer and essayist John Pinkerton noted how the recent geographical discoveries of "Australasia and Polynesia" together constituted a new "fifth part of the world."[12] The ancients considered the earth to have three divisions: Asia, Europe, and Africa. Columbus and others added a fourth, the Americas. But a note of caution is necessary here. Discovery—the moment, place, and fact of contact—does not at once become a fact of content, a secure geography. These new "maritime divisions" to the world extended knowledge more about the shape of the continents, and thus the surface of the globe, than they did about continental interiors and the geography within. This point was clear to contemporaries. "Nothing worthy of research by Sea, the Poles themselves excepted, remains to be examined," observed the founders in 1788 of the Association for Promoting the Discovery of the Interior Parts of Africa. "But, by Land [they noted], . . . the objects of discovery are still so vast, as to include at least a third of the habitable surface of the earth: for much of Asia, a still larger proportion of America, and almost the whole of Africa, are unvisited and unknown."[13]

Such evidence strengthens my arguments not just about the importance of geographical knowledge in knowing the Enlightenment world but about the role of geographical difference in the Enlightenment. The polar world and coastal northwestern America were imprecisely known (see fig. 8, for example, for knowledge by the later 1770s). Africa was the poor relation among the Enlightenment world's emergent geographies. North Africa was known, to European traders at least, but sub-Saharan Africa remained hidden from European view excepting what little was known around coastal trade centers run by the British, Portuguese, Dutch, French, and Danish. And even what was known of northern Africa was tainted with suspicion. The descriptions of the Ottoman Empire that figure, for example, in *Travels in Syria and Egypt* (1787) by Constantin François de Chasseboeuf, comte de Volney, reaffirmed existing stereotypes rather more than provided new insight into the North African and Near Eastern worlds.[14] There is evidence that some eighteenth-century European ideas—mainly in medicine—found a receptive audience in Islamic circles.[15] But the picture more generally was one in which a profoundly unsure "Orientalist" gaze from the West was based on ambivalent "imaginative geographies," "anticipatory geographies" even, which were distinguished not by fact but by notions of distrust, fear, and the exotic allure of the East.[16] Exploration of the sub-Saharan African interior is a late Enlightenment matter. Mungo Park's efforts to solve the two-thousand-year-old "Niger problem"—Which way did the river run? Where did it end?—began in 1794; the attempt was partially successful but ended with his death in 1805–6. (The problem was not finally solved by geographical fieldwork until 1830.)[17] So, too, China and Japan were hardly known at all. Even when the commercial and scientific worlds of Britain and China did meet in the late Enlightenment, the encounter failed: each, thinking itself the center of the world, could not reconcile the other's authority with their own worldview, the Chinese having little conception of the geography beyond their "Western Ocean."[18] Only India was known in some detail—and only then mainly through the efforts of the East India Company and for some districts more than others.[19]

On land and at sea, successful travel and exploration depended on knowing how to do it. Encountering new places demanded that the effects of geographical strangeness be regulated—that, in short, new facts should not overwhelm for want of method. Ensuring a consistency of approach was important if the results of geographical scrutiny in one place were to be made sense of by others elsewhere. Recognizing this, proponents of the new science from the late seventeenth century wrote what were, effectively, guides to geographical practice in the field. Enlightenment natural knowledge began with regulating geographical diversity. Robert Boyle's posthumously published *General Heads of the Natural History of a Country, Great or Small; Drawn Out for the Use of Travellers and Navigators* (1692) was one such attempt at regulation. John Woodward's *Brief Instructions for Making Observations in All Parts of the World* (1696) was even divided into geographical headings—"At Sea," "Upon the Sea-Shores," and "At Land." Throughout, his language steers the traveler over what to do and how to do it—"Keep an Journal of the *Ship's Course:* of the *Latitude, . . .* Of the *Variation of the Compass,*" "Take an account of the more observable and peculiar *Diseases* of the Country," and so on. The peoples encountered were to be observed and enumerated: "observe the *features, shapes* and *proportions* of them," their "*Tempers, Genius's, Inclinations, Virtues,* and *Vices.*"[20] This is a language of methodological direction and of moral instruction. What underlies these guides on "how to" travel—what, for geography and natural history, we may think of as the "scientization of travel"—are concerns about "why" travel, about the instrumentation of travel, and about the status of the reporter. Boyle's own instructions were, he said, "the only sure Foundation of Natural Philosophy." Observe, collect, classify, and systematize: in these ways, the world will be revealed, and revealed, moreover, by persons following a method whose use together with their own reliable status made them credible witnesses.[21]

Not everyone who traveled was a geographer or an explorer. "Scientific traveler" is not an Enlightenment term. It is a modern—and imprecise—one. In French, the term "explorer," meaning "one who is sent to discover a land in order to find out its extent, its situation," appears only in 1718. Even after that date, it was never common as a term denoting persons engaged in the geographies of global encounter. Like Cook and Bougainville and others, Lapérouse was described as a "navigator." The term *voyageur naturaliste* was commonly used in a French context. Exploration was undertaken by men who, like the Forsters, trained and worked as botanists, naturalists, hydrographers, or mineralogists. Specialization in natural philosophy was an exchangeable commodity, a passport to participation in international science as well as a marker of individual and national prestige. "National" expeditions often relied on scientific colleagues from other countries to provide

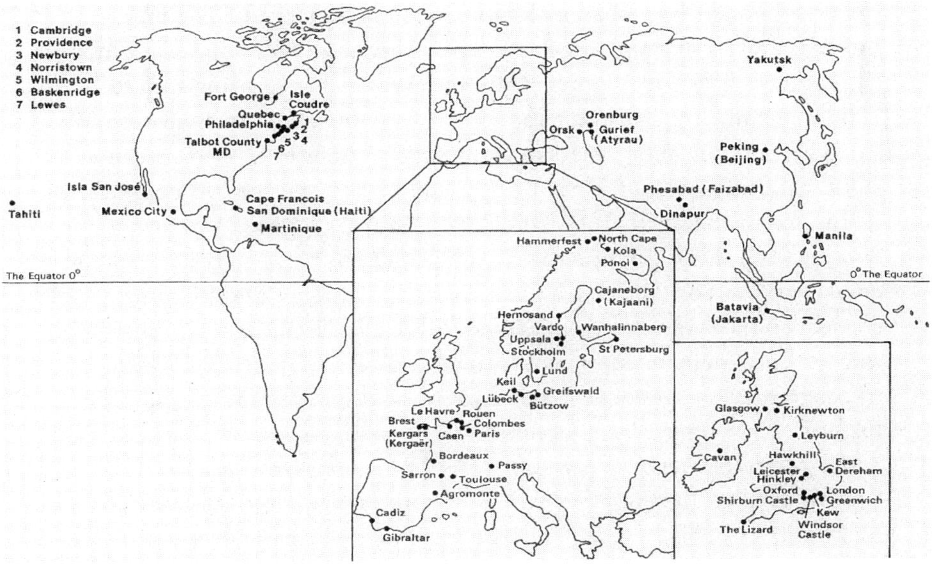

FIGURE 9

Planetary geography. The principal recording sites used in observing the transit of Venus, 1769. Although by distribution a cooperative global geography of planetary measurement, it was by coordination and observers a largely European enterprise. The map is based on listings in Harry Woolf, *The Transits of Venus* (Princeton, 1959), 182–87.

key expeditionary skills. The French geographer Joseph-Nicholas Delisle, for example, worked for over twenty years on behalf of the Russian court, and his brother died on the 1741 expedition to Kamchatka of the Dane, Vitus Bering, who was working for the Russians.[22] The Danish expedition of 1761–67 to Arabia was prompted by a German and staffed by Frisians and Swedes as well as Danes. On the French Pacific voyage of 1761—the first to observe the transit of Venus—over one-quarter of the 120 participating scientists were French, with 21 Swedes and 21 Germans, 18 British, and a handful each from Italy, Portugal, Denmark, and Russia. In the 1769 British voyage for the same purpose, 80 of the 150 scientists were British; others came from France, Germany, Spain, Sweden, and Russia.[23] The geography of planetary observation was at once a locally sited yet cosmopolitan affair (fig. 9).

Enlightenment exploration was "big science" in the sense of it being government sponsored, well equipped, an instrument of shared international interests in the nature of the earth. But it was not simply international or just scientific. As we have seen in part 1, what might be taken as "national" matters may also be seen as "transnational" exchanges between individuals and between institutions located throughout Europe. Underlying the cosmopolitanism of global exploration were national, usually commercial, interests and strategic political issues to do with military empire and trade. Cook sailed with sealed orders from the Admiralty requiring him to keep Britain's commercial advantages in view. Lapérouse's expedition was motivated by the style and success of Cook's navigations and by nervousness—rightly as it turned out—that British exploratory science hid commercial intentions in the northern Pacific regarding the supposed Northwest Passage between the Pacific and the Atlantic, and thus between Britain and oriental markets. For the French king Louis XVI, science and its commercial consequences were a means to extend the glory of France, not a thing to be shared. Louis demanded secrecy over all "operations relative to astronomy, to geography, navigation, physics and to the different branches of natural history." The Pacific voyages on behalf of the Spanish crown between 1789 and 1794, by the Genoese Alejandro Malaspina, likewise aimed at restoring Spanish prestige through science and oceanic navigation. The intention that Malaspina's "mobile science" be a source of national prestige also prompted the refashioning of Enlightenment science in a number of Spanish colonial institutions in South America.[24]

Land travelers encountered the globe differently from seafarers. Unlike Lapérouse's ships, terrestrial exploration was small scale: instruments had to be portable, food had to be found as one went along, disguise was often necessary. Some travelers used their own body as an instrument. During the course of their South American travels between 1799 and 1804, for example, Alexander von Humboldt and Aimé Bonpland used their bodies as surrogate barometers—their noses bled at certain altitudes and barometric pressures—and as electrical devices (in conducting experiments with Amazonian electric eels).[25] Peter Simon Pallas used his body's temperatures to judge the habitability of harsh climates during his six-year Siberian expedition beginning in 1768. Everywhere explorers went, the facts of geography demanded that instruments be locally calibrated. For James Rennell, mapmaker, geographer to India, and, after 1798, a consultant engaged in filling in the blanks on the map of Africa's interior, this meant camels. Given the difficulty of obtaining a terrestrial and positional fix by celestial observations and "because few or none of the African travelers carry watches with them," time distance between known places in Africa, argued Rennell, could only be calculated using the speed of camels. He based his claims on the reports of European travelers in Arabia: "it appeared to me, that if the African caravans are composed of the same kind of camels, and are governed in their motions and economy by the same circumstances, as those which cross the Arabian deserts; there is no scale, of the computed kind, that can be more applicable to the African geography, than that formed on the camel's rate of travelling."[26] In such ways, Enlightenment travel involved bringing the world to light less by imposing a single universal standard—as is often argued—than by calibrating others' local standards with a view to ensuring, in time, commensurability over space.[27]

At sea, knowing time distances and getting directional fixes were much greater problems. Measuring latitude, one's position north or south of the equator, was reasonably straightforward. Knowing longitude, the distance east or west from a given or "prime" meridian, was not. Astronomical measurement was unreliable. The solution rested in knowing the time at the prime meridian, since longitude can be calculated from the difference between local time (found by in situ observation of the sun at its highest elevation) and the time at the prime meridian. What was needed was a timepiece that would retain its original time throughout long voyages: gaining or losing against Greenwich time was not a problem if the daily rate of change—the "going rate"—was known. The solving of this problem between the British government's Longitude Acts of 1714 and 1773, when the Board of Longitude finally rewarded the clock maker John Harrison for his marine chronometer, illustrates the high value placed on maritime instrumentation in Enlightenment exploration. It was several years, however, before more than a few explorers had access to chronometers.[28]

Longitude's solution is less an unproblematic narrative of scientific advance than a complex tale of social competition and local geographical credibility. Harrison's chronometer had to compete against the lunar-distance method (a means of calculation in which longitude could be found by measuring the angular distances between the moon and the sun or stars), advocated by James Bradley, the astronomer royal, and by the astronomer Nevil Maskelyne. Both methods were field-tested—Maskelyne's at Saint Helena during his observations of Venus in 1761, Harrison's on several occasions, notably in sea trials in 1761 and 1764 between Britain and the West Indies. Testing accurately depends on knowing particular geographical locations with certainty—the Royal Observatory at Greenwich, for example. Embarrassingly for all concerned in the 1761 sea trials between Portsmouth and Jamaica, the island "moved" according to different longitudinal readings: no one could confirm where Jamaica was, longitudinally speaking.[29]

My point is that the facts of instrumental measurement are always geographically contingent: location makes a difference to the production of knowledge and to the contexts of its reception and justification. This fact was plain enough to George Vancouver, the navigator who sailed with Cook.

Vancouver's objective in his 1790s survey of Alaskan waters was to test new technology, chiefly chronometers. But for such knowledge to "travel," to be useful in other places, at-sea longitudinal measurements demanded land-based corroboration. Vancouver's choice of Nootka Sound, on the west coast of the island that now bears his name, was politically charged: the site, of trading importance to the French and Russians, was fought over by the British and the Spanish in the Nootka Sound Crisis of 1790. Nootka became Vancouver's point of reference, a regional equivalent of Greenwich. Yet cold weather and fog prevented the land-based calibration of deck-based longitude. Instruments broke, and even when working they had different going rates. As he headed homeward from Cabo San Lucas on the California peninsula (whose longitude he accepted since it had been used in observing the 1769 transit of Venus), Vancouver noted only that "from St. Lucas in California to Cape Douglas in Cook's Inlet . . . the position of the western coast of America would be found *tolerably correct*."[30]

Whether by ship, camel, or marine chronometer, Enlightenment explorers' encounter with instrumentation and with method did not straightforwardly provide precision and accuracy in geographical terms, even though the motive was to render space measurable in consistent ways. It could not do so because such notions are always locally constituted. The fact that explorers and others exercised a tolerance in and of their instruments is echoed in their negotiations with other humans. One geographical and cartographic problem high on Lapérouse's list, for example, was Sakhalin in the Russian Far East: was it an island or a peninsula? Geographically imprecisely known, Sakhalin was a site of political and commercial tension between the Russians, the Chinese, and the Japanese, just as Vancouver Island was for the British, French, and Spanish. On one occasion, Lapérouse and members of the region's indigenous peoples met on the shore. Neither securely on land nor at sea, the Europeans quizzed the natives over the nature and extent of Sakhalin: an Ainu man drew lines in the sand to indicate that Sakhalin was, indeed, an island. Only later was his information recorded on paper. Only later still was the native's paper sketch combined with Europeans' field measurements and graphical conventions to become a fact of geography—as a map—in any "standard" sense.[31]

This moment of cross-cultural encounter on Sakhalin's shoreline in May 1787 illustrates several additional things about Enlightenment travel and exploration. Navigation meant not just crossing geographical space safely, guided by one form of instrument or another, encountering continental margins, and pressing beyond them but negotiating and accommodating cultural differences and meanings. Exploration, at least the oceangoing kind, was larger scale but in its contact with natives was more fleeting than land-based travel. Moving over land, some travelers learnt their hosts' language, not for security but to allow better understanding. Michel Adanson's 1757 *Histoire naturelle du Sénégal,* for example, is based on his own discussions with the Senegalese. Missionaries—foreigners but residents—often stayed longer: volume 9 of the *Lettres edifiantes et curieuses* (1780–83), the twenty-seven-volume account of Jesuit missions in the Amazon, records Jesuits grappling with 150 different native languages. Exploration and travel were everywhere exercises in translation.

Plotting moments of geographical encounter, charting them on the Enlightenment's "Great Map of Mankind" as Burke put it, or, in Lapérouse's case, determining ethnic differences within Sakhalin's native populations, necessitated different languages in order to make sense of geography: gestures, signs in the sand, spoken words, partial written traces. Maps and charts provided milestones—but not always securely named ones—in a process of "ethnographic navigation" (see fig. 10). In his *Discoveries of the French in 1768 and 1769* (1791), for example, Charles Fleurieu documented the Pacific worlds as, first, French space before considering, second, the subsequent visits of the English navigators "who gave them new names" and, third, the Spanish, who often gave yet different names to the same places. As Fleurieu put it, "geographical discoveries are a kind of property, less useful, without doubt, than territorial property, and forming only an imaginary wealth: but as they are connected with national self-love, unsubstantial as they are to the possessor, they have been at all times, envied and disputed."[32] Natives thought Enlightenment explorers no less strange than Europeans did them, as each tried to fit the other into different worldviews. The Maoris, for example, who initially thought Cook's *Endeavour* a floating island, later referred to European navigators as "goblins."[33]

Mapping and Inscribing

Lapérouse's moment of enlightenment and "ethnographic navigation" on Sakhalin's shore speaks to broader questions of Enlightenment mapping and inscription. Despite the presumed certainty of its language of lines and symbols, a map is not an immediate and a static accomplishment so much as a process aimed at achieving some sort of commensurability: between different claims to knowledge, and between the map and the world it portrays. Maps are only scaled representations of the world, not mirrors of it. Of necessity, maps distort, reduce, and symbolize and do so in different ways and places.

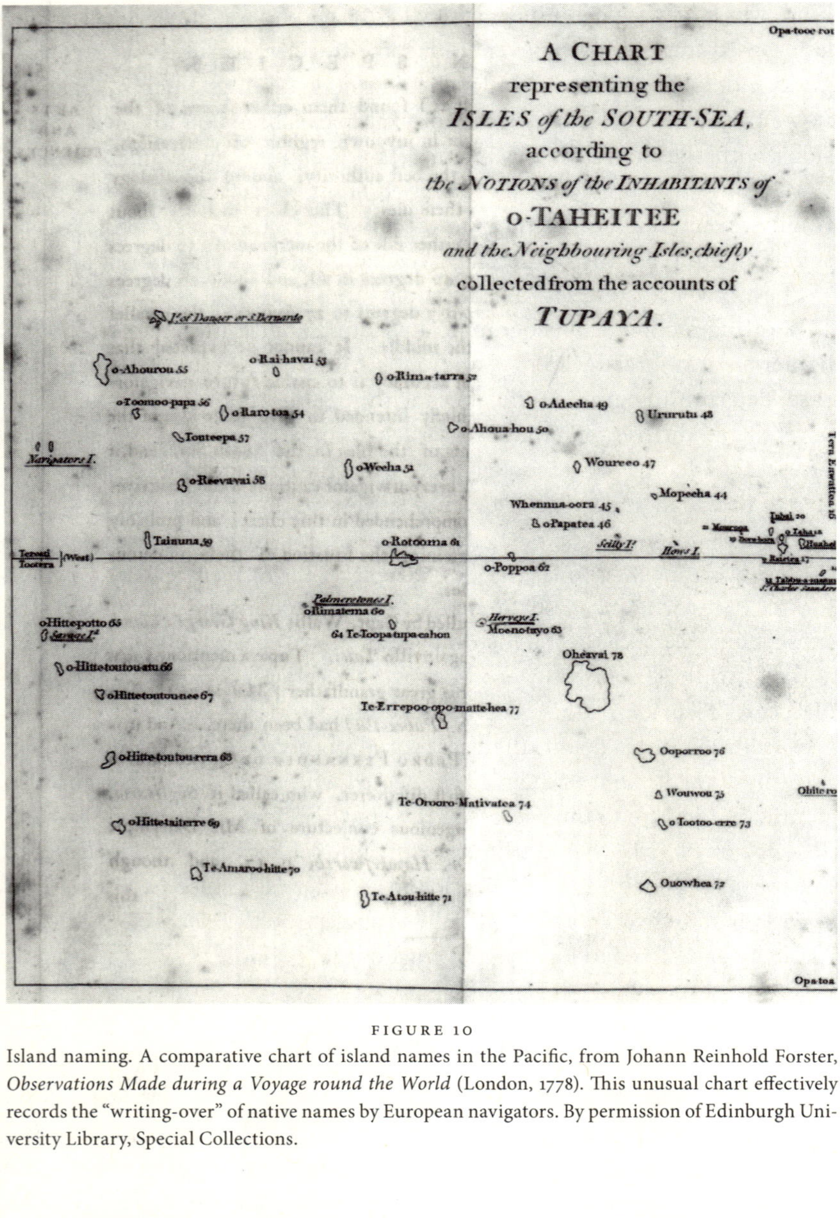

FIGURE 10

Island naming. A comparative chart of island names in the Pacific, from Johann Reinhold Forster, *Observations Made during a Voyage round the World* (London, 1778). This unusual chart effectively records the "writing-over" of native names by European navigators. By permission of Edinburgh University Library, Special Collections.

The Enlightenment brought new mapping projects and new types of mapmakers into being and ushered in new ways of looking at the world. By the end of the seventeenth century, four groups of "cartographic practitioners" were active in most European countries: mariners, geographers, commercial publishers, and land measurers. Their modes of mapping were

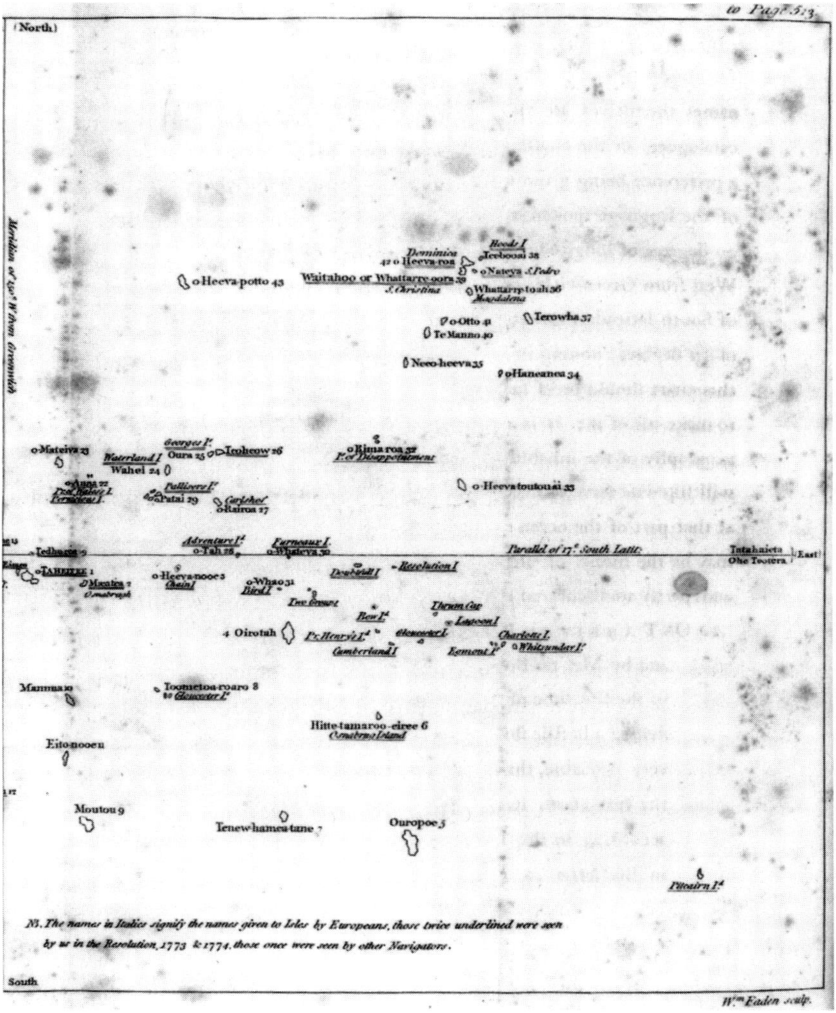

maritime charts, small-scale maps of the world or its regions, and large-scale land surveys. Interest in large-scale territorial survey—part of emergent systems of taxation and civic administration—prompted the appearance in the Enlightenment of a new fifth group, the professional military or "state surveyor."[34] With their base in academies, Enlightenment societies, or military schools and nourished by a faith in the usefulness of "mathematical cosmography," these men led the reordering of geographical space in the Enlightenment by using maps as a technology of relative spatial knowledge.[35]

Maps are geographical and political instruments. But as Fleurieu and others knew, they are not automatically documents of absolute truth. These claims may be illustrated with reference to France.[36] Jean-Baptiste Colbert, minister-in-chief to King Louis XIV of France, established the Paris Observatory and the Académie Royale des Sciences in Paris in 1666 specifically to advance mapmaking and to produce maps necessary for commerce, transport, and the military. Under the guidance of Colbert and the mapmaker and mathematician Jean-Dominique Cassini, "the gentlemen of the Academy" subjected France for the first time to coordinated mathematical measurement—and shrank it, by several hundred square miles, notably on the Mediterranean and the western coasts (fig. 11). Political knowledge about geographical space depends, in part at least, on the particular type of instrumentation employed and on its purpose. Colbert and others used triangulation: the measurement of a short distance—usually, in British terms, about five miles or so—from which, using a theodolite, a third point visible from both ends of the base line can be sited and distances measured using trigonometry. This process is repeated across the ground to be mapped, producing a series of triangles of known dimensions within which the topography was determined by eye and sketched in. This method of triangulation and topographic "in-filling"—the mathematization *and* artistic depiction of space—is a definitive feature of Enlightenment mapping.

France may not have been as big after 1682 (whatever its imagined reach), but its continued mapping in the Enlightenment illustrates how maps constitute geographical identity rather than reflecting it. With renewed emphasis after 1733 and working from a new set of triangles, the French under Jacques Cassini, Jean-Dominique Cassini's son, and Jacques's son, César-François Cassini de Thury, undertook the world's first scientific trigonometric survey of national space. The results were published in 1744 as *La carte générale des triangles de Cassini.* From 1747, an even grander scheme for the nation's mapping was hatched. This was Enlightenment geographical and national science at both small and large scale: small scale since greater levels of detail were expected (at a scale of 1 inch to 1.36 miles, or 1:86,400), but large scale given central coordination from Paris, the appointment of ten teams of surveyors, and a target of 180 map sheets to be published at a rate of ten sheets a year. Under the direction of Cassini de Thury until his death in 1784, all of France except Brittany was mapped. The project was the cartographic equivalent of the *Encyclopédie.*[37] France's geographical constitution through mapping continued throughout the Revolution: de Thury's son, Jacques-Dominique Cassini, known as Cassini IV, completed in 1793 what his namesake, Cassini I, began over a century earlier. And at

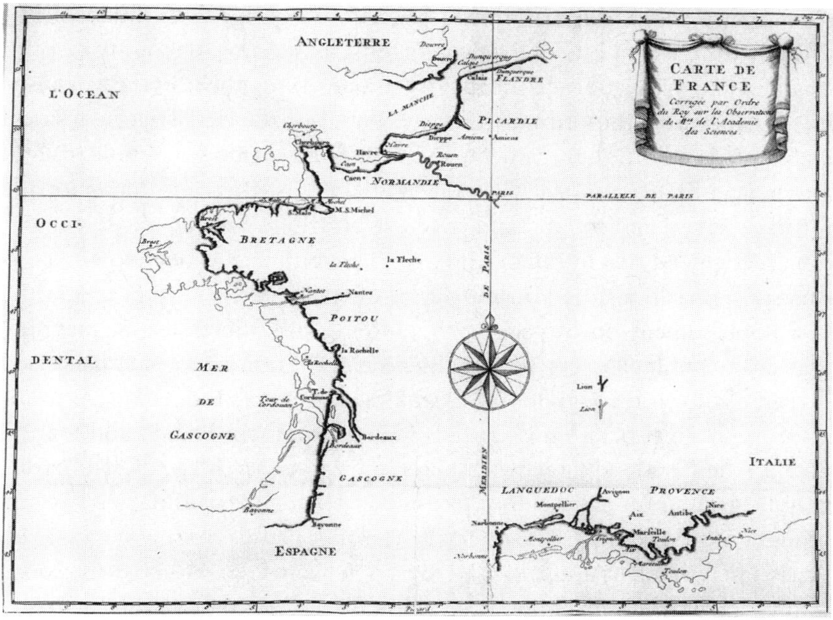

FIGURE 11

Map of France, with imagined and corrected outlines compared. The map, engraved in 1693, shows the imagined extent of France in thin lines, and, in thicker shading, the correct extent after geodetic measurement by Jean-Baptiste Colbert and others: "Carte de France corrigée par ordre du Roy sur les observations de Mrs de l'Académie des Sciences." By permission of the Trustees of the National Library of Scotland.

its completion, of course, because France's geography had changed in the intervening century, the project of national mapping began again.

In France particularly, national mapping was also natural mapping. Late seventeenth-century French geographical texts argued that France's frontiers had been marked out by nature—by its river systems and its mountains to the south and east. French geographers and others elaborated on this belief. For Rousseau, "we may fairly say that the political order of the Continent is in some sense the work of nature." Political arrangements furthered this notion of France's "natural space": the government signed thirty territorial agreements with its neighbors between 1738 and 1785. In the Enlightenment, the very idea of France as a geographical entity depended on belief in the naturalness of its political boundaries as well as on trust in the geometric power of the Cassinis and others.[38]

Other European nations followed France's lead. From 1747 onward, military engineers mapped Scotland, beginning with the Gaelic-speaking Highlands after the 1745–46 Jacobite Rebellion. This was mapping as political surveillance.[39] The original emphasis on military needs, "ordnance," together with the activities of the Society for Arts' scheme for county maps of England and Wales (see also chap. 9), was one prompt behind the establishment in Britain of the Ordnance Survey (initially called the Trigonometrical Survey).[40] In the Enlightenment, the military became map-minded, and mapping military-minded. General William Roy, the coordinator of the Military Survey of Scotland and advocate of Britain's systematic triangulation, recognized this in remarking that "if a country has not actually been surveyed, or is but little known, a state of warfare generally produces the first improvements in its geography."[41] Roy's work illustrates for mapping that cosmopolitan nature to Enlightenment natural knowledge discussed earlier, for it was through his work and Joseph Banks's influence that triangulation schemes begun by France connected southern Britain and France, notably the Paris and Greenwich observatories, in an international "cartographic grid" from 1787.[42] Elsewhere and at other times, military surveyors brought order through the map to the Hapsburg lands from 1763 to 1787, to Prussia between 1767 and 1787, and to Saxony from 1780. From 1793, responsibility for France's mapping, invested between 1747 and 1789 in the many *ingénieur-géographes militaires* of the second Cassini survey, was transferred to the Dépôt de la Guerre, the army's cartographic branch. And beyond France, French colonies were made to appear like France. The maps in the multivolume *Description de l'Egypte,* produced from trigonometrical survey, engraved in 1808, and published in 1818, are remarkable for their style and underlying ideology: "ancient" Egypt was cartographically colonized, made through mapping to appear a natural extension of "modern" France.[43]

What the French did to themselves after 1682, after 1747 and from 1791, and to Egypt from 1798, the British did in Bengal (1765–71), in Quebec (1760–61), elsewhere in North America (1764–70), and in Ireland (1778–90). In Russia, official instructions from 1721 on mapping (modeled on the French) emphasized geometrical accuracy over breadth of coverage. Such was Russia's size and diversity, the state so unsure of its own geographical frontiers, that mapping helped "fix" its peoples as much as its places. "In all the other states of Europe, the people who live under the same government are very little different one from the other, but this is not the case with all the people who compose the great Russian empire, where the inhabitants of some provinces differ as widely from those of the others as a Chinese does from a Hottentot."[44] As William Roy observed in 1787, mapping of the Russias was needed not just for its own sake but to open up a space for international geodetic collaboration and corroboration: "an Empress who commands so great a portion of the world . . . can scarcely be supposed to suffer a region . . . to expire, without directing something of this sort to be executed under the polar circle, or as near to the pole as the severity of the climate will permit, by way of confirmation or correction of the Lapland measurement, which stands single and by itself, without any collateral proof of its correctness" (see also chap. 6).[45]

Across the globe, large-scale Enlightenment mapping projects represented a new form of cartographic literacy—recognition of the utility of maps as geographical instruments of statecraft. Understanding precisely what Enlightenment mapping did in ordering space and quite how depends, however, on where one looks and at what scale. Other sorts of maps—enclosure maps of the agricultural landscape, property maps, road maps, town maps, maps of postal services—differently reflected local emphases on rationalizing space and place. Again, France provides an illustration, for there, despite moves to a unifying grid, regional differences in culture, language, and in the way maps were made underlie the idea of natural-national space—and fracture notions of an "essential" cartographic Enlightenment. In mapping Languedoc in the south, Paris-based mapmakers had, in theory, local equivalents to help coordinate their efforts as a result of the presence of astronomer-geographers within the Montpellier Société Royale des Sciences. In reality, local practices of instrumentation determined otherwise: because Montpellier academicians aligned to the tops of hill without reference to their base, and thus to their height, "in-built" local discrepancies did not accord with national maps centered on Paris. On-the-ground checking by local informants was necessary to make mapping practices meet: priests as trustworthy sources to verify academicians' figures, peasants for land names.[46] In Brittany, by contrast, no one at all could be relied upon. Parisian mapmakers—emblems of a French-speaking authority—were regarded with mistrust. For César-François Cassini de Thury, "the course of my travels in Brittany nearly cost me my life." His men were stoned, his father shot at, locals refused all hospitality: "The *seigneurs* knew neither how to read nor write, and not having the least notion of geography they denigrated my work and tried to incite others against it—whereas in foreign lands, my name alone draws respect."[47]

Cassini de Thury's problem with respect to Brittany—that his name traveled more easily than he could—happened everywhere in one way or another. Behind the map, "beneath" its surface languages of line and symbol, Enlightenment mapmakers, particularly those venturing beyond the edges of continental margins, depended upon accommodating others' knowledge, upon negotiating different views altogether about maps and geographical space. As French and British explorers pushed inland into the North American continental interior, for example, their activities are traceable by the maps they left. But unless used circumspectly, the term "mapmaker" is misleading: geographic and cartographic knowledge was provided by missionaries, traders, Hudson's Bay Company trappers, as well as by natives. In his *Geographical, Historical, Political, Philosophical and Mechanical Essays,* the American mapmaker Lewis Evans noted that because "the Country is yet a Wilderness," parts of the map accompanying his book showed rivers by their nature—"fast," "broad," and so on—not as surveyed features. One map of 1760 by Moses Norton, the Métis governor at Churchill, which showed the northern parts of Hudson's Bay, "laid down on Indn. Information" as Norton puts it, is by no means unusual.[48] In the Indian Subcontinent, European administration depended on collaboration with native intermediaries. Rennell's 1783 *Memoir of a Map of Hindoostan,* completed during his time as first surveyor general of Bengal, figuratively depicts this reliance—and Britain's expectant gratitude—on indigenous sources (fig. 12).[49] For Joseph Banks, president of London's Royal Society, which awarded Rennell its Copley Medal in 1791 for his surveying, the "margin" was better mapped than the "center": "Would I could say that England proud as she is of being esteemed by surrounding nations the Queen of Scientific Improvement, could boast of a general map as well executed."[50] What Rennell was doing for the mapped representation of British India, William Jones did in his work for the Asiatick Society of Bengal, whose scope was delimited by "the geographical limits of Asia" and whose work included correcting "the geography of *Asia* by new observations and discoveries."[51]

Other moments of cartographic exchange were, as noted, less formal and did not take place in institutions or on paper. Like Lapérouse's on the beach in 1787, Cook's attempt to solicit information about the coastline of New Zealand from Maoris on board HMS *Endeavour* in November 1769 began with him sketching—in charcoal on the deck—then pointing to the land. After a native traced the outline of North Island, Cook "took some white stuff [paper], on which he made a copy of what the old chief had made on the deck." Exploring New Zealand in 1773 on his second voyage, Cook named one unexplored inlet Doubtfull Harbour, another No Body Knows What. In 1791, Vancouver renamed it only slightly more definitively: Some Body Knows What. As maps traced out contact narratives and traced over native identity, even as they sought to record it (see fig. 10), so they helped inscribe different historical and geopolitical "traditions": Native Hawaiians' views of Cook, although far from uniform, give a different view based on their geographical experience than do many Western accounts.[52]

FIGURE 12

"Mutual Enlightenment," the title cartouche from James Rennell, *Memoir of a Map of Hindoostan* (London, 1783). Britannia accepts knowledge of the Indian subcontinent as military and commercial ties also bind Europe and Asia. By permission of the British Library (IOR X/223).

Enlightenment mapmaking thus emphasized empirical observation and quantifiable measurement. Yet as these examples show, Enlightenment mapping was more a complex series of exchanges specific to particular times and places—"islands of truth" if you will—than an immediate overlaying of "core" European ideas on passive people on the world's "margins."

And because European map producers worked to certain conventions, the consequences of these geographical encounters were always uneven. Enlightenment maps, products of moments of "cultural engagement" in particular geographical circumstances, were as commonly products of "cultural effacement"—that is, they "wrote out" because they "wrote over" the conditions of their making.[53]

Envisioning and Publicizing

Mapping was one form of visualizing knowledge about the world during the Enlightenment—a powerful and portable "geographicacy"—but it was not the only one. In travel accounts and in paintings, Enlightenment writers, commentators, and artist-naturalists actively envisioned and represented foreign places and peoples and the unknown closer to home. In Highland Scotland, for example, the mapping work of the Military Survey was accompanied by that of the landscape painter Paul Sandby, whose artistic depictions, including surveyors at work, helped establish a particular aesthetic view of an upland wild geography that was as foreign to most people as the Pacific world was.[54] In the South Seas themselves, painterly depictions of foreign geographies were increasingly concerned with accuracy. In that sense, the representation of Enlightenment exploration has itself been described as a voyage—a "voyage into substance"—in that the development of a "plain" style of landscape representation happened at the edges of Europe's Enlightenment world. Traveling, encountering, and exploring demanded and gave rise to a "neutral" style of image, a "scientific" image, aimed only at showing things as they really were.[55]

This view is open to criticism. In terms of eighteenth-century practices of topographic representation, the distinctions between art and science were never fixed and must be understood, moreover, in relation to geographical location and particular artists' concerns. Oceanic artist-naturalists like William Hodges and Sydney Parkinson of course aimed for accuracy of view. For the facts of foreign geographies to travel, pictures of them had to be realistic, had to be capable of "translating" into conventions and conceptions understood by their audiences. Yet such aesthetic practices were taking place in topographic depictions of the Alps and of other mountainous parts of Europe at much the same time: in and of Europe itself, not just at the frontiers of European intellectual expansion.[56] Questions of location matter even there, for in late eighteenth-century Brazil and in British Guiana, the "art of tropical travel" was different: the local effects of tropical heat and light undid attempts to establish general aesthetic conventions.[57]

Maps, paintings, and written accounts both aid exploration and are the result of it. For many people, such things allowed a form of sedentary travel. Even of those who could travel and did, not everyone was a *voyageur naturaliste,* mapmaker, or artist-naturalist. Men—and a few women—of taste and breeding traveled throughout Europe on the Grand Tour, particularly to France, Italy, and Greece, where the remains of classical civilization provided education and aesthetic refinement. For Goethe, Rome was "die Schule für alle Welt" (the school for all the world). Travel in this sense equated to education: broadening the mind through experience of others' geographies.[58] The Enlightenment was also characterized by works of imaginary travel—such as Montesquieu's *Lettres persanes* (1721), Jonathan Swift's *Gulliver's Travels* (1726), and Voltaire's *Lettres philosophiques* (1734)—that offered moralizing perspectives on other peoples and their customs and an opportunity to reflect on one's own. China, as we have seen, was unknown to most in the West. D'Anville's maps in his *Nouvel atlas de la Chine* (1737) incorporated partial information from a Swedish military officer and from Jesuit missionaries, but his depictions of much of the country were not based on a detailed firsthand encounter. Yet that did not prevent Rousseau—given the absence of reliable sources for his refutation it may even have encouraged him—from commenting critically on science's place there as a basis to enlightenment: "There is in Asia a vast country in which the sciences are a passport to the highest positions in the State. If the sciences really purified morals, . . . if they really inspired courage, then the people of China would assuredly be wise, free, and invincible. But as a matter of fact, there is no sin to which they are not prone, no crime which is not common amongst them."[59] Rousseau in *Discours sur l'origine et les fondements de l'inégalité* and Voltaire in *Candide* (1759) likewise used the idea of geographical travel as a narrative device without either the author or the reader having to travel. The poet William Cowper recognized just this point in *The Task,* based on Cook's voyages:

> Thus sitting and surveying thus at ease
> The globe and its concerns, I seem advanced
> To some secure and more than mortal height,
> That lib'rates me and exempts me from them all
> It turns submitted to my view, turns round
> With all its generations, I behold the tumult and am still.

And further, how

> He travels and I too. I tread his deck,
> Ascend his topmast, through his peering eyes

Discover countries, with a kindred heart
Suffer his woes and share in his escapes,
While fancy, like the finger of a clock,
Runs the great circuit, and is still at home.[60]

Because philosophers and poets wrote about places without traveling to them and explorers who did travel had to write, for Europeans, about the geographically unknown in ways their audience would understand, distinguishing in the Enlightenment between the philosophical traveler or sentimental traveler and the scientific traveler/explorer is not easy.[61] Usually, the last named is marked by the larger scale of the undertaking and by claims to authority that rested either in having seen the phenomena in question or in their instrumental measurement and assessment by others. Even so, securing new geographical information, even the information itself, could be dangerous. In Malaspina's case, the new facts he brought back and his claims about them were at first neglected and later judged seditious: Malaspina was jailed and then exiled by the Spanish authorities. He at least returned. Cook did not, killed by islanders after unwittingly transgressing local customs, a fact that has prompted opposing interpretations as to how Cook should be understood in the Enlightenment.[62] Lapérouse and his ships did not. Sailing west from North America in 1788, his expedition vanished without trace. His journal survives only by an act of foresight: Lapérouse had dispatched it to France some days before.[63]

All this is to observe that the facts of exploration, travel, and mapping added to the sum of Enlightenment knowledge only when (and if) they made a final move—into print. The litmus test of this "voyage into narration" was utility: did it increase knowledge?[64] For Pallas, the German edition of his Russian travels followed the exhortations of the director of the Saint Petersburg Academy of Sciences to publish "in order to respond to the eagerness . . . of the scholars of Europe." It was not just scholars who were eager. Among popular audiences in London in 1772, "the people who are most talk'd of at present are mr Banks and doctor Solander"—the naturalists on the *Endeavour,* Cook's first voyage (see plate 6). Between 1770 and 1800, over a hundred editions and impressions of works giving details of the new discoveries were published in Britain alone. Library borrowings testify to public interest in tales of Enlightenment exploration: works by Bougainville, Forster, and Cook were borrowed dozens of times from Bristol Library between 1773 and 1784. The work most frequently borrowed was John Hawkesworth's 1773 three-volume compilation, *Account of the Voyages Undertaken by the Order of His Present Majesty for Making Discoveries in the Southern Hemisphere.*[65] Cashing in on public interest, Hawkesworth's is also a cautionary tale: he was vilified as a compiler for altering others' narratives and making too much for public sensibility of those scenes of public sexuality witnessed in Tahiti by Wallis, Banks, and Cook.[66] As James Bruce, author of *Travels to Discover the Source of the Nile* (1790) also found to his cost, travel did not necessarily make truth in the eyes of one's readers. His work, belittled by Samuel Johnson and parodied by Erich von Raspe as the *Adventures of Baron Munchausen,* was not confirmed as truth until twenty years later.[67] Being "out there," the Enlightenment *voyageur naturaliste* had to control his ship, his crew, his instruments, his words—and himself. "Back home," he had to control his knowledge, his sources, and his reputation: the aim of securing geographical knowledge in the Enlightenment had to negotiate the temptation to pander to public demand, the claims of rivals, and the skepticism of one's peers.[68]

Explorers, mapmakers, and writers encountered, mapped, pictured, and represented the Enlightenment globe, but they did so, at least initially, locally. The world was not at once brought into view. Geographical knowledge was central to how the world came to understand itself in the Enlightenment—to how, indeed, the earth came to be known as a world—but such knowledge was gathered piecemeal, in particular ways and places before bigger claims were made about it. In effect, geographical knowledge in the Enlightenment made and left a network of uneven traces—as ships' wakes, hesitant steps on the shoreline, and in mapped but broken lines of national spaces and continental edges (as, for example, in fig. 2). Encountering the globe in such ways established a program for later discovery and a contemporary basis for understanding the physical world and the diversity of its human occupants.

6

Encountering the Physical World

In the Enlightenment, the earth was the subject of scientific study as never before. The previous chapter disclosed how and where Enlightenment explorers, mapmakers, and travel writers went about their work and with what geographically uneven consequences. Here, I examine what George Vancouver succinctly termed the "ardour of the present age, to discover and delineate the true geography of the earth."[1] In the undertaking of this "true geography," the encounter with the physical globe took the form of primary survey, that is, it involved the "in-the-field" enumeration of natural knowledge, its classification, and the role of system. To those undertaking in various ways what contemporaries understood as physical geography, *naturbeschreibung,* or *la physique du globe,* the task was huge, the object seemingly without order. As Georges-Louis Leclerc, comte de Buffon, noted in his *Histoire naturelle* (1749–89), "on the surface of this immense globe we are presented with heights and depths, prairies, seas, marshlands, rivers, caves, chasms and volcanoes and, on first inspection, we find no regularity, no order in all of that." Buffon exactly captures here two things about natural philosophy and Enlightenment study of the terraqueous globe: the sheer scale of what there was to know, and the role of systematizing what was known and how it was known.[2]

Several other themes mark Enlightenment encounters with the physical world. One was what we now may think of as the secularization of the world, consequent upon new ideas about the earth's size, richness, and age. As the botanical sciences scrutinized associations between plants and geographical proxies such as altitude and latitude with a viewing to determining both continental patterns and local causal links that might be harnessed to promote greater return, the geological sciences revealed a dynamic object with a longer history than earlier supposed. Pre-Enlightenment interpretations of earth history were exercises in unequal reconciliation: between observable features—what we now take for granted as fossils, sedimentary deposition, stratigraphic superposition, igneous intrusions, and so on—and biblical accounts about the Flood and the earth's age, with the scriptures being presumed true. It was commonly accepted that the earth was no older than 4004 BC. Late seventeenth-century natural history and natural theology were closely associated, each recognizing that scientific reason, correctly applied to the world, would lead to knowledge of God. The natural world, it was claimed, was a divine production and science a revelatory activity that served the God whose products and role as Nature's creator it disclosed. The naturalist John Ray illustrated this idea of the argument from design in his 1681 *The Wisdom of God Manifested in the Works of the Creation*. In one form or another, his views were repeated by numerous other "physico-theologists"—men like John Woodward in his *An Essay toward a Natural History of the Earth* (1695) and Thomas Burnet in his 1684 *Sacred Theory of the Earth*.

Yet the new science meant empirical science—exploring and seeing for one's self, not trusting to scriptural or ancient authority. "Let us not suffice us to be Book-learn'd," wrote Ray, "but let us ourselves examine Things as we have Opportunity, and converse with Nature as well as Books. Let us endeavour to promote and increase this Knowledge, and make new Discoveries. . . . Let us not think that the Bounds of Science are fix'd."[3] During the Enlightenment, facts resulting from geographical exploration of the physical world raised doubts about the argument from design and the age of the earth. New modes of thinking, which sought explanation in rational physical cause, began to appear as different "theories of the earth." The title of Burnet's 1684 work and that of James Hutton's 1795 *Theory of the Earth* are not dissimilar. The omission from the latter of the word "sacred" is illustrative nevertheless of that wider acceptance in the later eighteenth century of scientific explanation and theorization based not on theology but on observation and empirical encounter.

In another sense, studying the natural world in the Enlightenment was less "natural" than for earlier periods. Enlightenment thinkers "invented" or, at the least, laid the intellectual and institutional foundations for what we now see as disciplines or scientific fields: botany, geodesy, geology, meteorology, hydrography, even oceanography.[4] Natural history, for example, changed its meaning in the Enlightenment, away from a rationale of holistic gathering and the study of symbolic associations in nature toward more systematic and specialist languages of classification and measurement.[5] Such geographical in-the-field work helped establish new fields of learning because "thinking spatially" was an important element of this "new" natural history.[6] Further, geographically encountering Buffon's "immense globe" prompted new questions about meaning: How was the globe's "geographical architecture" to be explained if not as the work of God? Why did plant and animal species have the geographical variation they did? Could nature be explained, not just catalogued?

Emergent specializations threw up particular questions—about the age and size of the earth, the cause of earthquakes, about ocean currents, the chemistry of the air, and so on. Because natural knowledge of the globe went hand in hand with an interest in the political consequences of nature's economy, natural knowledge had strong associations with economic utility. Nature's exploration was often a precursor to nature's exploitation. It is for these reasons that the Enlightenment was also distinguished by questions of environmentalism as a philosophical concern: Was the earth in decay? What moral responsibility, if any, do humans have to the environment? What were the connections between human variability, between the distribution of humans across the globe, and the earth's physical environments?

In giving purchase to these remarks, this chapter addresses four themes in order to illustrate the ways in which Enlightenment thinkers made geographical sense of the earth as a physical object. Three of the themes are the geographies of the plant world, the role of flood and of fire concerning the earth's internal dynamism and age, and meteorology and oceanography in the Enlightenment. I begin, however, with that deceptively simple question posed of the physical globe in the Enlightenment: What shape is the earth?

Putting the Earth to Shape

The shape-of-the-earth debate was a question central to Enlightenment interest in what we might now think of as terrestrial physics.[7] The answer was far from clear. The solution depended greatly upon geographical fieldwork. And as I shall also show, the implications of this problem and of its solution illustrate several of the themes of this book, for in considering the question and its solution, we can discern something of the varying geographical reception of Newtonianism in the Enlightenment and of the way local social interests were played out in different locations.

In Proposition III.19 of his *Principia Mathematica* (1687), Newton argued that the earth, originally a homogenous fluid mass, had been acted upon by universal gravitation and centrifugal force due to the earth's rotation. The result of such actions, he reasoned, was that the earth would be ellipsoid in shape: the earth, as a result of these physical principles, should bulge at the equator and be flattened at the poles. For Newton, the world was pumpkin shaped. Empirical evidence seemed to support him. Measurements undertaken in 1672 (repeated in 1682) by the French mathematician Jean Richer—who was sent by the Parisian Académie Royale des Sciences to French Guiana to test the theory—showed that pendulums of the same length swung more slowly in Cayenne than they did in Paris, France. For

Newton, arguing as he was for a law of universal gravitation, this was proof that points on the equator were farther from the earth's center. A further element of the earth's oblate shape was that, in a Newtonian world, a degree of latitude would be lengthened as one moved away from the equator either north or south toward the poles.

Not everyone agreed with this view of the world's shape or with these underlying mathematical principles. For the French mathematician and philosopher René Descartes, the effects of gravitational forces—what he considered "vortices" swirling about the earth—would be otherwise: in a Cartesian world, the earth was flattened at the equator and bulged at the poles. Like Newton, Descartes had evidence to "prove" his case, provided chiefly by the French astronomer-geographers and mapmakers Jean-Dominique Cassini and his son Jacques. Building on the work of Jean Picard, who in the 1670s had measured a degree of latitude along the meridian running through the Paris Observatory, the Cassinis extended their latitudinal measurements and noted a decrease in the length of a degree toward the poles. To some extent, these different views crystallized along national lines: British mathematicians and natural philosophers supported Newton; the French were generally Cartesian in supporting the Cassinis, not least given their role in establishing those features of French geographic authority, the Paris Meridian and the triangulation of French space. Debate on the issue took a further turn in 1732 with the publication of *Discours sur les différentes figures des astres* (Discourse on the different shapes of the heavenly bodies) by the leading French academician, mathematician, and physicist Pierre-Louis Moreau de Maupertuis, in which Newton's case was strongly supported.

In order to know the shape of the earth precisely, argued Maupertuis, it was vital to measure the earth's shape in the places where it was presumed to differ most—at the equator and at the poles. At Maupertuis's prompting, two expeditions were established to field-test the contrasting theoretical claims. One, under the command of Maupertuis and including the Swede Anders Celsius and the French mathematician Alexis Clairault, went north, into Lapland, in 1733. The other, nominally led by the mathematician Louis Godin but actually run by the geographer Charles-Marie de la Condamine, headed in 1735 for equatorial locations in Peru (which then included modern-day Ecuador). Maupertuis's expedition used English instruments operated by mathematically trained observers who understood Newtonian physics. La Condamine's expedition, staffed by mapmakers using techniques and instruments favored by the Cassinis, undertook detailed surveys in South America (see figs. 13 and 14), despite being troubled by internal squabbling,

FIGURE 13

Triangulation and the Quito meridian, by Pierre Bouguer, Louis Godin, and Charles-Marie de La Condamine, from Pierre Bouguer, *La figure du terre* (Paris, 1749). Note the hachuring in place of contour lines and the stylized representation of woodland cover outside the area being surveyed. By permission of Edinburgh University Library, Special Collections.

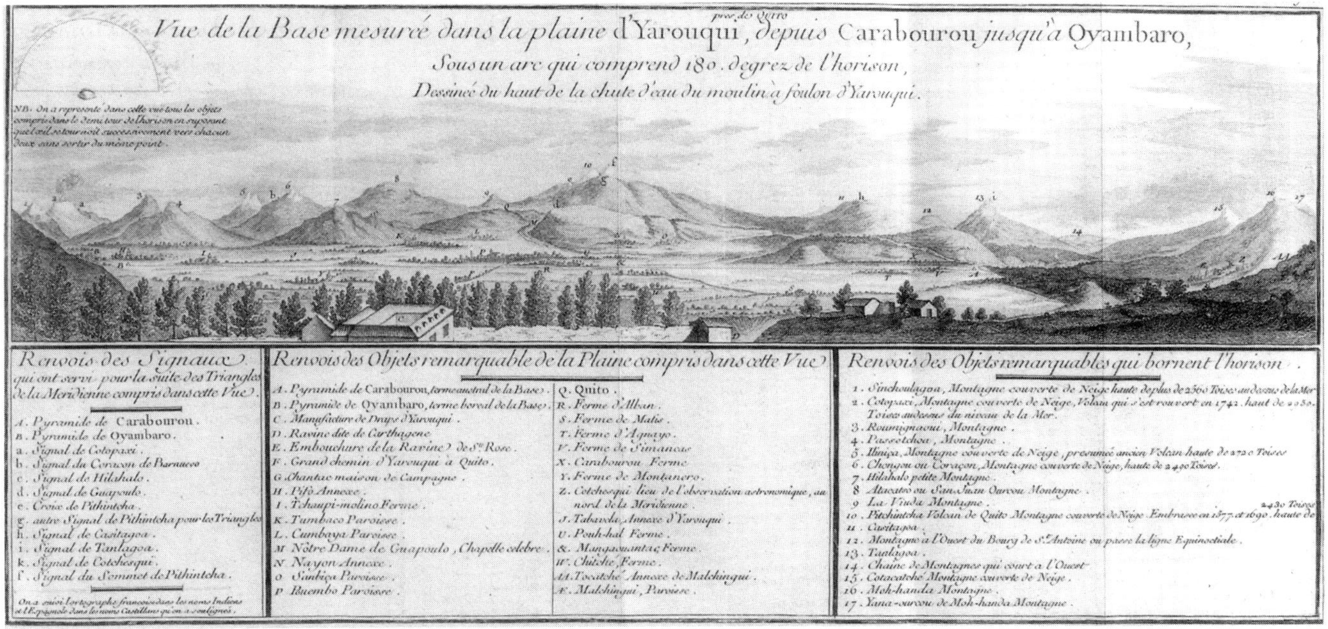

FIGURE 14

Topographic views of the landscape near Quito, from Charles-Marie de La Condamine, *Journal du voyage* (Paris, 1751). This form of landscape depiction helped to fill in the spaces formed by triangulation survey (cf. fig. 13). By permission of Edinburgh University Library, Special Collections.

civil war, earthquakes and volcanic eruptions, and the vicissitudes of colonial bureaucracies. The Lapland expedition returned home by 1737. La Condamine took over nine years to return. He spent much of his time mapping the Amazon, lost much of his work en route, and returned to find that virtually all his findings had been rendered useless: his fellow expedition member, the mathematician Pierre Bouguer, had got home sooner and already presented his work to the Académie. La Condamine was lucky: Joseph de Jussieu, the expedition botanist, returned to France only in 1771. Even then, the results of the Lapland expedition were not immediately agreed upon. Cassini II disputed the reliability and accuracy of its English-made instruments precisely because they were English and associated with a different, more practical natural philosophy. But, in time, the Newtonian worldview prevailed. Maupertuis is remembered, because he so saw himself, as "the man who flattened the earth"—and, of course, the Cassinis (see plate 3b).[8]

In one sense, this question of geodesy, the science of the earth's shape, had a precisely locational dimension: depending on where you stood, the world was a different shape. As Voltaire put it in his 1734 *Lettres philosophiques,* "in Paris you see the earth shaped like a melon, in London it is flattened on two sides."[9] The shape-of-the-earth debate was not simply a matter of national difference. It was also—and more so—a local disciplinary and social contest within Paris and in the Académie Royale des Sciences between Maupertuis and the Cassinis over the nature and political purpose of mathematical inquiry. In yet another sense, it illustrates the geographical variation in the reception of Newtonianism evident in the first fifty or so years of the eighteenth century, which bridges the Scientific Revolution and the Enlightenment. In a different period, Darwinism varied in its different local and social contexts—his evolutionary ideas had, in short, divergent social and spatial geographies of production and of reception.[10] So too Newtonianism has recently been studied in its contemporary scientific and popular forms, and in terms of the modern commemoration of Newton.[11] One element within this contextual reappraisal is the variation over space in engagement with Newtonian ideas. Initially, of course, Newton's theories were not understood by his peers in Cambridge, and not everyone there or elsewhere in England agreed with him even when they did comprehend his work. In Gascoigne's terms, "the mighty vessel of the *Principia* required humbler intellectual tugs to tow it out of harbor."[12] The supportive activities of the Scottish mathematician David Gregory, first in Edinburgh and then in Oxford, of Colin MacLaurin in Edinburgh in the 1740s, and the contrasting resistance of many French theorists suggest differing "contours of Newtonianism" in the Enlightenment—variations in what Newtonianism was held to be in the natural and in the moral sciences, and in regard to the significance placed on his work by different people in different places. And so it was in regard to Newtonian precepts underlying the question of the earth's shape: this was Enlightenment mathematics and geodesy as matters in competing intellectual worlds. The Lapland expedition succeeded because key institutions, personnel, and instruments were mobilized and calibrated *before* the expedition even took place as well as long afterward, as William Roy recognized in 1787 in calling for corroboration of the Lapland work. Maupertuis—or Sir Isaac Maupertuis, as Émilie du Châtelet termed him—negotiated a social space for the defense of precision along particular strategic lines by involving numerous European astronomers and the use of particular techniques in his schemes.[13]

Putting the earth to shape was not simply a matter of global variations in how the world was seen. It also meant standardizing—or attempting to anyway—the units used in earth measurement in very local scales. Visiting France in 1789, the English agricultural commentator Arthur Young raged against "the infinite perplexity of the measures." No place was like another: "They differ not only in every province, but in every district and almost in every town." His dissent was rooted in geographical variation. In Paris a *pinte* (a unit of volume) came to 0.93 liter, but to 1.46 liters in Saint-Denis. In Seine-en-Montage it was 1.99 liters, and in Précy-sous-Thil 3.33 liters. The geography of the *aune*, a unit of length commonly used of cloth, was even more varied: Paris alone had three, each relating to a different sort of cloth; Rouen had two. France as a whole had seventeen, each different from the other. Rennell, whose mapping work in India we have noted, faced just such difficulties in accommodating different linear scales. In places where no surveys had been undertaken by British mapmakers, the usual measure employed in Hindoostan was "the coss, or crores, commonly estimated at two British statute miles." But, continued Rennell, "I have not been able to get to the true length of the coss, as fixed by Acbar, and other Emperors; and, even if I had, it would be no use in the present enquiry, as all my Hindoostany itineraries and tables are in computed cosses." We should recall, too, that in the Revolution, the French adopted—if only for a time—standard "Revolutionary time" based on divisions of the year into twelve parts of thirty days each.[14]

A further illustration of the geographical basis to Enlightenment calculations about the size and shape of the earth concerns the meter. The debate over degrees of latitude did not go away with the gradual acceptance of Newtonianism or with Maupertuis's claims about a pumpkin-shaped world. And as we have seen, in France especially, mapmaking that demanded uniformity in order to work underlay the move to equate the country's national measurement with its natural and rational measurement, even as such concerns commonly accommodated and reflected local differences (fig. 15).

In 1792, the French astronomers Jean-Baptiste-Joseph Delambre and Pierre-François-André Mechain undertook to produce a standard meter at the behest of the Académie des Sciences in order to further standardize France's sense of itself. The plan was to measure enough of a north–south meridian through Paris that they would be able to calculate the distance from the North Pole to the equator: one ten-millionth part of that would be the meter. Almost as a mirror of the La Condamine and Maupertuis continental expeditions sixty years before, Delambre headed north, Mechain south and into Spain. And exactly like their global counterparts, measurement proved difficult because of the local facts of geography. Delambre was taken for an aristocrat (dangerous in Revolutionary times), and the fact that he used church towers—where he could find ones that had not been burned down—merely added to locals' sense that his triangulation schemes were in reality a Royalist ruse. Mechain faced worse since by the time he was at work near Barcelona, Spain was at war with France. His chief difficulty, however, lay in the fact that errors made in longitude as a result of atmospheric refraction could not be corrected. In 1799, when plans for the metrication of French life and French space were laid before the authorities, a "definitive" platinum meter bar was presented. But because of triangulation errors and the fact of local geographical variation, the bar, and the unit it signified, was not a perfectly precise device capable of being used everywhere: the world was simply too irregular. As Enlightenment contemporaries observed, "the more the earth was measured, the more uncertain its shape became."[15]

In sum, calculating the earth's shape had global scientific consequences but differing local expression and political significance. The events of the 1680s, 1730s, and 1790s in France, Lapland, and South America inaugurated new kinds of geodetical practice involving long-distance travel, international diplomacy, variability in instrumentation, and the "on-the-ground" local working-out of social and intellectual authority. As in mapping, intellectual authority over the shape of the earth in the Enlightenment rested in general terms on principles that aimed at the mathematization of space. Yet as Picard and Richer found out in the late seventeenth century, as Delambre and Mechain found out in the 1790s—and as Harrison, Maskelyne, and Robison understood from their testing of longitude in the 1760s and Vancouver found in the 1790s—attempts to rationalize space had always to overcome the facts of local difference in order to come to an "accurate" view of the whole. Geography got in the way of plans for order.

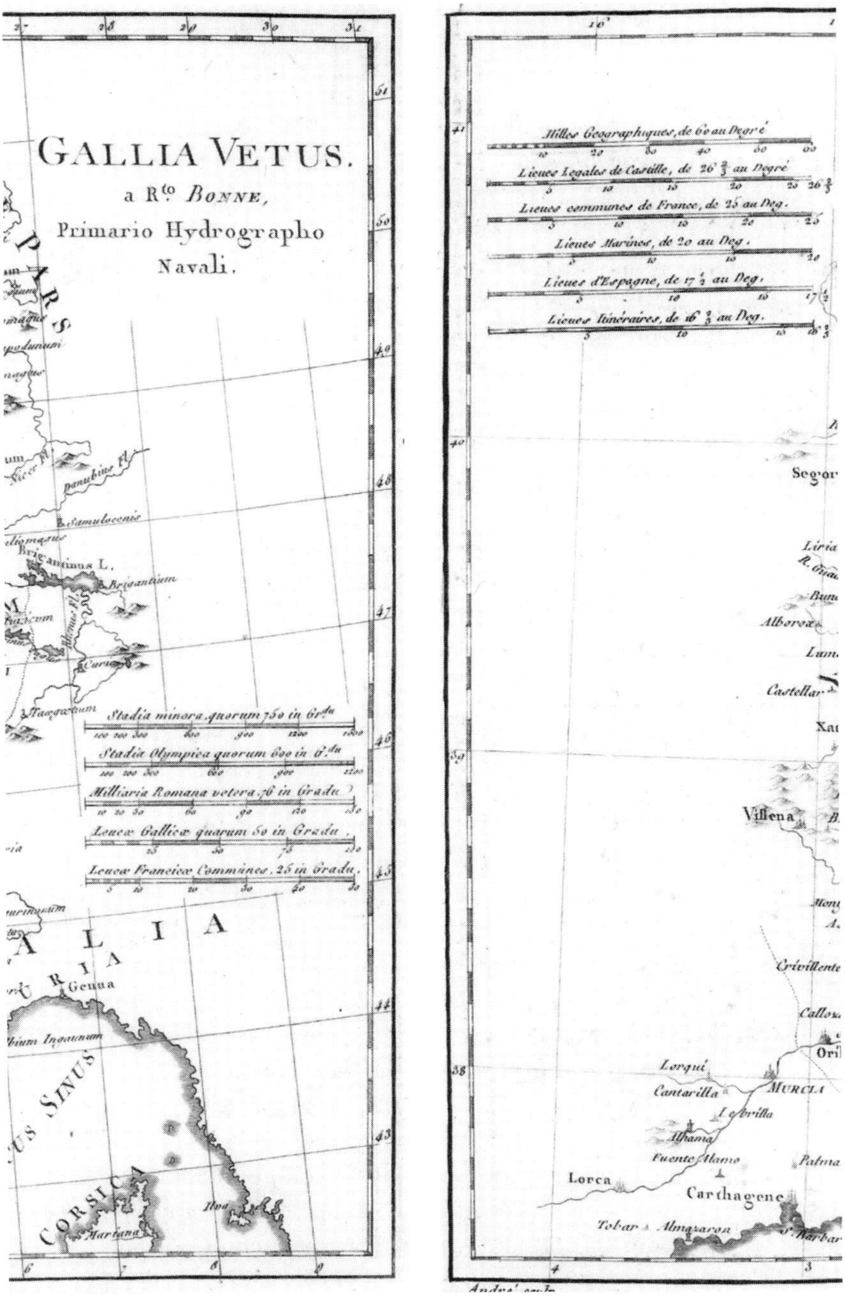

FIGURE 15

Accommodating different scales. Examples from maps of eastern Gaul and the Mediterranean coast of Spain, from Rigobert Bonne and Nicolas Desmarest, *Atlas encyclopédique* (Paris, 1787–88), vol. 2, inserts from plates 12 and 50. By permission of Edinburgh University Library, Special Collections.

Ordering the World of Plants

The mid-1730s is a key moment in Enlightenment physical science for another reason, for as geodesists were arguing over the world's shape, botanists were putting it to order. In his 1735 *Systema Naturae* (The system of nature), the Swedish naturalist Carl Linné (known usually as Linnaeus) advanced a classificatory system aimed at categorizing all plant forms, known and yet to be known, on the planet. Linnaeus's work, like Maupertuis's, was initially based on fieldwork in Lapland, also in 1732. Linnaeus's schema provided a systematic charting—Linnaeus called it "a map"—by which to order the plant and later the animal world. His *Geographia Naturae* (A geography of nature) as he also termed it was for plants a descriptive and to an important degree an artificial system, based as it was on the characteristics of plants' reproductive parts. Earlier botanists had proposed classificatory systems of one sort of another, but in its use of simple Latin and the binomial system, Linnaeus's had a simplicity not found in that of others.[16]

For these reasons, Linnaeus's 1735 *Systema* is a foundational text of Enlightenment science, concerned as it was to place the worlds of plants and of animals. Yet in its time it was neither unchanging nor unopposed. Linnaeus continually revised his system, bringing it into final shape in his 1751 *Philosophia Botanica* and his 1753 *Species Plantarum*. These works established the standard botanical nomenclature that assigned, first, a generic and, second, a species name to plants. It is a global system, capable of fitting all plant life within its schema. What made it work—and what underlay its production at all—was the fact that botanical information traveled: from "the field" that was, initially, Lapland and, latterly, the world itself, to Linnaeus in his Uppsala garden in Sweden, and out again into the world. This botanical disciplining of seeming geographical chaos relied in its later manifestations less upon Linnaeus moving and more upon his "disciples" doing so (Linnaeus called them his apostles)—his students—traveling the world, collecting, enumerating, and classifying. Between 1745 and 1792, nineteen of Linnaeus's students went on voyages of geographical discovery: Pehr Kalm, for example, in North America in 1747, Pehr Löfling in South America in 1754, Pehr Forsskål with the Danish Arabian expedition of 1761–63, Daniel Solander with Cook in 1768, and Anders Sparrman with Cook and the Forsters in 1772.

Rooted as it was in the facts of fieldwork and travel, botanical inquiry was of practical benefit. For Linnaeus, nature's riches were divinely disposed—"the earth then is nothing but a museum of the all-wise Creator's masterpieces," as he put it in 1754—but they were useful and could be improved on.[17] "Utility" for him had national value: domesticating native wild flora in order to reduce reliance on foreign imports, overcoming the facts of geography—soil and latitude—in order to benefit Sweden economically. Linnaeus proposed five climate zones: Australian (which ran from Ethiopia to southern Africa), Oriental (Siberia to Syria), Occidental (Canada to Virginia, China, and Japan), Mediterranean, and Boreal (Lapland to Paris). Later, he added an Alpine zone (which included all mountains). This geographical ordering of the world on the basis of its plants was part of his ideas on the adaptability of plants to different places—how could patterns in the botanical world be explained?—and also of his acclimatization experiments, which included (unsuccessfully) the growing of tea in Sweden (but not coffee, which he regarded as too fancifully French). Linnaeus was of course not unique in emphasizing economic utility. But in his horticultural experimentation—which we might liken to monitoring the effects of present geography, in climate and soil type, in order to manage its potential—he was driven by visions of future geographies, of Sweden (and the world) transformed by Enlightenment science. For Linnaeus, "the sciences are thus the light that will lead the people who wander in darkness."[18]

For all its universalizing appeal, the Linnaean system was not the only one at work in ordering nature's diversity. Depending on where in the world one went collecting, and where one resided in making sense of it all, different schemas emerged. Largely on the basis of his West African travels, Michel Adanson, for example, emphasized "natural" systems and outlined over sixty classificatory schemes in his 1763 *Familles des plantes*. In Paris, the keeper of the Jardin du Roi, Antoine-Laurent de Jussieu, likewise offered a classification based on natural families in his *Genera Plantarum* (1789). Also in Paris, Buffon never warmed to Linnaeus's proposals, seeing them, as did others in Saint Petersburg, as too overwhelmingly sexual, even immoral. In London, botanists did not think his taxonomy accurate. In Edinburgh, some botanical theorists even doubted the fundamental notion of plants having two sexes. In Berlin, botanists like Gleditsch used a modified version of Linnaeus's schema. In Göttingen, Albrecht von Haller based his classificatory schema on questions of geographical distribution, not plants' sexuality. For the plants of the Philippines and elsewhere in the Indian Ocean in the mid-1780s, the Spanish botanist Francisco Noroña used a mixture of Adanson's and Linnaeus's schemes. To Rousseau, because Linnaeus's work was simple and could be used by nonexperts—Linnaeus, as it were, "democratized botany"—Linnaean thinking had preeminent moral qualities. Like others' if more so, Linnaeus's classifications helped make the Enlightenment's botanical world "visible." But his global vision was differently interpreted in different places.[19] Buffon in particular saw the world quite differently. His view was of nature as a creative force, not the design of a Christian God. Buffon was more concerned to understand diversity—in, for example, his writings on the geographical distribution of plants and animals—than to classify it.[20]

Different locations tended to produce different conceptions of the connections between the plant world and the physical environment. In his work in the Vivarais region of France in the 1780s, Abbé Jean-Louis Giraud Soulavie studied "nature's geography" as he called it, the associations between plants, geography, altitude, and climate in ways we would now consider more holistic, more regionally ecological, than Linnaean distributional and taxonomic concerns.[21] Humboldt's South American botanizing led to his 1807 *Essai sur la géographie des plantes* (in which he terms Giraud-Soulavie the founding father of plant geography). Humboldt's own vision of a different form of plant geography, one that emphasized dynamic interrelationships in contrast to the Linnaean artificial description of nature, was first publicly laid out in his 1793 *Flora Fribergensis Specimen*. It was not "tested" in the field, however, until 1802 and his continental botanical field excursion in South America with Bonpland. Linnaean systems of botanical classification were retained, of course, but so too was the more dynamic Humboldtian science, notably in early nineteenth-century German-speaking and Scandinavian botanical communities.[22]

Despite such national differences, the empirical science of botany was everywhere also an imperial and cosmopolitan science. Associations between the empire of nature and political empire were apparent in the role of Enlightenment botanical gardens as local sites of classificatory power and "hubs" of national control over global natural knowledge. Buffon's role as director of the Jardin des Plantes was mirrored by Casimiro Ortega in Madrid's Real Jardin Botánico, in Pierre Poivre's curatorship in Mauritius, and Sir Joseph Banks's at Kew Gardens. As chapter 3 showed, men such as Banks in London, Linnaeus in Uppsala, Haller in Göttingen, and Thouin in Paris stood at the center of networks of botanical knowledge: controlling, directing, exchanging, displaying. Banks saw his own encounter with the globe as pioneering. As he immodestly noted, "I may flatter myself that being the first man of scientific education who undertook a voyage of discovery and that voyage of discovery being the first which turned out satisfactorily in this enlightened age, I was in some measure the first who gave that turn to such voyages."[23] In truth, Banks's authority rested not only in his own scientific aptitude but in his being in a position to assist others and advise governments. For Banks, Enlightenment natural science and enlightened politics both necessitated geographical propagation of the right sort. In overseeing Matthew Flinders's 1801–4 *Investigator* voyage, Banks ensured that the ship carried fruit bushes, grain, and other seeds of civilization to Britain's Australian colony. Instilling Enlightenment ideals of improvement and cultivation in the fertile soils of a new continent demanded the right climate, morally and politically as well as geographically.[24] Yet it is noteworthy that Enlightenment interests in the spatial distribution of plants occurred alongside the destruction of forests for their lumber, in eastern Australia, as on the American coasts and throughout Europe. By the late Enlightenment, the imperatives of economic botany tended to outweigh the imperatives of taxonomic classification and geographical distribution—they did so especially on the margins of European empires.[25]

It was not just plants that were the object of natural explanation and exploitation. Both Linnaeus and Buffon wrestled with and refuted the biblical account of the origins and dispersal of animals on the earth's surface. But the death knell for the religious explanation of zoo-geographic diversity came not from Uppsala nor from Paris but from Brunswick in Germany and from Eberhardt Zimmermann, a professor of mathematics, whose *Specimen Zoologiae Geographicae Quadrupedum* (1777) was the first to describe in detail the distribution of animals by invoking what we might now think of as concepts of natural habitat based on associations among climate, food supply, and physical environment.[26] For Zimmermann (whose work I return to in chap. 9, in looking at medical geography), species were "fixed" not by a single God but by the diversity occasioned by geography.

Of Flood, Fire, and a Dynamic Earth

For Enlightenment earth scientists and others, one event in particular helped recast current thinking about God, Nature, and the earth's "physical geography." On 1 November 1755, All Saints' Day, between 9:30 and 9:40 a.m. local time—in a world then knowing only local time—one of the largest earthquakes in human history hit Lisbon, the Portuguese capital. Tens of thousands of people were killed, many by the tsunami that followed. Physical aftershocks continued for weeks. Coastal towns elsewhere in Portugal and Spain and northwest Africa were devastated. The effect was felt across Europe. In southern Ireland, "sudden and surprising reflexes of the sea" at three in the afternoon tore boats from their moorings; in Scotland, "Loch Lomond was agitated in a very surprising manner." Reports from Sweden,

Holland, and southern England tell similar stories, notably of the sudden rising and falling of well water. In the West Indies, a smaller version of the tidal wave that swept Lisbon's quayside clean arrived nine hours later.

Charting the news about Lisbon captures one "chrono-geography" of knowledge reception in the Enlightenment, a time geography of the event's "mental aftershock": the news took four days to travel from Lisbon to Madrid, fourteen days to reach Paris, from where it traveled to London and to Dublin. News of the event took twenty-five days to appear in public in newspapers in The Hague but in private correspondence had reached the business community there several days before, seriously affecting the financial markets. It reached Uppsala and Boston on the same day, 8 December, New York on 19 December, and South Carolina not until early February 1756. Yet in a world in which natural circumstances had theological significance, how was the news received? What did it mean?[27]

For some, churchmen mainly, the Lisbon earthquake was divine retribution, the town's destruction inevitable given its inhabitants' wickedness. Across Europe, "earthquake sermons" moralized upon Lisbon as simply a hardship to be endured: God was the primary agent, even though He worked through secondary natural causes. Others such as Voltaire, Rousseau, and Kant publicly disputed whether God could have meant it to happen and how humans should behave in the face of such uncertainty. For Voltaire, signaling his disquiet in his *Candide* and 1756 *Poème sur le désastre de Lisbonne,* it was a turning point, away from an "all is well" enlightened optimism toward a profound pessimism. In Italy, the initial reaction of some commentators was incomprehensibility at the nature of the event, their later reactions disquisitions upon the nature and significance of ruins—as the result of natural catastrophe and as moral comment upon the decline of civilizations. In Holland, Dutch commentators interpreted Lisbon's fate overtly in religious terms and covertly in economic ones: Lisbon had been slow to see the true light of northern religion, Protestantism, and Portugal was not modern enough in contemporary European terms. From the Dutch perspective, "the destruction of Lisbon 'makes sense,' in that it signals Lisbon's radical 'unmoderity.'"[28] What shook people overall was not the event but its seeming lack of purpose. Earthquakes were anyway familiar things: major quakes had struck Lima in Peru in 1736—as La Condamine testified—and Port Royal, Jamaica, in 1748. Earthquake "swarms" were reported throughout Europe and North America in 1737. London shook in 1750. The established explanation of earthquakes in the eighteenth century was that they resulted from the subterraneous explosion of deposits of sulfur, niter, or bitumen. For some, including Kant, Buffon, Pierre Bouguer, the Swiss philosopher Elie Bertrand, and John Michell, the English natural historian latter dubbed the father of seismology for his view in 1760 that earthquake energy was released in waves, earthquakes were essentially chemical phenomena. For William Stukeley, the English antiquarian and theologian, and in America, Benjamin Franklin and Thomas Prince, the cause was electrical discharge.[29]

However understood, the 1755 Lisbon earthquake's more lasting effect was to accelerate existing trends toward materialistic and historical philosophies of nature. The 1783 Calabrian earthquake did likewise. There, the Naples Academy of Sciences sent observers to sketch the destruction and to report on it at firsthand—and to feel for themselves the tremors that lasted on and off for three further years. Because of the sketches made and the accounts produced on the spot, Calabria 1783 is perhaps more important than Lisbon 1755 as a key moment in the sensory perception and visualization of nature's destructive power.[30]

At the same time, it is possible to chart different interests in "flood geology," in the role played by water in the earth's history, and to see how natural philosophers reconciled scriptural accounts of the flood with material evidence gleaned from geographical encounter. In general terms, the Noachian account had three central features: the flood was universal, submerging even the highest mountains; it lasted forty days and involved waters from inside the earth as well as from the heavens; and its duration was less than one year. What is also clear is that there were national and local differences in view regarding the literal belief in Genesis—from theological and scientific commentators alike—and that departures from a literal interpretation were dependent on arguments derived from different sites of stratigraphic and fossil evidence. As a result of such geographical variation, the flood was reduced in status during the Enlightenment, from a universal myth to a series of locally determinable and even recurrent natural events.[31]

Perhaps the most spectacular playing-out of nature's magnificence—not least because of their visibility in the present—appeared in volcanic eruptions. For the fashionable, volcanoes were artistic objects, sites of aesthetic wonder as well as of natural process. The eruptions of Mount Vesuvius, which was more or less continually active for about sixty years from 1750, feature in European art—by Jacob More, Joseph Wright of Derby, and Claude-Joseph Vernet among others—as a sublime natural spectacle (plate 8). William Hamilton, the British envoy in Naples—one of the first foreigners to be on the scene in Calabria in 1783—documented and depicted Vesuvius's eruptions in his reports to the *Philosophical Transactions;* they later were incorporated in his 1776 *Campi Phlegraei: Observations on the Volcanos of the Two Sicilies*.[32]

By the later Enlightenment, and in association with debates in "flood geology," volcanoes were increasingly used as test sites for contrasting theories of the earth rather than as sublime spectacles. Among those who held that the flood was universal, one theory held that all the rocks of the earth had formed aqueously, through precipitation and deposition. For the Neptunists as they became known, the sea was the origin of all the earth's solid matter. The chief exponent of this view was Abraham Gottlob Werner, superintendent of mines in Freiburg, who held that the earth's geological features were the result of one universal ocean. In contrast, the Vulcanists took subterraneous heat—fire, not water—to be the crucial factor in the earth's formation. A leading theorist in this regard was James Hutton, the Scottish natural philosopher and agriculturalist. The views of these men and of their followers were not, strictly, opposites. Seeing the two views about the physical earth as simply between a Huttonian Edinburgh and a Neptunian Freiburg underplays a complex geography of ideas. The important differences concerned the relative emphasis accorded to fire and to water and whether heat or fire existed in the earth's core or was limited to its crust. Establishing as Hutton did in his 1795 *Theory of the Earth* that heat was a force deep within the globe did not disprove Wernerian thinking about the role of water, and theorists continued to be exercised about the deluge for decades after. Yet the sort of reasoning from nature that Hutton and others employed, and that Buffon brought to bear in his *Histoire naturelle* in arguing for a 75,000-year-old earth, helped lay the basis for a "deeper time" than the scriptures claimed. As Hutton and Buffon and others thus deepened time, and as men like Johann von Charpentier, Werner's colleague in Freiburg, began to see surface distribution as clues to the nonvisible world (see plate 9), other conceptions of the "earth sciences" were formed through geography: for the French "mineralogists" (geologists) Jean-Étienne Guettard and Antoine Monnet, "mineralogists had the honor of making Geographers aware that they could considerably extend the sphere of their science and that this sphere should present the physical as well as the political on our planet. Thus united, these two sciences would give rise to a third science, Geographical Mineralogy."[33] And so it did, as a physical geography of earth description in terms as various as "mineralogische geographie," "géographie souterraine" (as Soulavie called it), "anatomia della terra" in Italy, or, commonly, "geognosie." This physical geography was not, to put it in modern terms, a "geohistorical science" seeking explanation of the features discerned; it was in Enlightenment terms a science of spatial order in which the emergent earth sciences were informed and shaped by the languages and practices of geography.[34]

Belief in one or another earth theory did, however, have different contours of meaning and different locational expression. British thinkers generally saw volcanoes as constructive agents and equated volcanism with a cyclical model of dynamism and stasis and, indeed, political progressivism. In Germany, it was otherwise. So when Leopold III Friedrich Franz of Anhalt-Dessau built a (working) model volcano on his Wörlitz estate as part of its remodeling as an "English" landscape between 1763 and 1790, he knowingly associated volcanism with enlightened improvement.[35] In France, the work of Nicolas Desmarest and Jean-Étienne Guettard in the Auvergne region provided evidence of the earth's past volcanic activity: from their work as well as Hutton's, the earth could be inferred to have an ancient history.[36] Kant's own 1785 work on volcanism and his Königsberg lectures in physical geography were based on the mountains of Bohemia and Moravia.[37] Werner looked to the mountains between Bohemia and Saxony for his empirical evidence. In Sweden, Anders Celsius took recent sea-level change in the Baltic as evidence to support Werner. Hutton's advocacy of vulcanism, or Plutonism as it was also known, was molded by fieldwork at particular sites in Scotland: Arran, Jedburgh, Glen Tilt in Perthshire, Siccar Point on the Berwickshire coast, and Salisbury Crags in Edinburgh.[38]

In explaining the earth's present physical geography and its history through the remains of its past geographies, Enlightenment earth scientists moved away from supernatural causes toward natural ones. They did so by engaging in fieldwork, observation, and deductive reasoning. Just as in botany, the different theories advanced of the earth depended significantly on where people went looking for evidence to "prove" them, on specific localities and places of knowledge, and on how others elsewhere interpreted them.[39]

On Oceans, Climate, and Meteorology

In the Enlightenment, the world's oceans were subject to attention in several ways. One was renewed interest in terrestrial magnetism. Magnetic phenomena had been investigated ever since compasses became navigational aids: the English physician William Gilbert had promoted new ideas about magnetism, including its importance for latitude, in his *De Magnete* (1600). Like earthquakes, magnetism's cause was mysterious. Unlike earthquakes, magnetism was useful. Given that oceanic navigation in the Enlightenment had enlarged the world, improved navigation was a necessity.[40] To navigate accurately in poor visibility necessitates knowing the variation in magnetic north in different geographical locations. Edmund Halley, the English astronomer and philosopher, undertook several voyages between 1698 and 1700 in the *Paramore* in order to plot such variations—voyages also concerned with astronomical measurement and with the testing of Newtonian theories on the earth's shape. Later work in geomagnetism was undertaken by the Germans Tobias Mayer and Johann Wilke, and in Britain, by John Robison in Edinburgh and by James Rennell in London. Rennell's work includes a chart of the magnetic variations around Africa in order to correct land-based measurements recorded in Mungo Park's travels in West Africa. The first nation to rule the seas, geomagnetically speaking, was France: Feuillée in the Pacific in 1707–12, Pierre Bouguer in the 1740s, Lacaille in South Africa in 1750, Le Gentil in 1761 and 1769 during the "Venus voyages," Bougainville in 1766, and Bellin over longer periods all measured inclination and declination in the earth's magnetic field. By 1795, when the French established their Bureau des Longitudes, their institutional site for "la physique du globe"—geomagnetism, physical astronomy, and geodesy—doing such "big science" was too expensive for single nations to pursue. The story of terrestrial magnetism from the early nineteenth century is one of international collaboration, not national science.[41]

Oceanography—the study of the world's seas, their chemistry, and contents—was likewise unevenly developed in the Enlightenment.[42] Despite the impetus of Luigi Marsagli's 1725 *Histoire physique de la mer,* little systematic work was done on oceanic chemistry until the late eighteenth century. In America, Benjamin Franklin and Thomas Pownall published on ocean currents, as did Rennell in Britain, with Franklin making repeated on-ship thermometric observations during his many crossings of the Atlantic on political and literary matters. In hydrography, the science of marine mapping, the French led Enlightenment Europe, as they did in terrestrial triangulation. Notable in this regard is the work of Jacques Nicolas Bellin, *ingénieur-hydrographe,* whose two-volume *Hydrographie française* (1792) collected French marine maps of studies undertaken between 1737 and about 1772—on ocean currents, magnetic variation, and coastal soundings, for example—and whose earlier *Recueil des mémoires* (1751) discusses the work involved in many of the maps produced between 1737 and 1751. Through the work of Bellin and others, the Dépôt des Cartes et Plans de la Marine, established in 1720 within France's Navy Ministry, acted as a collecting center for maritime cartographic information and as a model for the Observatorio de la Armada in Spain (1753) and the Hydrographic Office in Britain (1795).[43]

As was the case for oceanic navigators and mapmakers, hydrographers everywhere came, in time, to place reliance on their chronometers. Generally, however, hydrographers everywhere relied on sailors for their information. This posed a problem. There was, of course, nothing wrong with relying on facts borne of experience. But for those crossing the world's oceans or mapping their borders, accepting others' practical knowledge—in this case, that of ships' captains and common seamen—meant treating as equal people who, in other circumstances, would not be deemed legitimate sources of rational knowledge. This point has been earlier alluded to with reference to Franklin's attempt to map and explain the Gulf Stream from the late 1760s. Let me here illustrate this point with reference to Murdo Mackenzie, Britain's leading hydrographer in the Enlightenment and author, for the British Admiralty, of the *Maritime Survey of Ireland and the West of Great Britain* (1776). Mackenzie's work was subject to sustained public attack by the Scottish philosopher James Anderson, who accused Mackenzie of poor instrumentation in his charts, and worse—of relying "upon the eye only, and that even in a hasty and superficial manner." Mackenzie did not deny his dependence on direct observation: how could he? To delimit land from sea you have to see where the two meet. Mackenzie defended his coastal charts (and his own credibility) by noting that their usefulness was warranted by the very persons he had spoken with to produce them—"seamen, . . . who after the continued use of the said surveys, from ten to thirty years, have never once found fault."[44]

The cases of Franklin's whalers and Mackenzie's sailors hint at bigger issues over the social differences underlying geographical knowledge about the physical world in the Enlightenment. One relates to that commonplace reliance on natives' knowledge—what we have seen of Lapérouse at Sakhalin, Cook in New Zealand, or of Linnaeus's students as they sought specimens for transplanting—even if the final form, such as the map or herbarium, often obscured the conditions of its making. Another, similarly a matter of "native" knowledge, concerns socially inferior "locals" among one's own countrymen (and women). Such people, "experience-based experts" if you will, were not members of any academy or its equivalent. Their knowledge was not recognized in any formal sense. Nor did it reduce to a system or, in Enlightenment terms, "theory." Yet it cannot be ignored. The experiences of Cassini de Thury in Brittany, of MacKenzie in Hebridean waters, of Cook and Vancouver in Pacific ones, and of mapmakers in Montpellier illustrate this. A further distinction embraces the sense in which people reacted differently to what others construed as "moments" of, or opportunities for, Enlightenment. To those Neapolitan philosophers and military officials sent to observe and to restore order in Calabria in 1783, the earthquake there was literally a heaven-sent opportunity to enlighten and modernize a devastated region and a backward society. But the Calabrians, peasants and landowners alike, resisted the proposed geographies of modernity: their future was their past.[45]

Collectively, such distinctions in the making and reception of natural knowledge mean we can no more talk of a socially uniform Enlightenment than we can of a geographically uniform one. And there is a further observation to be made given the presence in the Enlightenment—as now—of "folk knowledge," belief systems rather than science, through which ordinary people understood their world. One important illustration of the way scientific knowledge of the physical world in the Enlightenment distanced itself from "lay" natural knowledge—what has been elsewhere termed "rustic science"—was meteorology, the science of the atmosphere. In classical terms, meteorology meant "reading the skies" for natural portents, curious or wondrous events that foretold social upheaval, poor harvests, or the like.[46] Strictly, meteorology in the Enlightenment meant more than systematic observation and prediction of the weather. Both terrestrial magnetism and hydrography fell within its compass, and the influence of weather on agriculture and human health meant that it was as much a life science as an earth science. But what is undeniable is that in the Enlightenment, weather became more and more an object of scientific study and not of folklore. Individual clerical meteorologists and natural historians recorded it; societies charted rainfall and temperature in relation to harvest yields; medically minded philosophers debated connections between disease and weather; wind, rain, and temperature series figured as objects of systematic attention using instruments specifically designed for the purpose.[47]

Enlightened study of the weather was about the importance of place. This was true in a qualitative sense: observers noted the "warmth" of local airs, the "goodness" of waters, and so on. It was also true, if more so in the later Enlightenment, in quantitative terms. In Britain, the mathematician James Jurin and philosopher William Derham used the Royal Society to coordinate and record weather data from a handful of sites in the 1720s and 1730s, as had the Breslau physician Johann Kanold. But interest in weather recording really only took off after about 1770. What prompted it was both a strongly utilitarian impulse—making weather recording useful by linking it, notably, to agriculture and to public health—and emergent concerns about what we now would term the environment. In France, the agrarian writer and chemist Duhamel du Monceau coordinated a program of "observations botanico-météorologiques," annual tables on weather, crop returns, and public health information, for forty years beginning in 1741. In 1776, the politician Anne-Robert-Jacques Turgot proposed a formal body, initially in response to public unrest over cattle disease, to oversee the management of "les maladies epidémiques." The Société Royale de Médecine, established in 1778, undertook a program of "medical topographies" linking local physicians' reports on the airs and waters of places with the moral qualities of their inhabitants. Over 220 such topographies—and the sketch of a plan for the whole of France—were completed by the time the Société was suppressed in 1793. In the Netherlands, the medical-meteorological society the Natuuren Geneeskundige Correspondentie Sociëteit worked likewise through its networks of weather watchers and correspondents.[48] Across Germany in particular, meteorological societies sprang up, and agricultural ones were revived to record the weather in "naturalist calendars" as part of useful state knowledge. Such were the standards of the Palatine Meteorological Society in Mannheim that its times and practices of observation—the so-called Mannheim Hours, and Mannheim Cloud Cover and Wind Scale—were widely adopted.[49] Europe had a series of extreme climate years in the 1780s—at the end of what is now understood as the Little Ice Age.[50] The hailstorm of 13 July 1788 that devastated harvests across France is even thought to have played a part in fomenting Revolutionary sentiment.[51] Elsewhere, the Swiss natural philosopher Jean-André Deluc established new instrumental standards for barometers and other instruments in his 1772 *Recherches sur les modifications de l'atmosphere*. That year, the English chemist and radical dissenter Joseph Priestley published his influential essay "Observations on Different Kinds of Air." From then on, eudiometry, investigations of the oxygen content of air in relation to soil type, became a widespread practice.

Although not globally connected in its findings and organizing institutions, climate study was becoming a global pursuit. Officers in colonial field stations and botanical gardens recorded weather and drew conclusions between single events and longer-run observational series. In 1791, for example, weather watchers in India, and on Saint Helena and Saint Vincent in the Caribbean, noted extremely high mean temperatures and drought conditions in Australia, southern Africa, Mexico, and several islands in the Southern Hemisphere. They did not know it then, of course, but they were recording what we now understand as El Nino oceanic current and southern oscillation events, elements within global climate circulation models. In these colonial observatories—in the work of men like William Roxburgh, superintendent of the Botanic Garden in Calcutta, and Alexander Beatson, governor of the East India Company lands on Saint Helena, for example—lie some of the origins to modern environmentalism.[52]

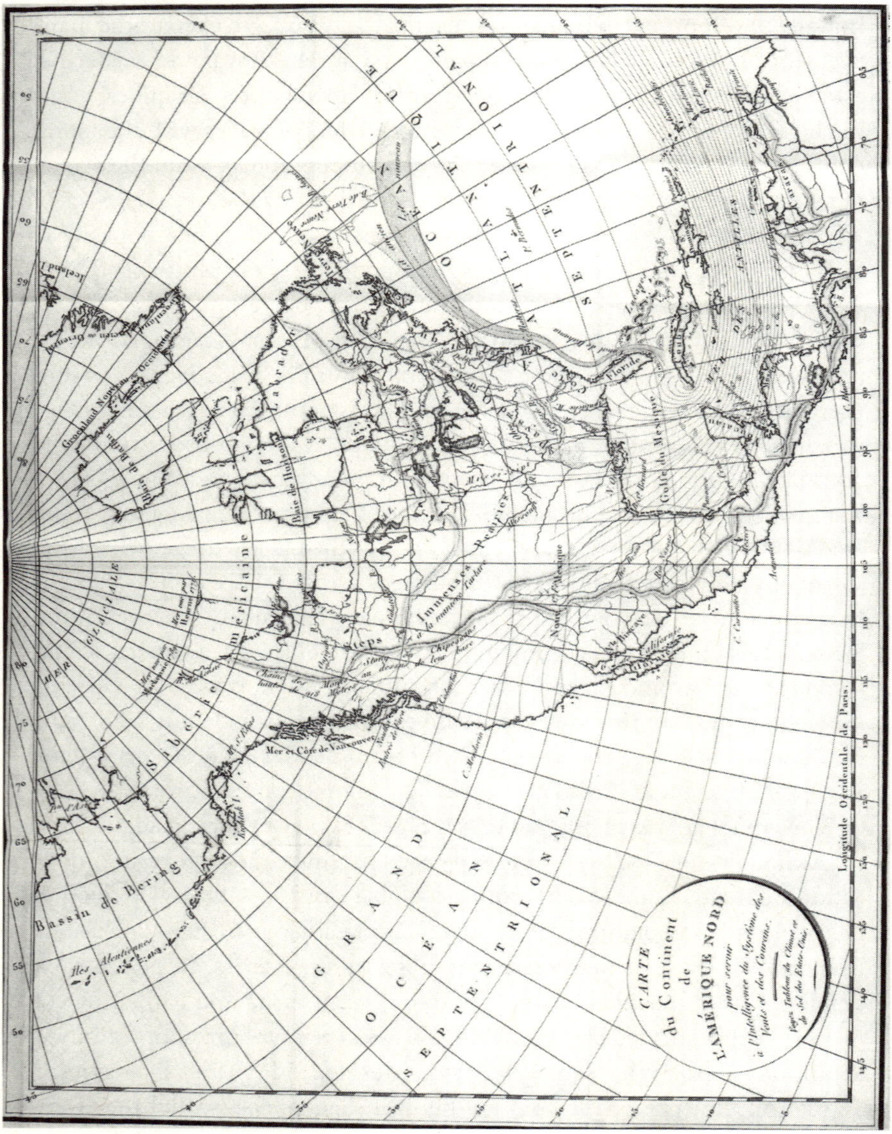

FIGURE 16

Map of North America, from Constantin-François de Chasseboeuf, comte de Volney, *Tableau du climat et du sol des États-Unis Amerique* (Paris, 1803 ed.), vol. 2. In here combining soil and climatic description, wind directions and an outline of the Gulf Stream, Volney is providing a physical atlas of the continent's earth processes. By permission of Edinburgh University Library, Special Collections.

Why all the interest in meteorology? One answer was utility—this was natural knowledge to human ends. Another lay in the fact that, as for Linnaean botany and geodetic measurement, the Enlightenment world was being everywhere subject to the sciences of system. Issues of instrumental method, quantification, standardization, and theorization coalesced in the nascent earth sciences. It is anachronistic to speak in the Enlightenment of climatology and of environmentalism. Yet after about 1770, knowledge of the terraqueous globe was characterized by preoccupations that had the environment—and its intrinsic connectedness—at their center. This is so whether we look at Enlightenment emphases in analysis, moves toward historical explanation, or mapped attempts to connect natural processes (see fig. 16), at aesthetic engagements with wild and cultivated nature, or at the recognition that the globe was more dynamic and much older than was once thought.

There is, however, a further, more complex explanation—and it concerns the place of humans themselves. Encounters with the physical world in the Enlightenment always had humans in view and were always looking to understand the connections between humans and their physical setting and to test long-held beliefs in doing so. Was Man determined in his nature, even governed in moral character, by the facts of his geography? For Montesquieu, whose *L'espirit des lois* is partly an important work of climate theory, the answer was yes. He was not alone in thinking so. Explaining national differences and the contours of moral virtue and physical capacity by reference to geology and climate—Montesquieu held, for example, that basalt was conducive to piety—might work for Europe, although there were many who held it did not. But how to make sense of North American peoples, Pacific islanders, Siberian natives? How were differences in human culture to be explained? Answering these and other questions in the Enlightenment required a Science of Man.

7

Geographies of Human Difference

Enlightenment encounters with the globe provided new and radical visions of humankind. From *voyageurs naturalistes* and others, reports poured in: of pygmies, cannibals, "wild men," "hairy men," of Hottentots and Patagonian giants, this last an age-old belief held as late as the 1760s by Commodore Byron, the British navigator, but refuted by Bougainville and by Cook.[1] Understandings of human origin and of human difference as derived from classical learning and the Bible were challenged by the human diversity that so struck those oceanic navigators, philosophical travelers, mapmakers, and shipboard naturalists and artists whose work we have considered. For in many parts of the world revealed through such encounters, there were humans different from Europeans by skin color, custom, culture, language, and, in Enlightenment terms, their "stages" of social development. The evidence of human diversity and discussion about what it meant provides an important part of what I take to be the connections between geographical knowledge and Enlightenment thinking. What emerged in the Enlightenment from this "shock of the human 'new,'" as we might put it, certainly the shock of the human Other, and in the theoretical and speculative writings that followed, were new ideas about the human condition and about human differences understood as a matter of geography.

These ideas laid the foundations for the "human sciences." As with counterpart developments in the natural world, what contemporaries termed the Science of Man was distinguished by attempts to bring questions of science as method and matters of utility as outcome to bear upon the moral and human world. David Hume (who proclaimed a desire to be the "Newton of the Moral Sciences") posed it thus in 1740: "'tis at least worthwhile to try if the Sciences of man will not admit of the same accuracy which several parts of natural philosophy are found susceptible of."[2] In this regard, the Enlightenment gives us anthropology (social and physical), psychology, economics, sociology, critical history (in the sense of methodologically aware narrative practices), and even moral philosophy and linguistic studies. Evidence in answer to the question of whether it gave us modern geography, touched on in chapter 1, is a central element of part 3. Although we cannot for reasons of historical accuracy strictly define such things in the eighteenth century as "disciplines" in modern terms, the Enlightenment was the birthplace of the modern in regard to the scientific study of humankind.[3]

Notions of what was "scientific" about the human sciences centered on two related matters. The first was the emphasis on experiment, or, more properly, on observational method. So, for Hume, "we must glean up our experiments in this science from a cautious observation of human life, and take them as they appear in the common course of the world, by men's behaviour in company, in affairs, and in their pleasures." And, he continued, "where experiments of this kind are judiciously collected and compared, we may hope to establish on them a science."[4] The second was in the attention paid to method, to the analysis of human phenomena, including human thought, but also to the material conditions necessary to social existence and how geographical variations in human society might be explained.

With a few exceptions, contemporaries likewise identified and shared several assumptions about what distinguished the "human" in the human sciences. These included the power of rational choice, language, the importance of individual and societal self-preservation and of personal and social well-being—"happiness" as many theorists put it. Attention to the process of mental analysis—to the nature of human thought—also characterized a distinctively Enlightenment interest in "the anatomy of the mind."[5]

Discussion of these issues by modern scholars has emphasized the historical development of the human sciences and the place of historical reasoning in the Enlightenment. Such factors have also influenced modern understanding of the Enlightenment as a historical entity. It is, after all, precisely because of such late seventeenth-century intellectual precursors—in the work of John Locke, for example, and because of the variety of Newtonian legacies from the 1680s—that the Enlightenment can be broadened as a historical period.[6] One assessment of the origin and histories of the human sciences has considered their reduction to just two categories, history and anthropology, with, in the Enlightenment, the first category shorn of its modern historicist emphases and the second far greater in its Science of Man period than in later periods.[7] This pairing certainly does justice to features of the human sciences in ways contemporaries would have understood. History and anthropology were united in Enlightenment theorizing about human society, namely, in stadial theory—sometimes also termed "conjectural history," or even the "Natural History of Man." This was the idea—which had variants around its central theme—that human society developed over time in stages: from innocent rudeness to mannered refinement (with the cultures of Enlightenment Europe at the more "developed" end of the spectrum).

Such questions to do with method and with the origin and history of the human sciences cannot be neglected in thinking about the Enlightenment study of Man. They are, however, implicit in what follows since the arguments central to this chapter are geographical rather more than historiographical or methodological. What follows considers in particular the part played by geographical knowledge in formulating the central questions posed in the Enlightenment to do with human society, human difference, and human distribution.

Simply, my claim is that the Enlightenment Science of Man was in several crucial ways a Geography of Man. I mean a number of things by this. There can be no doubt that the human sciences in the Enlightenment depended on geographical knowledge for their origin and later refinement. Stadial thinking as a peculiarly Enlightenment form of historical explanation was dependent on geographical descriptions of human difference. Geographical information, in the form of travel accounts and "philosophico-physical geography," as Johann Herder put it of Johann Reinhold Forster's work, provided "the reservoir of human 'experiments'" in the human sciences.[8] Geographical information—even admitting of the presence of the erroneous and the credulous—was the raw material "from which discerning philosophical historians drew much of their "experimental" knowledge.[9]

Yet it was not simply so. There is, to reiterate a point made earlier, a difference between understanding based on secondhand geographical accounts and understanding based on firsthand experimental method and observation. In terms of how geographical information was used in stadial theory, this difference itself varied over time and in space. For example, in that ethnographic laboratory that was the Pacific, one "crucial turning point in the European understanding of the lands and peoples of the South Seas" was the publication in Paris in 1756 of Charles de Brosses's *Histoire des navigations aux Terres Australes*. This work was the distillation of 250 years of written accounts by European voyagers to the Southern Hemisphere. Cook took it with him on his voyages—as an "enquire-within," a portable guide to his travels, whose maps directed his thinking even as he contradicted them.[10] Yet Brosse himself did not travel beyond France farther than Italy. By contrast, the 1778 work of Cook's fellow voyager Johann Reinhold Forster, his *Observations Made during a Voyage round the World, on Physical Geography, Natural History, and Ethic Philosophy*, provided an account of human difference rooted in the immediacy of personal observation. In surveying linguistic and other differences among island groups, Forster provides a proto-anthropology of the Pacific world and its peoples. The same point—but *not the same timing*—may be made of the Americas. Lafitau's

Moeurs des sauvages amériquains comparées aux moeurs des premiers temps (1724) was based on personal knowledge of the native peoples he described. William Robertson's influential *History of America* (1777), for some the locus classicus of stadial theory, was written without visiting America.[11]

In a crucial sense then, the Enlightenment was characterized by geographies of human difference *in time*. Information on native peoples in the "fifth part of the world," in the Americas (the fourth part), and regarding the foreign and exotic close to home—the Laplander and the Scottish Highlander to name but two—underpinned the theoretical explanation by Europeans of human society over time. Speculations about human conjectural history as a whole were predicated on geographical evidence of actual human difference. These ideas are elaborated on here with reference to work by Rousseau, Turgot, Smith, Millar, Robertson, and Forster among others, and to stadial theory's engagement with American natives, the peoples of the Pacific world, southern Africa, and those thought of as the "marginal" cultures of Europe.

In a further vital sense, revealed differences in human society amounted to geographies of human difference *in space*. Why were the peoples charted on what Burke called the Great Map of Mankind at such seemingly different stages of social and political development? Was human nature everywhere the same? Why were human organizational systems so different geographically? Geographical encounters with human diversity threw up these and other questions that centered not just on human origins and social progress over time but also on distributions over space. Why were the different peoples encountered located where they were? Could social and cultural differences and, notably, physical differences in human types (skin color, hair, beardlessness, stature, even fertility) be geographically explained? Was there a correlation among geographical location, physical difference, and moral capacity? At the same time as Enlightenment theorists posited models that explained non-European social formations as "before Europe" by virtue of their geographical discovery "beyond Europe," so, in the writings of men such as Montesquieu, Hume, Linnaeus, Buffon, and others, related ideas emerged to explain differences in human distribution within and beyond Europe. These cohered around questions of climate, race, nation, and environment, and it is to them I turn first.

Physical, Moral, Natural? Explaining the World's Human Geography

Montesquieu's 1748 major work, *L'esprit des lois* (*The Spirit of Laws*), provides a comprehensive understanding of the laws and moral codes governing human conduct, at individual and societal levels.[12] His attention centered on the principles that informed the ways in which humans organized their collective lives. Montesquieu looked for explanation to factors such as popular custom, differences in political constitution, and population—and, importantly in discussing physical causes, to climate. For Montesquieu, climate was "the first and most powerful of all empires." Montesquieu was not the only Enlightenment writer to consider climate an influence on human society, and many of his ideas drew on contemporaries' as well as on longer-running environmental theories. Nevertheless, the attention paid in *L'esprit des lois* to climate's role in shaping human culture, particularly in books 14–18, is significant, even if it is underplayed in some discussions of Montesquieu's role in developing an Enlightenment science of society.[13] Yet we also must be circumspect. Not everyone then agreed with him. And the claim by one modern scholar that the book overall "marks a turning point in the prehistory of the social sciences, laying out a route that ultimately leads to human geography," places too great an emphasis on this work alone and on a falsely linear sense of disciplinary history.[14]

Montesquieu's theories of climate have forebears in classical thought, notably in Hippocrates' *On Airs, Waters, and Places,* and in the *Methodus* and the *Republic* of the early modern French political theorist Jean Bodin. Like these and other writers, Montesquieu held to the views that climate (by which was meant, essentially, variations in temperature with latitude) influenced the human body's physical state. In turn, this physical state influenced mental states, both individually and of societies as a whole. Simply put, where you were in the world, climatically speaking, determined how you were in the world. Unlike earlier theorists who saw climatic factors either as divinely disposed or as fixed by planetary influence—in bodily "humors" for example—Montesquieu sought to discern climate's influence in order to reveal the relationship between climate, geographical difference, custom, and law. As he put it, "if it be true that the temper of the mind and the passions of the heart are extremely different in different climates, the laws ought to be in relation both to the variety of those passions and to the variety of those tempers." He further extended existing arguments in seeing an association among climate, geography, and moral capacity. "If we travel towards the North," Montesquieu argued, "we meet with people who have few vices, many virtues, and a great share of frankness and sincerity. If we draw near the South, we fancy ourselves entirely removed from the verge of morality; here the strongest passions are productive of all manner of crimes, each man endeavouring, let the means be what they will, to indulge his inordinate desires." What geographers and others knew as the "temperate zone" was for Montesquieu a climatic zone without extremes, a "space" of moral equability: "In temperate climates we find the inhabitants inconstant in their manners, as well as in their vices and virtues: the climate has not a quality determinate enough to fix them."[15]

Montesquieu additionally saw climate as the explanation of cultural stability, not just of cultural difference—in interpreting Oriental despotism, for example. He invoked climate's influence to explain differences in religion, in the nature of laws and customs, even in patterns of suicide (where the "British type" differed from the "Roman" in being the consequence not of education but of "distempers" ultimately owing to the climate). For Montesquieu, geography in the shape of climate had a constitutive if not also a determining role in the nature of political systems, not least in explaining the contrasts between Asia, a land of climatic extremes and despotic government, and Europe, whose temperate nature underpinned "the liberty of Europe."[16]

To the modern reader, Montesquieu's emphasis on climate—even what he held "climate" to be—appears odd, out of tune with contemporary sensibilities. This view, however, ignores the context in which he was writing; it emphasizes what he wrote rather than explaining why. To be sure, Montesquieu was in part advancing environmental explanation from "old and well-tested theories," and he was heavily reliant on others' earlier geographical accounts. But he was doing so as part of widespread Enlightenment speculation about the nature of the human condition. In the history of medicine, his attention to climate in relation to health and disease was firmly rooted in classical concerns. But medical theorists were then debating what we would know now as human physiology and the body's responsiveness to environmental stimuli (see also chap. 9 for evidence of this in medical geography). Montesquieu was much influenced, for example, by John Arbuthnot's 1733 *An Essay Concerning the Effects of Air on Human Bodies,* with its claim that different temperaments—local "manners" if you will—predominated in different regions because of the body's differing response to temperature variations: cold promotes sloth, heat liveliness. And there was anyway in the eighteenth century a renewed emphasis on "environmental medicine," a "neo-Hippocratic" impetus to medical thinking.[17] The *Encyclopédie* treated Montesquieu as an authority for elements of its definitions of "climate." Even oceanic navigators made practical decisions based on perceived connections between geography and physical capacity: Malaspina chose men from the north of Spain for his crew as they were more used to cold weather than were southerners.[18] In his writings about climate, Montesquieu was thinking geographically for clues as to how human society worked and why it worked differently in different places. In seeing climate as one component of the "general spirit" of a society, Montesquieu was, in modern parlance, using geography to understand sociology—to explain what he and others saw as geographical differences in human society.

For contemporaries like Voltaire and Hume, Montesquieu's attention to climate as a physical cause was misplaced. For Voltaire, Montesquieu underplayed the role of political process: social agency, not temperature or latitude, shaped people and nations alike. For Hume, variations in "national characteristics" were not so easily explained. In his 1748 essay "On National Characters," Hume challenged the view that physical factors such as climate act to constitute social structures. "As to *physical* causes," he wrote, "I am inclined to doubt altogether of their operation in this particular; nor do I think, that men owe any thing of their temper or genius to the air, food or climate." "Physical causes," he continued, "have no discernible operation on the human mind. . . . If we run over the globe, or revolve the annals of history, we shall discover every where the signs of a sympathy or a contagion of manners, none of the influence of air or climate."[19]

For Hume, national characteristics and their geographical differences were to be explained by *moral* causes. They stemmed, that is, from such things as political constitutions, customs, what Hume commonly referred to as "manners." As he put it in listing reasons why moral not physical causes prevailed, "the same set of manners will follow a nation, and adhere to them over the same globe, as well as the same laws and language. The SPANISH, ENGLISH, FRENCH and DUTCH colonies are all distinguishable even between the tropics."[20] Hume was aware of geographical differences in what we might consider the *level,* the development status, as well as the nature of national character. "And indeed there is some reason to think, that all the nations, which live beyond the polar circles or between the tropics, are inferior to the rest of the species, and are incapable of all the higher attainments of the human mind." For him, such difference was the result of social—moral—rather than natural factors. As he further—if rather hesitatingly—put it, "the poverty and misery of the northern inhabitants of the globe, and the indolence of the southern, from their few necessities, may, perhaps, account for this remarkable difference, without our having recourse to *physical* causes." In only one respect was Hume prepared to countenance a physical explanation for social differences:

> The only observation, with regard to the difference of men in different climates, on which we can rest any weight, is the vulgar one, that people in the northern regions have a greater inclination to strong liquors, and those in the south to love and women. One can assign a very probable physical cause for this difference. Wine and distilled waters warm the frozen blood in the colder climates, and fortify men against the injuries of the weather: As the genial heat of the sun, in the countries exposed to his beams, inflames the blood, and exalts the passion between the sexes.[21]

The differences in these arguments—Hume's explanation resting in human agency, Montesquieu's on the determining role of climate—should not be taken as extremes. Montesquieu did not see climate as the only factor. Hume made assumptions about geographical differences in social capacity in his remarks about peoples in the polar and tropical regions, and he more than hints at climate's role in explaining the "cartography" of Europeans' predisposition to drink and passion. In some of his other writings, Hume drew on geographical fact in historical explanation: in his *History of England*, for example, the discovery of the New World is cited as a prompt to Europe's science as well as to its commerce.[22] And elsewhere in his *Essays* he wrote of employing the method of the natural historian in order to draw out what he termed a "mental geography, or delineation of the distinct powers and parts of the mind"—what we might call classificatory space for Hume's "the different operations of the mind."[23] More broadly, both men were attempting to explain the human geography of the Enlightenment as a matter of civil history.

For others such as Linnaeus and Buffon, explanation of human difference lay in the geography behind natural history. In their work and in that of others, we can also chart the place of geographical thinking to explanations of human difference that embraced the language of "race" as well as of "nation."

Linnaeus's 1735 *Systema Naturae*, significant as we have seen for the world of plants, was just as revolutionary and provocative in the emergent human sciences. Linnaeus—with reference to earlier classificatory schemes—included there the genus *Homo*, with a single species, *sapiens*, differentiated into four varieties, *varietate*, on the basis of their continental geographical distribution. These were *Europaeus albus, Americanus rubens, Asiaticus fuscus*, and *Africanus niger*. In later editions, Linnaeus revised his taxonomies on the basis of new geographical information. As has been remarked, "Linnaeus's classifications were also in a historical position to reflect . . . the remarkable expansion of information through which the earth's 'geographical space' . . . was transformed in the eighteenth century into a 'human space.'"[24] In 1758, for example, he replaced the order Anthropomorpha with a new order, Primates. New varieties appear: *Homo sapiens ferus*, feral "wild man," included on the basis of discoveries in Europe (such as the Wild Boy of Hanover in 1724), and, following reports from the East Indies, *Homo troglodytes* (Nocturnal Man), said to live in caves there. In later editions, such distinctions were either placed in the genus *Simia*, the apes and chimpanzees, or dropped altogether.[25]

The fact of change in the Linnaean classification of the human species is arguably less important than its implications. Thinking taxonomically and geographically in terms of human difference was not new: earlier divisions to which Leibniz and Locke had subscribed worked with categories such as Europeans (including Persians and North Africans), Black Africans, Chinese, and Lapps, or Americans instead of Lapps. Several things make Linnaeus's work significant. Linnaeus classified the human species as part of nature—in effect, he naturalized humans into systems much as he did plants and other animals. Linnaean systematics was a major achievement of the Enlightenment's human sciences, but it blurred the human–animal distinction. Further, his was a static taxonomy in a God-given world. His emphasis on "variety" paid little attention to the reasons for the different forms. Explanation of the variation mattered less than description of difference—"to know these bodies . . . by those marks imprinted on them by nature" as Linnaeus wrote.[26] In these terms, the Linnaean human world was polygenetic, a world of different fixed varieties. And in his later classifications especially, Linnaeus underplayed reason—that feature of human life by which others distinguished humans (*Homo*) from their "natural" taxonomic bedfellows, apes and chimpanzees (*Simia*). In all these ways, the "view from Uppsala" about human diversity was very different from that version of the natural history of Man proposed in Paris by Buffon.[27]

Buffon's approach to the natural history of the human species stemmed both from distaste for Linnaean "abstract" classification and an interest in the "physical truth," the relationships between natural species, human beings included, and their surroundings. This *is* a matter of method—an interest in process, in the empirical forms, *and* an attention to natural relationships that placed the human being at their center.[28] Here it is sufficient to note that Buffon's 1749 essay "Variétés dans l'espèce humaine," which appears in volume 3 of his *Histoire naturelle*, offers an understanding of human difference in dynamic association with factors such as food, habits of life, climate, and geography. It does so on the basis of new meanings of the term "species."

Buffon's revision of the term "species" stressed natural relationships, a reproductive capacity that ensured biological continuity. For him, "la variété humaine" was a matter of "race," a lineage of traits passed down through generations. In paying attention to race as the successive generations of species—humans included—Buffon went beyond the deterministic tones of Montesquieu by emphasizing the relationships between race and environment over time. In believing that there was only one species of human being, Buffon was a monogenist. In explaining human diversity, Buffon saw variations in human races as dynamic, a result of geography and climate and of movement over space, where Linnaeus saw only fixity. In a Buffonian view, variation could, of course, mean degeneration when the effects of climate and access to food in new environments might alter the original stock. That is how Buffon explained the generally smaller, less robust, and less fertile humans and other animals of the New World in contrast to those of the Old World. In discussing "The American," he cited low population density and great distances as additional reasons for the continent's "universal want of civilisation"—"I am persuaded that there are more men in Paris than all the natives of North America." He further distinguished between "degenerating" and "isolated species" on the basis of the degree of geographical variation between humans. And everywhere climate was a major influence: "nothing can be proved more clearly that the climate is the principal cause of the varieties of mankind, than this colour of the Hottentots"; "the heat of the climate is the chief cause of blackness among the human species."[29] In discussing Geneviève, a white Negress from Dominica born of black parents who had originated from Africa's Gold Coast (fig. 17)—he also examined her physically to take various physiometric measurements, mainly of skull size and shape—Buffon saw her not as a "race apart" but as a "one-off," the result of "unhealthy air" in the region of her birth.

As with Linnaeus's, Buffon's work on human difference is neither unambiguous nor unchanging. Neither man could ignore the new geographical evidence emerging from the later 1760s. That is why, in Buffon, there are differences—on the "degeneracy" of the "American aborigine," for example, and on the question of human giantism—between his 1749 essay and his 1788 *Époques de la nature,* Buffon's final published discussion on the human species. There he stresses an animated nature, with humans, like other species, changing in relation to their local environment over long periods of time. Buffon's method of classification by "race" is thus inexact, his terminology "maddeningly inconsistent," as one modern reviewer has noted.[30] Yet his general hierarchy of species–race–nation influenced others. In the first edition of his *De Generis Humani Varietate Nativa,* for example, the German Johann Friedrich Blumenbach followed Linnaeus, who had, Blumenbach noted, been "following common geography" in finding four human races.

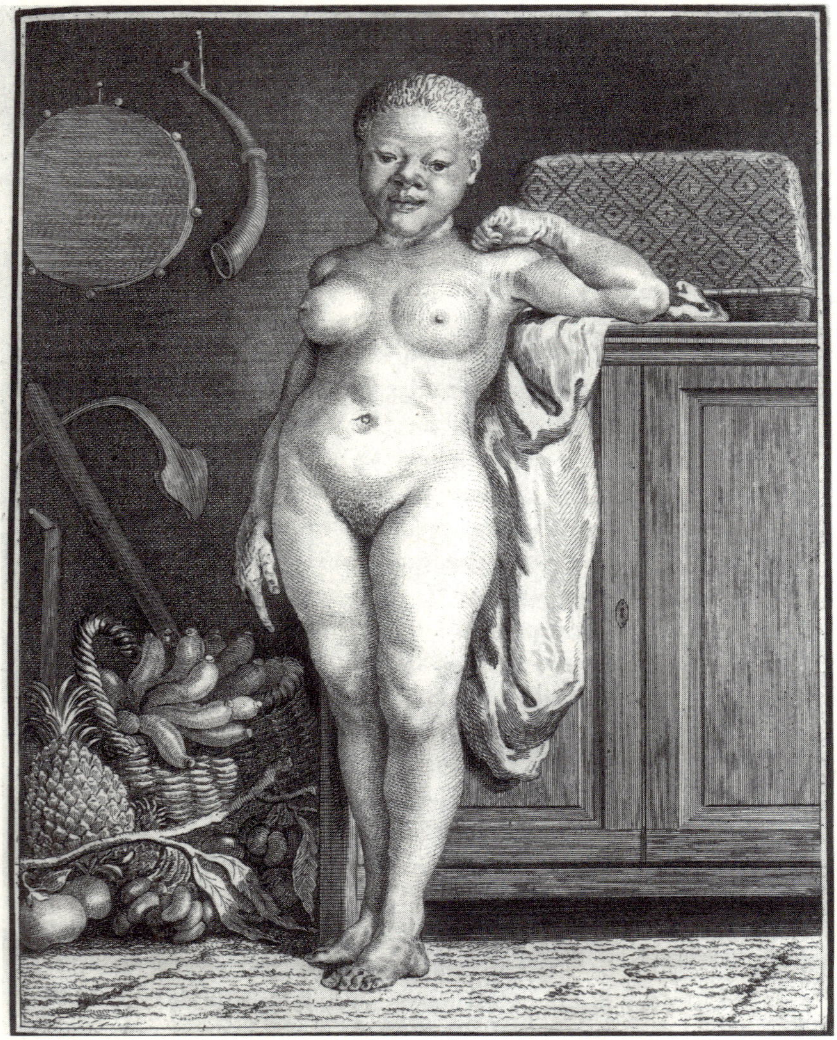

FIGURE 17
White Negress, from Georges Louis LeClerc, comte de Buffon, *Histoire naturelle, général et particulière* (Paris, 1778), vol. 4. The plate is one of several white and "piebald" figures that Buffon used to illustrate his ideas on race and climate. By permission of Edinburgh University Library, Special Collections.

By the 1795 edition of this work, Blumenbach had moved away from "varietate" to "gens," a term that translated in the German as "race." Blumenbach, sometimes seen as the founding figure of physical anthropology for his interest in the physiological differences in "racial type," identified "national varieties" below the racial level within his later division of humankind into five great races—Caucasian, Mongolian, Ethiopian, American, and Malay. By 1795 then, geographies of human diversity looked different from Göttingen, not just from Uppsala and Paris. And as the works of Buffon and others moved into translation, so differences appeared in the nature of what was said, even in how the books were laid out, in order that French and German scientific claims should appeal to the Anglophone public.[31]

The language of "race" and of "nation" was inexact more widely in Enlightenment writings, not just in Buffon's. Numerous authors used "race" and sought explanation in geography to remark on the separateness of peoples apparent in "national" language and "character."[32] More than that, associations were commonly made between race, nation, and moral capacity. Noted Hume in a 1753 version of his 1748 essay,

> I am apt to suspect the negroes and in general all other species of men (for there are four or five different kinds) to be naturally inferior to the whites. There never was a civilised nation of any complexion than white, nor even any individual eminent either in action or speculation. No ingenious manufactures amongst them, no arts, no sciences. On the other hand, the most rude and most barbarous of the whites, the ancient Germans, the present Tartars, have still something eminent about them.

Others thought similarly: for Buffon, "the most temperate climate [between the 40th and 50th degree of latitude] . . . produces the most handsome and beautiful men. It is from this climate that the ideas of the genuine colour of mankind, and of the various degrees of beauty, ought to be derived." Kant, writing in 1775 on the different races of man, was even more geographically particular. He argued that the "natural disposition" of the "white brunette" race—which had probably existed historically "between the 31st and 52nd parallels in the Old World"—was now best approximated by the inhabitants of northern Germany.[33]

Constructing immutable categories of human difference in terms of racial superiority, of physical differences and their presumed moral and social correlates, is rightly an anathema. In terms of Enlightenment thinking about race and racial geography as elements in the emergent human sciences, Hume, Kant, and others cannot be judged in such ways. True, Hume's is "not a fleeting observation"; nor should we take Kant's or Buffon's to be.[34] But neither are they indicative of a full-fledged "modern" conception of race as the determinant of human difference and of unequal political and intellectual positions in consequence. It makes no more sense to place the origins of modern racism as a politicized ideology in the Enlightenment than it does to see its later refinements as simply a nineteenth- or a twentieth-century phenomenon. On the one hand, what we are dealing with are speculations about distributional differences in human social structure—were they "naturally determined" or rooted in "moral" cause? What role did climate play? What was the effect over time of local setting? On the other, these questions signaled ambivalent even contradictory moves toward languages and methods that sought to explain the human world.

Conjectural Histories, Actual Geographies: Stadial Theory and Human "Progress"

For Forster in the Pacific, one obvious way to answer such questions was direct comparison. Reflecting upon the peoples encountered, Forster conjectured in his 1778 *Observations* on the two great "varieties of people in the South Seas" and the differences and relationships between them. The "one more fair, well limbed, athletic, of a fine size, and a kind benevolent temper" inhabited "O-Taheitee, and the Society Isles, the Marquesas, the Friendly Isles, Easter Island and New Zeeland." The other, "blacker, . . . the body more slender and low," was to be found in "New-Caledonia, Tanna, and the New-Hebrides." Each was divided into "several varieties, which form the gradations towards the other race." Noting that "it seems from thence to follow, that the Malay tribes still endeavour to spread, and to subdue the aboriginal tribes in the various South-sea isles," Forster concluded: "These hints, it is to be hoped, may induce some future navigators more carefully to examine the languages, manners, customs, temper, habit and colour of body of the inhabitants of the various South-Sea isles, in order still better to trace the origin and migrations of these nations; and to throw a still greater light on this interesting part of the History of Man."[35]

Forster's comments juxtapose a "savage-unfriendly-black" population in the western Pacific islands and a "barbarian-friendly-brown" race in the eastern islands, and hint too at that comparative ethnography apparent in later decades with the invention of terms such as Micronesia and Melanesia (the latter meaning, literally, "black islands"). Forster's comparative remarks can even be considered an attempt to establish a common ground between Linnaean and Buffonian views. They should also be understood as part of that longer-run Enlightenment concern to understand "the Natural History of Man" in comparative terms or, put another way, to explain the "Geography of Man" as a question of historical development. Language was a crucial element in such thinking. As writers like John Millar knew in his *Origin of the Distinction of Ranks* (1779) and William Marsden in his 1783

History of Sumatra, geographical discovery was a matter of linguistic exploration, the object being to show the close similarity between, for example, the Polynesian languages of the South Seas and of New Zealand peoples (fig. 18). And from words and phrases, even island names (see fig. 10), a more dynamic geography would emerge of the movements of peoples across the Pacific. Even other more material artifacts were noted with the same view in mind (fig. 19). Being geographically out of Europe afforded writers the chance to observe others' worlds *and* to look back at Europe historically. As James Dunbar put it in his *Essays on the History of Mankind in Rude and Cultivated Ages* (1780), "the history . . . of some of the South Sea isles, which the late voyages of discovery have tended to disclose, enables us to glance at society in some of its earlier forms."[36]

These ideas lie behind one of the Enlightenment's most distinctive intellectual features—its engagement with the idea of "progress," the development of human society over time. Theories of progress revolved around ideas of the passage of humankind from a "rude" or a "savage" to a "civilized" state. This was often portrayed as taking place in four stages—hunting, pastoralism, agrarianism, and commercialism—each determined by a particular conjunction of economic activities and social capacities. Conjectural history was the study of the features distinguishing both the stages typologically and the sequence of human development overall—the passage, in effect, from the natural to the cultural. Which social and intellectual characteristics marked each stage? How did they act to change societies from one stage to another?

The first major Enlightenment writer to ask why human society was so varied—indeed, for him, so distinctively local in its customs and laws—was Giambattista Vico in his *Principi de una scienza nuovo d'intorno alla natura delle natzione* (1725). He saw the development of human society over time as the working out of God's providential plans through humans' creation of their own languages and institutions. Other leading figures echoed Vico's concerns if not his argument. In *L'esprit des lois,* Montesquieu offered a fourfold classification of societies in supposing that each nation's laws would be different, depending—in addition to climatic factors—on which economic activity predominated. Turgot speculated likewise by using three stages, "hunters, shepherds, husbandmen." In his 1755 *Discours sur l'origine et les fondements de l'inégalité,* a response to Montesquieu, Rousseau considered the "progress" of human society from savagery to civility, seeing in some "savage" peoples a primitive virtue not apparent in the civilized societies of western Europe. Johann Herder's *Ideen zur Philosophie der Geschichte der*

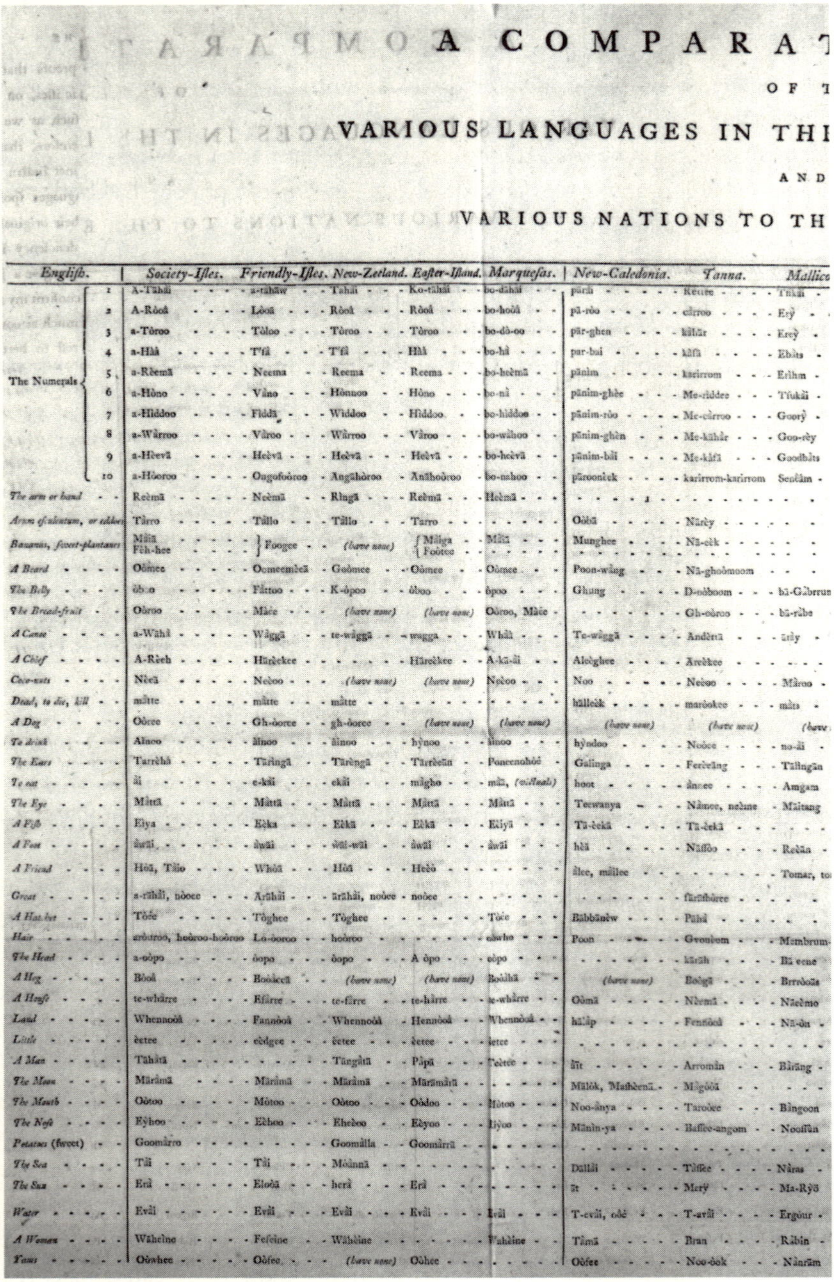

A COMPARAT

OF T

VARIOUS LANGUAGES IN THE

AND

VARIOUS NATIONS TO TH

Engliſh.		Society-Iſles.	Friendly-Iſles.	New-Zeland.	Eaſter-Iſland.	Marqueſas.	New-Caledonia.	Tanna.	Mallico
The Numerals	1	A-Tāhāi	a-tāhāw	Tahāi	Ko-tāhāi	bo-dāhāi	pārīh	Rettee	Tthāi
	2	A-Ròoā	Lòoā	Ròoā	Ròoā	bo-hoòā	pā-ròo	càrroo	Erỳ
	3	a-Tòroo	Tòloo	Tòroo	Tòroo	bo-dò-oo	pār-ghen	kāhār	Ereỳ
	4	a-Hàà	T'fā	T'fā	Hàà	bo-hà	par-bai	kāfā	Ebàts
	5	a-Rèemā	Neema	Reema	Reema	bo-heèmā	pānim	karirrom	Erihm
	6	a-Hòno	Vàno	Hònnoo	Hòno	bo-nà	pānim-ghèe	Me-riddee	Tſukāi
	7	a-Hiddoo	Fiddā	Widdoo	Hiddoo	bo-hiddoo	pānim-ròo	Me-càrroo	Goorỳ
	8	a-Wàrroo	Vàroo	Wàrroo	Vàroo	bo-wàhoo	pānim-ghèn	Me-kāhàr	Goo-rèy
	9	a-Hèevā	Heèvā	Heèvā	Heèvā	bo-heèvā	pānim-bāi	Me-kāfā	Goodbàts
	10	a-Hòoroo	Ongofoòroo	Angāhòroo	Anāhoòroo	bo-nahoo	pārooneek	karirrom-karirrom	Senēām
The arm or hand		Reèmā	Noèmā	Ringā	Reèmā	Heèmā			
Arum eſculentum, or eddoes		Tàrro	Tàllo	Tàllo	Tarro		Oòbā	Nārèy	
Bananas, ſweet-plantanes		Māiā Fèh-hee	Foogee	(have none)	Màiga Footee	Māiā	Munghee	Nā-eèk	
A Beard		Oòmee	Oomeemèeā	Goòmee	Oòmee	Oòmee	Poon-wāng	Nā-ghoòmoom	
The Belly		òboo	Fàttoo	K-òpoo	òboo	òpoo	Ghung	D-oòboom	bā-Gābrrun
The Bread-fruit		Oòroo	Màee	(have none)	(have none)	Oòroo, Màee		Gh-eòroo	bā-rùbe
A Canoe		a-Wāhā	Wàggā	te-wàggā	wagga	Whāā	Te-wàggā	Andètā	āthy
A Chief		A-Rèeh	Hārèekee		Hāreèkee	A-kā-āi	Aleèghee	Āreèkee	
Coco-nuts		Nèeā	Neèoo	(have none)	(have none)	Neèoo	Noo	Neèoo	Màroo
Dead, to die, kill		màtte	màtte	màtte			hālleèk	maròokee	màts
A Dog		Oòree	Gh-òoree	gh-òoree	(have none)	(have none)	(have none)	(have none)	(have
To drink		Ainoo	àinoo	āinoo	hỳnoo	àinoo	hỳndoo	Noòee	no-āi
The Ears		Tarrèhā	Tāringā	Tārèngā	Tārrèeān	Poneenohòè	Galinga	Ferèeāng	Tālingān
To eat		āi	e-kāi	ekāi	màgho	māā, (victuals)	hoot	ànnee	Amgam
The Eye		Màttā	Màttā	Màttā	Màttā	Màttā	Teewanya	Nàmee, neàme	Māitang
A Fiſh		Eiya	Eèka	Eèkā	Eèkā	Eèiyā	Tā-eèkā	Tā-eèkā	
A Foot		āwāi	àwāi	wāi-wāi	àwāi	àwāi	hèā	Nāffòo	Reèān
A Friend		Hoā, Tāio	Whoā	Hoā	Heèò		ālee, māilee		Tomar, to
Great		a-rāhāi, nòoee	Arāhāi	ārāhāi, nooee	nooee			fārāfhòrre	
A Hatchet		Tòée	Tòghee	Tòghee		Tòée	Bābbānèw	Pāhā	
Hair		aròuroo, hoòroo-hoòroo	Lo-òoroo	hoòroo		oòwho	Poon	Gvonèum	Membrum-
The Head		a-oòpo	òopo	òopo	À òpo	oòpo		kārāh	Bā eene
A Hog		Bòoā	Boòàeeā	(have none)	(have none)	Boòāhā	(have none)	Boògā	Brrròoās
A Houſe		te-whàrre	Efàrre	te-fàrre	te-hàrre	te-whàrre	Oòmā	Nèemā	Nāeèmo
Land		Whennoòā	Fannòoā	Whennoòā	Hennòoā	Whennoòā	hālàp	Fennòoā	Nā-òn
Little		èetee	eèdgee	èetee	èetee	èetee			
A Man		Tāhàtā		Tāngàtā	Pàpā	Teètee	āit	Arromàn	Bàrāng
The Moon		Mārāmā	Mārāmā	Mārāmā	Mārāmàrā		Mālòk, Mathèenā	Māgòoā	
The Mouth		Oòtoo	Mòtoo	Oòtoo	Oòdoo	Mòtoo	Noo-ànya	Taroèee	Bàngoon
The Noſe		Eỳhoo	Eèhoo	Eheèoo	Eèyoo	Iiyoo	Mānin-ya	Baffee-angom	Nooffān
Potatoes (ſweet)		Goomàrro		Goomàlla	Goomàrrā				
The Sea		Tāi	Tāi	Mòannā			Dāllāi	Tàffee	Nàras
The Sun		Erā	Eloòā	herā	Erā		āt	Merỳ	Ma-Ryò
Water		Evàì	Evàì	Evàì	Evàì	Evàì	T-evāi, oòè	T-avāì	Ergòur
A Woman		Wāhèine	Fefeine	Wāhèine		Wahèine	Tàmā	Bran	Ràbin
Yams		Oòwhee	Oòfee	(have none)	Oòhee		Oòfee	Noo-òok	Nànrām

FIGURE 18

Word charts. The comparative language geography of the Pacific peoples, from Johann Reinhold Forster, *Observations Made during a Voyage round the World* (London, 1778). Such work on languages reflected Enlightenment concerns with human origins and geographical diversity. By permission of Edinburgh University Library, Special Collections.

TIVE TABLE

THE

E ISLES OF THE SOUTH-SEA,

D OF

HE EAST AND WEST OF IT.

…icollo.	Malay.	Tagala.	Pampangos.	New-Holland.	Mexican.	Peruvian.	Chileese.
	Sa	ysa	Isa, Metong		ce		Quya
	Dua	dalava, dalova	Ad-dua		ome		Eppo
	Teega	tatl, ytlo	at-lo		Jei		Quila
	Ampat	apat	apat		Nahuy		Meli
	Leema	Lima	Lima				Quechu
	nam, a-nam	anim	anam				Cuyn
	toojoo	pito	pitu				Relgi
r	delapan	valo	valo				Pura
ts	fambelan	Siyam	Siam				Ailla
	fapooloo	Polo, pobo	Apalo			Chunca	
	Tangan, Lingan	camas	camavo			Maqui	{ Puilpa *(brachium)* Cue *(manus)*
	Kelady Bira } *Rumph.* { Tallo *Javanice*	Tarac, *(camote gande)*	Tugui				
	Pefang *(Rumph.)*						
	Jangoot	gumi, Baba	Gumi *(barbado)* Baba	Waller			Pajun
rrun	Proot, Pooroot	Tiyan	Attian, Butad				Pue
	Soccun, *(Rumpf.)*						
	Praw	Berai *(navio pequeño)*	Pangga, Lunday	Maragan			{ Butawampa *(navis)* Pichampo *(navicula)* Dolio, *(lembus, canoa)*
	Raja	Hari	Ari			Apu *(dominus)*	{ Apò Curaca *(dominus)*
	Necor	Niog, Niyog	Ungot				
	{ ca-matayan patay, mâttee	{ ca-matayan, hallimolan patay *(to kill)* }	matai				{ Lay *(mori)* Langawyn *(necare)*
ve none)	Anjing	Darapova	Dapur	Rotta			Tewa
	Minnum	Yuom	Inum	chuchala			butun
in	Telinga	Taynga	Balug-bug, Talinga, *(bored ears)*	Melea			Pilum
a	Macan	cain	can, afan, putat	bootina			Jen
g	Mata	Mata	Mata	Meul			Ne
	Eekan	Ysdâ	afan				Challua
	Cak-kee	calis *(foot of an animal)*	Bitis	Edamal			Namon
tomarro	Sobat, Toolan	{ Tiap, Lagoyo, Lagoma, Cutoto, Calagoma, Calagoya, Caffi }	{ Lugud, Paffyag Cutang, Dugo }				
	befarr	Daquila	Dagul			Iitun	buta
	baliong, kapak campa	Daras, Pan-daras	Daras				
m-baltung	Ramboot	Boboc	Buac, Cavad	Morye			Lonco
	Capala	Olo	Bantuc	Waggheegee	Tzontecontli		{ Lonco Towongen
s	Babbi, Bobbee	Babuy	Babi	*(have none)*			Cuchy
	Rooma	Bahay	Balay		Calli		
	Darat	Lupa	Labuad, Gabon		Lan	Lafta	Tue
	{ Kecheel } *in quantity* fedekit } *in magnitude*	onti, bali-balian	intak, lati				Pichi
	Manufia, Orang	Tavo	Tavo	Bāmmâ		Runa	Wento
	Boolan		Bulan			Cilla	
	Mooloot	bunga	Bunganga, Afboc	Yembe, Jembi	Chal		Oun
	Eoong	Ylong	Arung	Bonjoo			Ju
	Oby kecheel						
	Laoot	{ Daggat Laot, *(mar alta)*	} Dayat Laut *(mar alta)* }				Lonquen
	Mata-harree	Arao	Aldao	Galan		Inti	Ante
	Ayr	{ Tubig Tambang *(freshwater)* Taifik *(salt water)*	{ Sabug, Danum Tabang *(fresh water)* Alat *(salt water)* }	Poorai	Ael	Unty	ko
	Pampooang	Babaye	Babai	Mootjel			Domo
n	Oby befarr	Obi	Ubi				

FIGURE 19

Pacific islanders' canoes, from Louis-Antoine Bougainville, *Voyage autour du monde* (Paris, 1771). Alongside linguistic difference (cf. fig. 18), Enlightenment navigators used variations in material culture to develop typologies of social and geographical difference in the Pacific world. By permission of Edinburgh University Library, Special Collections.

Menschkeit (1784–91) offered a stadial interpretation informed by Buffon's empirical enquiries. In his posthumously published *Esquisse d'un tableau historique des progrès de l'esprit humain* (1795)—written while on the run from the Revolutionaries—the mathematician Jean-Antoine-Nicolas Caritat de Condorcet argued for an almost indefinite human progress to be undertaken through rational laws and universal education: enlightenment as never ending.

Thoughts about human progress and its connections with the natural world found particularly fruitful expression in Enlightenment Scotland. There, numerous thinkers debated the empirical analysis of human society, past and present—producing an intellectual climate that has since been very largely taken (not always rightly) to "define" the Scottish Enlightenment as a whole. The economist and philosopher Adam Smith drew upon a four-stage model for his influential *Wealth of Nations* (1776). Others did likewise, such as John Millar, whose *The Origin of the Distinction of Ranks* (first published in 1771, and in fuller form in 1779) explored the various structures of political authority in relation to stages of economic and social advancement. In his 1767 *An Essay on the History of Civil Society*, Adam Ferguson considered the stages of human history to be "savage," "barbarian," "commercial," and "polite." William Robertson, churchman, historian, and university principal, made much of depictions of "the savage state" in theorizing about American "primitives" in his *History of America* (1777). The methodological term "conjectural history" is even a peculiarly Scottish invention, coined by the philosopher Dugald Stewart. Others, like Lord Monboddo, who speculated at length about orangutans (and much else besides) in his multivolume *The Origin and Progress of Languages and Antient Metaphysics* (1773–92)—considering them to be wild humans, not animals—worked with stadial theory, but in reverse as it were. For him, modern (enlightened) culture was a pale reflection of classical Greek and Egyptian culture.[37]

For all that belief in and "modeling" of the idea of progress was in these and others' writings a core element in the Enlightenment's exploration of human society, historians since have been more ambivalent in considering the evidence.[38] Whether it is strictly the case that "there is no longer consensus on the centrality of progress to the Enlightenment,"[39] it is certainly true that the ideas of progress, stadial theory, and conjectural history become more complex if we think about them geographically. Hume's attention to what may be thought of as conjectural history's basic problem—"Whence . . . the different forms which civilized society has assumed in different ages of the world?"—can be posed as a matter of "conjectural geography," as philosophers turned to the different *places* of the world. Consider again Dunbar's

Essays on the History of Mankind in Rude and Cultivated Ages (1780). Dunbar takes European human history to be a geographical history. Dunbar argues for what is, effectively, the geographical march of civilization from tropical climates to temperate ones—from nurturing origins to "mature" regions. Frugal environments, the result of a Providential design, spurred human ingenuity. For Dunbar, "there is no need to recur to the positive and direct influence of the outward elements on the human mind"—by which he means climate and latitude. "The series of events [that is, human history] is governed more perhaps by moral than by physical causes: and this propensity of genius and temper may owe its original to the primary direction of the sciences, and their early alliance with theology and civil government."[40] We have seen how new geographical discoveries were key to the conception of human difference. What Forster and others saw as the human geography of the Pacific present was, in stadial terms, the human geography of the European past. Difference was not simply a matter of distance, for in the peasant societies of Alpine Europe, in Scotland's clan-based and Gaelic-speaking Highlands, in Ireland as in Lapland, the European past was on the philosophes' doorstep.[41] For those disposed to see it so, in such geographical difference lay all human history.

For Turgot, stadial thinking depended on geography as a distinctive intellectual practice, as a way of thinking rather more than as New World knowledge, or as matters of human distribution or climate-cum-latitude. In his 1751 *Plan d'un ouvrage sur la géographie politique,* Turgot used the language of political geography to describe the human world in a series of *mappemondes politiques.* The first of these concerned the racial geography of the world; others centered on the distributions of peoples around the globe and changes from the period of "the most ancient nations" (China and Egypt) through the Greek and Roman worlds into the modern period. He took political geography to have four elements: historical study of the relationships among the physical world, population distribution, and the formation of nations; the study of resources, industries, and the state of commerce within countries; transport systems; and variations in forms of political organization around the globe. His further distinction between "theoretical" and "historical" political geography—the first, the relationships between the art of government; the second a historical analysis of how present political structures had emerged—emphasized political geography as an intellectual and a practical administrative exercise.[42]

Where Turgot thus used geography to think about universal history, many writers and philosophers equated the development of human society with the social and natural geographical state of particular countries. Recall John Locke's remark that "in the beginning all the world was America." What Locke meant by these words was that, to him, North American "Indians" were the vestiges of a primordial civilization, American "Nature" the remnant of a once pristine world. The indigenous peoples of America thus represented the original "stage" of human development, and in the first half of the eighteenth century, Locke's America and not Forster's Australasia was the focus of attention for these reasons.

Many of the ideas circulating in early Enlightenment Europe about freethinking American natives living in a state of nature came from the writings of Louis-Armand de Lom D'Arce, Baron de Lahontan, and the Jesuit Joseph-François Lafitau. Lahontan's *Nouveaux voyages de Mr. le Baron de La Hontan dans l'Amerique septentrionale* (1702–3) is based on his life in North America from 1683 to 1694 and draws greatly on natives' testimony. So too does Lafitau, whose *Moeurs des sauvages amériquains comparées aux moeurs des premier temps* (1724) is an engagement in comparative geographical history based on living among the Mohawk Iroquois (see figs. 20 and 21). Perhaps because of the relative absence of firsthand accounts and for their incorporation of the natives' voice, both works were widely read, with Lahontan portraying native Americans as the intellectual and physical superiors of the corrupt Europeans, and Lafitau providing for the Americas "perhaps the first text to attempt to combine a substantial collection of data which the author himself had collected 'in the field' with [a] significant measure of 'universal speculation' as to its meaning." Lafitau's writings helped establish the Americas as Europe's historical field site, the Enlightenment's pre-Pacific ethnographic laboratory: as contemporaries noted, Lafitau's method made "distances in time analogous to distances in space."[43] And his writings did more: his comparative method demanded that readers relate the manners and customs of the American "Indian" not just to the cultures of Enlightenment Europe but to the classical cultures of Greece and Rome and those even farther east. Adam Ferguson, for example, seeking not conjecture about human difference but hard evidence based on credible claims, turned to Lafitau—and to Buffon—to do so.

It is possible that Lafitau was read more carefully by Scottish philosophers keen to understand the natural history of Man than he was by the French.[44] It may even be that the representation of "primitive man" and the equation of primitivism with indigenous Americans often considered to have its origins in Rousseau's 1755 *Discours* owe more to other earlier connections. One scholar has detected an "alternative geography of intellectual influence," one that identifies "an earlier tradition associated with figures in Glasgow, Aberdeen and Philadelphia rather than Geneva."[45] This

FIGURE 20

American peoples, from Joseph-François Lafitau, *Moeurs des sauvages amériquains, comparées aux moeurs des premiers temps* (Paris, 1724), vol. 1. (1) Hurons and Iroquois; (2) Algonquins; (3) Eskimos; (4) peoples of Greenland. By permission of Edinburgh University Library, Special Collections.

FIGURE 21

Native American husbandry, from Joseph-François Lafitau, *Moeurs des sauvages amériquains, comparées aux moeurs des premier temps* (Paris, 1724), vol. 2. The picture, albeit stylized, is consistent with Lafitau's concern to document realistically the innate capacities of native North American peoples. By permission of Edinburgh University Library, Special Collections.

geography connects the philosopher Francis Hutcheson, Irish born but Glasgow resident—in whose writings and teaching some see the "origin" of the Scottish Enlightenment altogether—with William Smith, born near Aberdeen, whose *Some Account of the North-American Indians* published in Philadelphia in 1754 is sympathetic to Hutcheson's sentiments. But the truth is that Enlightenment sensitivities toward America were more widely shaped by the fact that different books on the human geographies of the New World were read differently in different places: Abbé Cornelius de Pauw's *Recherches philosophiques sur les américains ou mémoires interessants pour server à l'histoire de l'espèce humaine,* for example, more radical than Buffon's work in seeing American society and nature as degenerate, was received differently in Berlin than in Paris and engaged with more directly by those who had traveled to America than by those who had not.[46] So too William Robertson's historical writings—notable for their consideration of Europe's "progressive" rise as a republic of states out of the precivil feudal system as well as for his conjectures concerning the seeming inability of Americans to civilize themselves—likewise had a variable geography of influence. In France particularly he was highly regarded, among Parisian philosophes especially so thanks to the on-the-spot advocacy of David Hume. In Germany, Robertson was read no less favorably—in a translation by Georg Forster before German editions appeared—but there, some took different views about commercialism in Europe and in the Americas.[47]

It is of course true, albeit simply put, that much shared interest in explaining America and American peoples in the Enlightenment was undertaken by "exporting" theoretical models from Europe—models themselves based on "imported" ethnographic data. But in time, sweeping European portraits of "savage Indians" prompted a reaction from American intellectuals. In his 1791 *Travels through North and South Carolina, Georgia, East and West Florida, the Cherokee Country . . . and the Country of the Choctaws,* William Bartram offers a sympathetic reading of native American society—not least, perhaps, because he trained in Philadelphia under William Smith. The most notable American rebuttal of European theorizing is Thomas Jefferson's *Notes on the State of Virginia* (1787), which takes Buffon to task for his depiction of the degeneracy of American natives and their environment. In such work, and more evidently in Jefferson's instructions in 1804 to Captain Lewis over his expedition's need for "useful knowledge" about the extent of the emergent nation—a geographical endeavor that exposed the true scale of the continent's human diversity—Americans later came to know themselves better by themselves.[48]

This was never true of Africa. More accurately, it was not true of Europeans' conception of Africa. Seen from "the outside," the African continent was geographically little more than a coastline, its peoples either points of contact for commercial networks or, in West Africa especially, a raw material as slaves. Although for the French, North and West Africa became a subject of renewed attention after the collapse of their interests in North America following the Seven Years' War, this interest focused hardly at all on African natives. When it did, sweeping judgments were made based on climate, Africans' perceived cultural inferiority, and the moral dominance of "European Man." Montesquieu was blunt on this: the "greatest part of the people on the coast of Africa are savages and barbarians." Bougainville was blunter still: Africa was the "mother of monsters."[49]

Different geographies afford different perspectives. For the British and the Dutch, the establishment of trading colonies and of strategic interests on Africa's southern cape allowed close scrutiny of the region's indigenous peoples. Ethnographically diverse, the Khoekhoe peoples (sometimes Khoikhoi) were commonly seen as "Hottentots." There are, however, marked differences in their portrayal over time and in the Hottentots' place as objects of "racial" scrutiny in the Enlightenment human sciences and in their popular representation, in Britain especially. In his *The Present State of the Cape of Good Hope* (first published, in German, in 1719), Peter Kolb, a Prussian sent to the cape in 1706 to undertake meteorological research, offers a sympathetic account of the Khoekhoe—consistent with Lafitau's of native Americans (fig. 22). For the Linnaean-minded Anders Sparrman, however, writing in his *Voyage to the Cape of Good Hope* (published first in Swedish in 1783, in English in 1785), Kolb's sensitivity was misplaced. To Sparrman, the Khoekhoe were not human beings but cultural specimens to be inspected, described, classified. By the time John Barrow's *Travels into the Interior of Southern Africa in the Years 1797 and 1798* was published in 1801, the Khoekhoe no longer had even that dignity, being "written out" as impediments to Britain's imperial interests.[50]

Questions surrounding the claim that the Enlightenment Science of Man was a Geography of Man thus have several different answers. For those encountering human difference across the globe or writing about it on others' evidence, geography—that is, a truly human geography—was a matter of location and distribution. Who was where? Explaining human distribution as a matter of history and of geography—why were peoples as and where they were?—elicited geographical answers: climate, latitude, the natural environment. In a related sense, geography was the determining

FIGURE 22

Hottentots and European traders, from Peter Kolb, *The Present State of the Cape of Good Hope* (London, 1738 ed.), frontispiece to vol. 1. Kolb's portrayal of southern African peoples is more sympathetic than that of many other commentators. By permission of Edinburgh University Library, Special Collections.

natural "base" against which human varieties, societies, races, and nations were molded. To one degree or another, Montesquieu, Linnaeus, Buffon, Hume, Kant, Dunbar, and others all understood geography as "physical cause" to explain the Enlightenment's "moral" geography, the differences in manners between people.

At the same time, there is in Enlightenment thinking a strong sense of geography not as a fixed natural base but as something always in process. Far from being determined by "nature," the environment could be transformed by the power of human agency to produce "culture." This is what Buffon meant when he wrote of humans "seconding Nature"—changing one's world by reforming agriculture, clearing woodlands, draining the marsh, and so on. That is, in one sense, why works on economic and agrarian improvement are so distinctively an Enlightenment genre. They were, in effect, prescriptions for future geographies—how the farm, the countryside, the nation might—should—look. That is why, given interests in stadial theory and conjectural history, economist-philosophers like Adam Smith and agrarian-philosophers such as Adam Dickson in his 1788 *Husbandry of the Ancients* looked to human society in the past to explain present human circumstance and to imagine the future. It is one reason why humans in the aggregate—population—became an object of Enlightenment scrutiny: in Germany in the work of Johann Süssmilch; in Scotland in the inquiries of Robert Wallace, Alexander Webster, and Sir John Sinclair; and in England, of William Godwin and Thomas Malthus.[51] And it helps explain why interest in "national character" and human ethnography as a series of *mappemondes politiques* was paralleled at the level of the human individual in artistic depiction and in physiognomy, the "science" that equated "mapping faces" with mapping nations, human looks with cultural prediction.[52]

If these ideas may be thought of as the geography *in* or *behind* human difference, we can note too a geography *of* ideas about that difference. Different parts of the world were used at different times: the Americas before the 1760s before being overtaken by Oceanic data; North America east of the Mississippi until 1804; Patagonia and southern Africa always in view as local ethnographic sites but differently interpreted. Views about the human world differed by place—in Uppsala, Paris, and Göttingen, to name only a few. Because they did, we must be cautious about equating cosmopolitan debates about "national" character with national debates. Consider Scotland again. In its factual content as well as in its methodological languages, the Science of Man in Scotland was strongly a Natural History of Man. But it was not everywhere the same. Adam Smith drew on Linnaeus's *Systema Naturae* even as he criticized the work and praised Buffon for his style but dismissed him for his theories. In Aberdeen, Thomas Reid and others were openly critical of Buffon. In Germany, that essential Enlightenment tenet, progress, had similar local geographies.[53] We cannot claim to know the Enlightenment as a matter of human science and, in the several senses explored here, of human geography, unless we recognize such variable topographies to knowledge's production and reception.

Part 2 has shown that the earth as the home of mankind was revealed in a number of ways, made "modern" through intellectual practices that we can today collectively term "geographical knowledge." Hitherto ignored in studies of the Enlightenment that have considered it an intellectual phenomenon in time but not in space, processes of exploring, mapping, inscribing, and depicting show how the Enlightenment was "made" in space. In the Enlightenment, the earth was made older, more diverse, and given a "precise" shape through geographical endeavor. With the exception of the polar regions, the world's continental margins were delineated. The shape of the earth was revealed through geography—and revealed differently, depending on where one stood, philosophically and geographically speaking.

To Europeans, encounters with the human cultures and natural riches of the Pacific world in particular added a new "fifth part" to the world. There, and in the Americas—the Enlightenment's two major geographical laboratories—distant "margins" provided an essential "core" of material for theoretical speculation on the nature of Man, on human difference, and on the bases to cultures and social change. "Voyaging" is a key term in understanding those processes of geographical knowledge that, in combination, made the Enlightenment world. The term embraces notions of voyager-naturalists leaving port, traveling, and returning home, there to "launch" their systematic observations, classifications, and comparative reflections in the form of print and picture and performance for eager publics. Yet geographical knowledge was not made and did not voyage easily. Mapping depended on processes of translation and negotiation. Instrumental results calibrated for one place did not accord with measurements elsewhere. Everywhere, the facts of local geography got in the way of schemes to rationalize the Enlightenment's worlds of difference.

What has been seen as the Science of Man or the Natural History of Man may be considered a Geography of Man. For as geographical examination of the Enlightenment world provided the raw data for nascent "disciplines," theorists of the facts and causes of human difference proceeded methodologically via observation, empirical testing, and a search for system. They did so, moreover, by considering the geographies of human difference *over time*—in which context explanation of the human world followed the same logical precepts as explanation of the physical globe: by looking for evidence of the past in the present. And they did so by studying geographies of human difference *in space*—examining that map of human variation that explorers, voyagers, and proto-ethnographers were bringing to light. In looking for explanation in issues of moral superiority, economic capacity, and importantly, climate, Enlightenment thinkers turned to the facts of geography to explain human society.

PART THREE

Geography in the Enlightenment

8

Geography and the Book

For the anonymous reviewer of John Pinkerton's *Modern Geography* (1802), geography's usefulness, in book form anyway, lay in putting the world's diversity into portable order:

> There is no science so attractive as geography. It requires scarcely any preparation of previous study; and deals in a sort of information so popular and various, as to recommend itself even to those who have but little relish for literary occupations. It is indeed a kind of condensation of books of travels, and exhibits the most captivating collection of marvellous truths, that ever yet were assembled, to excite or to gratify curiosity. Of its substantial utility, it is unnecessary to speak.[1]

Pinkerton's subtitle—*Including the Most Recent Discoveries, and Political Alterations*—further indicated its up-to-date value. Enduring geographical enigmas of the Enlightenment world—the existence of a southern continent, the course of the Niger, Baja California an island or a peninsula, even the world's shape—had by then been more or less answered. Pinkerton's book synthesized the findings of *voyageurs naturalistes* and travel writers, presented a systematic description of the world, country by country, and so made distant geographies accessible. But what of others' books? Since, earlier, the Enlightenment was "modernizing" rather than "modern" in its geography, how did books of geography before Pinkerton's present the world? What, indeed, was a geography book? Who wrote them? Who read them? What did authors and audiences take geography in the Enlightenment to be about? What forms did geography's "substantial utility" take?

Geography in the Enlightenment was far from being a textual enterprise alone. Nor was it a single thing. General distinctions were made between mathematical geography, descriptive geography, and chorography, regional description. More specialist differences were also understood. In one form or another, the subject was widely taught: in universities by mathematicians, moral philosophers, and natural historians; in schools and in the coffeehouses, taverns, and other spaces of the public sphere by itinerant lecturers and self-taught "geographers." Geography was readily domesticated—learned at home by women as a genteel accomplishment, by men as a basis to commerce, by children as an educational accomplishment. Globes and atlases appeared in considerable numbers and in different sizes and purposes, both as scientific instruments and as objects of cultural status and ornament. Geographer-mapmakers, some working in teams for national armies, others for scientific societies or individual patrons, plotted new shapes to countries. Geographical parlor games with cards and jigsaw maps were commonplace. So too were works of geography quite different from Pinkerton's—books for children, works for women, texts to be read aloud. Painters turned their attention to "The Geography Lesson" as a subject. Pornographers worked with geography's words and works, using metaphors of "virgin territory" and mapping the landscape of the female body as part of their stock in trade. And in other ways—in enumerating population, in assessing natural and fiscal wealth, in understanding the connections among health, climate, and place—people likewise thought geographically in and about the Enlightenment. Geography, in short, was a subject and a diverse cultural commodity.

Part 3 examines geography's epistemic cultures and differences, of genre, place, practice and practitioner, social space, and textual tradition. To an extent, doing so for the "discipline" of geography and for geography's discourses in the Enlightenment is an act of recovery, even one of restitution. I say this because geography with its books, practices, audiences, and intellectual substance has too often been overlooked by Enlightenment scholars. And not only by them. Until recently, the subject's place in the Enlightenment was either marginal or ill served even among geography's historians, its texts derided as arid narratives, its place in the universities unexamined, its presence in the Enlightenment's public sphere unimagined. Work from a variety of quarters, testimony to a wider interest in the geographies of the eighteenth century, has now begun to address these and other issues and so to recover geography's presence in the Enlightenment.[2]

The central concern of this chapter is with geography and the book. I examine geography's textual traditions, the different genres that went to make up the "geography book," the nature of geographical gazetteers and dictionaries, geography's authors and readers, and the purposes that geography's books served. Far from being few and far between or being of interest only to experts, books of geography were numerous, varied, and popular in the Enlightenment. Take English-language works in "special geography," for example. In their simplest sense, books of special geography purported to describe all the countries in the world. They did so by working with sets of "particulars"—terrestrial, celestial, and human—and, unlike geographical dictionaries and gazetteers, presented information in the form of continuous prose. Of necessity selective and ordered, special geographies aimed to be useful and always up-to-date: "the 'discovery' of new lands was an intellectual constant." Dozens of such works were produced during the Enlightenment, most for adults, some for children, with several works going through many editions. For example, Patrick Gordon's *Geography Anatomiz'd; or, The Compleat Geographical Grammar* (1693), which was written for the children of the gentry and claimed to have "reduc'd the whole body of modern geography to a true grammatical method," was in its twentieth edition, "corrected and inlarged," by 1754. William Guthrie's *New Geographical, Historical and Commercial Grammar* (1770), intended by its author to be the successor to Thomas Salmon's *New Geographical and Historical Grammar* (1749), went through forty-six editions: one work among about 650, it was the most popular geography book in eighteenth-century Britain.[3]

Geography's textual traditions are considered in the second part of this chapter. In the third and final part, I show how geography books made the Enlightenment work, with specific reference to Jedidiah Morse's *American Geography* (1789). In Morse's text as perhaps no other, we can note both the rejection of others' "imported geographies" and witness an attempt, in thinking geographically about America's national identity, to provide a geographical blueprint for Enlightenment and for the American Republic. By "geography and the book," however, I do not mean to concentrate only on books of geography. I also consider the geography *in* and the geography *of* books. Let me begin by returning to the *Encyclopédie*, the book that many commentators then and now regard as *the* embodiment of the Enlightenment *mentalité*. In considering how and where that work was produced and received as well as how it defined "geography," we can trace the many geographies of the *Encyclopédie* and of the Enlightenment.

Geographies of the *Encyclopédie*

The *Encyclopédie: ou, Dictionnaire raisonée des sciences, des arts, et des métiers, par une Société des Gens des Lettres* was first published in Paris between 1751 and 1765 in seventeen volumes of text under the editorship of Denis Diderot and Jean d'Alembert, with eleven volumes of accompanying plates appearing between 1762 and 1772. In being seen as a crucial *machine de guerre* in the campaign against ignorance and unwarranted authority, the *Encyclopédie* has been considered "the embodiment of the Enlightenment," the Enlightenment's "manifesto" and "epitome," even its "central document." It certainly visualized the power of secular reason: the work's frontispiece illustrates the figure of Reason (with some help from Philosophy) lifting the veil from Truth and being cast in light in the process.

FIGURE 23

The Enlightenment personified, from Rigobert Bonne and Nicolas Desmarest, *Atlas encyclopédique* (Paris, 1787–88), frontispiece to vol. 1. The image is from the frontispiece to Denis Diderot and Jean d'Alembert's *Encyclopédie, ou dictionnaire raisonné des sciences, des arts, et des métiers*. By permission of Edinburgh University Library, Special Collections.

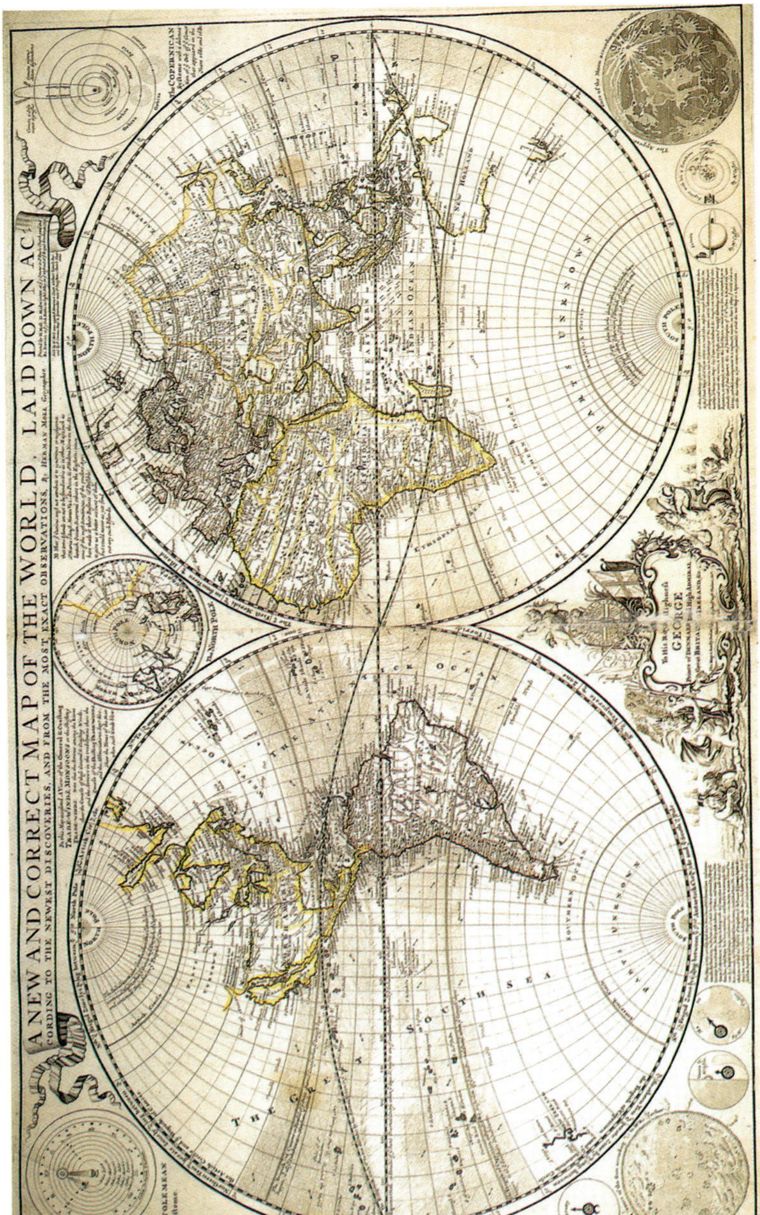

PLATE 1

The world in the early Enlightenment, from Herman Moll, "A New and Correct Map of the World" (London, 1709). Note the depiction of California as an island, the absence of information on northwest America, the uncertainties over the Arctic regions and northeast Asia, and the assumption that New Guinea and New Holland (Australia) are a connected landmass. By permission of the Trustees of the National Library of Scotland.

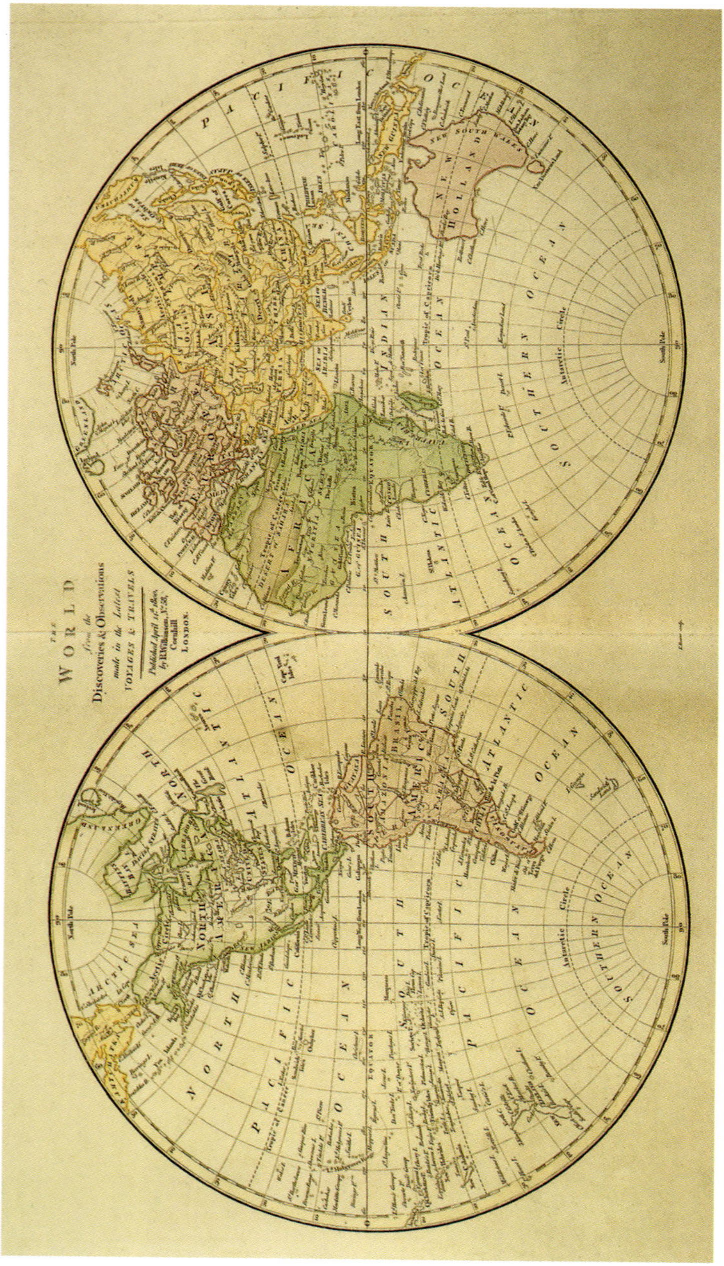

PLATE 2

The world in the late Enlightenment, from R. Wilkinson, "The World from the Discoveries & Observations Made in the Latest Voyages and Travels" (London, 1800). In contrast with Moll's map of 1709 (plate 1), northwest America, northeast Asia, and Australia are all given "modern" continental shapes. By permission of the Trustees of the National Library of Scotland.

PLATE 3

Enlightenment author: Denis Diderot, *encyclopédiste* and Enlightenment correspondent. Courtesy of the Louvre.

PLATE 4

Enlightenment calculator: Émilie du Châtelet, *hôtesse de salon* and Newtonian author. Courtesy of the Marquis and Marquise de Breteuil.

PLATE 5

Enlightenment measurer: Pierre-Louis Moreau de Maupertuis, geodesist and mathematician. Courtesy of Musée de Saint-Malo.

PLATE 6

Enlightenment personalities. Portrait of Omai, Sir Joseph Banks, and Daniel Solander, by William Parry. Courtesy of National Portrait Gallery, London.

PLATE 7

The ship as Enlightenment instrument. HMS *Endeavour* off the coast of New Holland, by Samuel Adkins. Watercolor, 1794? By permission of the National Library of Australia.

PLATE 8

The Enlightenment Earth observed: *Vesuvius in Eruption, with a View over the Islands in the Bay of Naples,* by Joseph Wright of Derby, ca. 1776–80. By permission of the Tate Gallery, London.

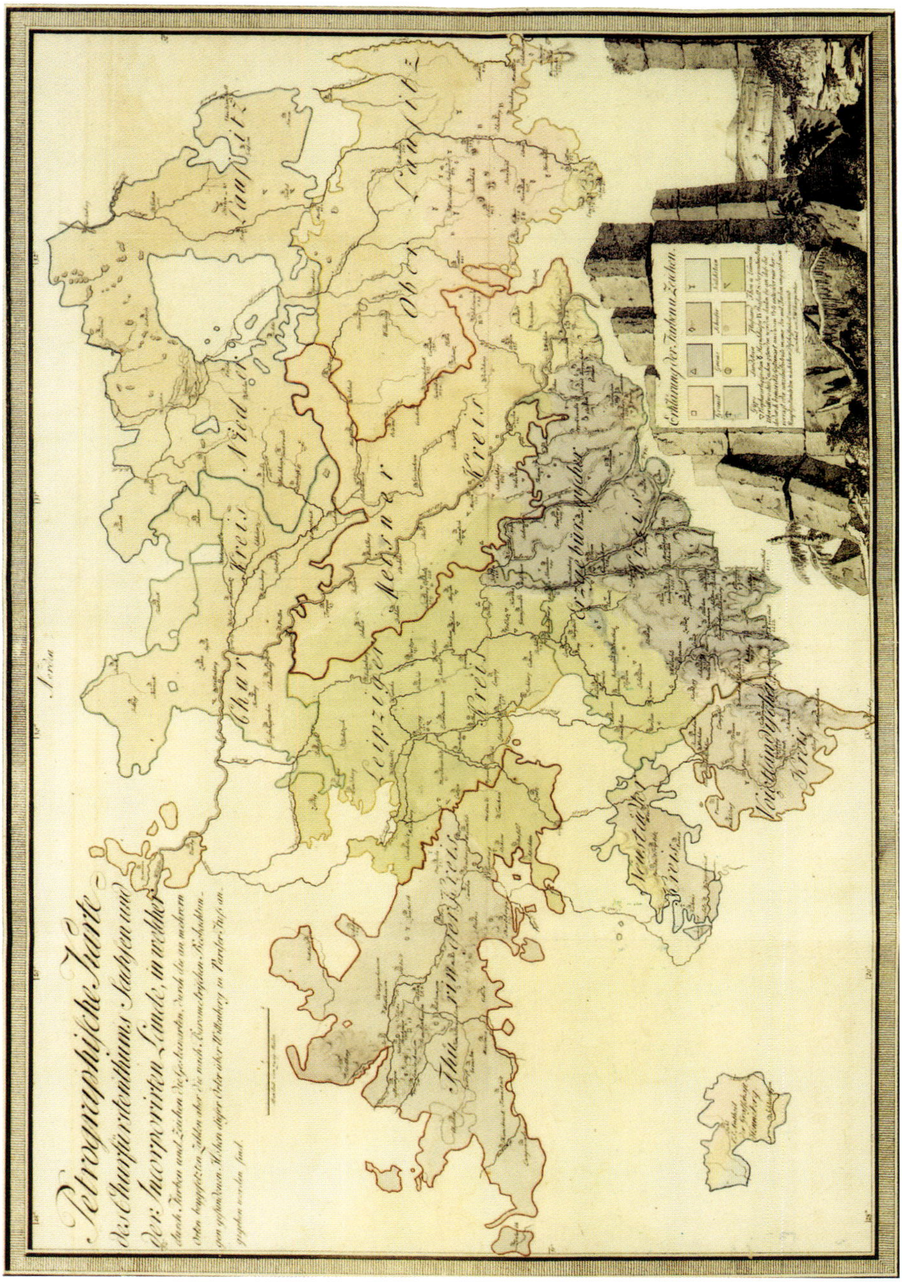

PLATE 9

The Enlightenment Earth mapped and inferred: Mineralogical map, from Johann Friedrich W. von Charpentier, *Mineralogische Geographie der chursächischen Lande* (Leipzig, 1778): 4° Sigma 407. By permission of the Bodleian Library, University of Oxford.

PLATE 10

Enlightenment geography as instructional games: *Portrait of Masters Thomas and John Quicke Assembling a Geographical Dissection,* by William Hoare, ca. 1770. The portrait, which is in a private collection, is reproduced here by courtesy of the Paul Mellon Centre for Studies of British Art, London.

PLATE 11

Late Enlightenment political geography: J. Wallis's jigsaw "New Map of Europe after the 1815 Congress of Vienna" (London, 1815). By permission of the British Library (Maps 162.p. 4).

PLATE 12

Geography, enlightenment, sociability and display: *The Geography Lesson,* by Pietro Longhi. Oil, ca. 1750–52. Courtesy of Museo Querini Stampalia, Venice.

Farther to the side, Memory and Imagination sit in association with their respective arts and sciences. In the foreground, Theology kneels (fig. 23).[4] In the hands of Diderot, the self-taught son of an artisan and the mathematician d'Alembert, the work was certainly used as a tool to oppose religious doctrine and the authority of the state. The *Encyclopédie* embodied a cosmopolitanism—160 contributors produced its 72,000 entries—whose author-practitioners were concerned to unite the arts and the sciences, order the world of knowledge, and provide a storehouse of useful human learning continually being replenished through inquiry into the truth. Importantly too, the *Encyclopédie* extended earlier encyclopedic traditions, in Britain in works such as John Harris's *Lexicon Technicum* (1704, 1710) and Ephraim Chambers's *Cyclopaedia* (1728), and in France, in the *Grand dictionnaire historique* (1674) of Louis Morieri.[5]

Diderot and d'Alembert used the *Encyclopédie* to map human knowledge. In his *Discours préliminaire,* d'Alembert made clear that the work was a dictionary of the arts and sciences giving the basic principles and content of each, and an encyclopedia exhibiting the connectedness of all human knowledge. It was

> a kind of world map which is to show the principal countries, their position and their mutual dependence, the road that leads directly from one to the other. This road is often cut by a thousand obstacles, which are known in each country only to the inhabitants or to travellers, and which cannot be represented except in individual, highly detailed maps. These individual maps will be the different articles of the *Encyclopedie* and the Tree or Systematic Chart will be its world map.[6]

Providing metaphorical-taxonomic maps of knowledge in such ways—what Francis Bacon 150 years before had called "A Description of the Intellectual Globe"—was a commonplace of the encyclopedic tradition. The use of map imagery was at once explicitly rhetorical and epistemological, and because this was so, reading the *Encyclopédie* as a map can help illustrate the conceptions of order and the search for "truth."[7] But in signaling as they did to the central place of Reason and Philosophy in the frontispiece and there marginalizing Theology, Diderot and d'Alembert did more than depart from established practice in a new search for order. They offered a certain iconography and a novel topography for their scheme for learning—and, of course, in so placing religion, a dangerous one. The source of human enlightenment was Reason, not Revelation, empirical inquiry not scriptural dogma. New places were found for the emergent sciences, geography included, under the overall label "Reason" in their "Detailed

System of Human Knowledge." Later in the chapter I return to this issue of the classificatory spaces for subjects, a new "geography to knowledge" in looking at geography's content and at its epistemic space and place in the *Encyclopédie*'s map of learning. Here, let me briefly review how in other ways the work illustrates the geographical dimensions of the Enlightenment.

For those publishing, selling, shipping, and reading the *Encyclopédie*, Enlightenment was a business—and had a geography—in several senses. In addition to the volumes of the Paris edition, the *Encyclopédie* was published in Geneva (1771–76), in Lucca (1758–76), in Leghorn (1770–78), in a quarto edition in Geneva and Neuchâtel (1777–79), and in Lausanne and Berne (between 1778 and 1782). In total, over 25,000 copies were produced before 1789. Paris was the initial center of production in publication terms and was the leading center of production in terms of authorship: in terms of residence, about 59 percent of all contributors lived in the city. In terms of birthplace, the figure is lower—many encyclopedists were born in the provinces, nine in Switzerland, and a few in Germany and elsewhere. Paris was the *Encyclopédie*'s creative core. Within the city, social spaces such as Baron Paul-Henri Thiry d'Holbach's convivial salon in the rue Royale or at his Grandval estate were more important sites for encyclopedism than either the Académie des Sciences or the Académie Française. This is an irony in respect of the first named since that body absorbed the Société Académique des Beaux-Arts, original home of the *Encyclopédié*'s conception by that "Société des Gens des Lettres." Beyond Paris, there were four "regional clusters of Encyclopedists," centered in Lorraine, Languedoc, Versailles, and Switzerland (in effect, Berne, Geneva, and Lausanne). In contrast with those in Paris, encyclopedists in Languedoc did cohere around local academic institutions—the Société Royale des Sciences de Montpellier, the Université de Médecine de Montpellier, and in Beziers, the Académie Royale des Sciences et Belle-Lettres. Behind the textual world represented in the *Encyclopédie* were different regional and local geographies of intellectual affiliation, different social spaces for its conception, discussion, and production.[8]

The *Encyclopédie* may have been largely a Paris creation, but its reception, in terms of subscription sales anyway, was otherwise. Lyon dominated the geography of the *Encyclopédie*'s sales, with nearly twice as many subscriptions (1,079) as in the French capital (575), and for a population one-fifth the size. Sales were high in France's provincial capitals—places such as Toulouse, Bordeaux, Besançon, and Dijon, which played administrative and cultural roles in their local area—and were significant even in many smaller towns. Across France, sales were on the whole high in administrative and cultural centers, low in cities with a commercial and industrial focus. The *Encyclopédie* seems to have appealed more to a "heterogeneous public of noblemen, clerics, and a group sometimes identified as the *bourgeoisie d'Ancien Régime* . . . notables, rentiers, officials, and professional persons, as distinct from the modern industrial bourgeoisie." Again, however, national pictures of social reception demand local geographical and social qualification. The work sold poorly in port towns and in the textile towns of north France; Lyons's domination reflected the activities of the bookseller there; in conservative Besançon, military men, *parlementaires* and lawyers predominated among the town's subscribers. Beyond France, subscriptions were high in Geneva (284 subscribers), Turin (53), Liège (52), Brussels (46), Neuchâtel (39), Genoa (32), Warsaw (31), and Mannheim (27), with many other European cities having only a handful of subscribers. London had only 13, Saint Petersburg 8. Distance was not directly a determinant of access for potential audiences. The few subscribers in America often got their copies more quickly than their European counterparts: one bookseller in Lisbon waited eleven months for a consignment dispatched via Geneva, six months for one sent through Amsterdam.[9]

In thus tracing the book geographies of the *Encyclopédie,* Darnton, Kafker, and others have powerfully illuminated the national, transnational, and local geographies of a textual Enlightenment. Using booksellers' letters and subscription lists as the basis of their work has advanced methods useful in plotting other texts' geographies. What cannot be so easily charted is how—and where—the *Encyclopédie* and other books were read. Was the *Encyclopédie* read at all or just consulted? Was it argued over in company? It is not easy to know. Unlike leading scientific texts whose marginalia in successive editions have allowed others to trace the ownership, reading history, and reception of the work, or whose diaries permit insight into private readings, the reception geography of the *Encyclopédie* is less certain than its geographies of production and sale.[10]

What cannot be ignored in understanding the geographies of the *Encyclopédie* is its content. In different places, different interpretations of the content did influence sales—where it was read and by whom mattered as to how it was judged. This was particularly so for the geographical entries that, on the whole, were uneven in their quality. Diderot noted in 1769 that those entries—several of which he had written—required considerable revision for the later editions of the *Encyclopédie* being proposed by that leading "merchant of ideas," the bookseller and publisher Charles Joseph Pancoucke, and in what became Pancoucke's *Encyclopédie méthodique,* begun in 1782. Some geographical entries were poor: the Baron Louis de Jaucourt's "Italie" looked hardly at all at eighteenth-century Italy, sneered at its priests, reported without firsthand knowledge on the country's geography and history, and stood in marked contrast to d'Anville's *Analyse geographique d'Italie* (1774), for example. That, more than his irreligious tone, made it hard for Italian booksellers to sell the work locally. Other geographical articles written by distant "authorities" likewise prompted a critical reception on national grounds. But others did not: entries on the West Indies were written by a French engineer in Grenada; professional geographers wrote several articles.[11]

In recognizing that the reception geography *of* the *Encyclopédie* was influenced by the geography *in* the *Encyclopédie,* we should make a distinction between poor entries on countries whose reading there affected the work's reception and entries on geography's definition. The principal entry on geography, "La Géographie," reads

> Geography should be seen under three different period headings, 1st. Ancient Geography which is the description of the land from the time of the development of the Ancient Peoples up to the fall of the Roman Empire 2nd. Geography of the Middle Ages from the fall of the Empire up to the rebirth of letters 3rd. Modern Geography which is the current description of the Earth from the rebirth of letters to the present day. One distinguishes further the areas of Geography as Natural, Historical, Civil and Political, Sacred, Ecclesiastical, and Physical.

The development of geography is then traced through its three "ages." Attention is paid to the principal divisions making up "Géographie Physique," to geography's connections with astronomy, military surveying, even to geography's use in card games as a form of sociable instruction. Geography had both current credibility and significance in historical inquiry about the development of peoples and nations. "Ancient Geography" was an established and active form of scholarly geographical inquiry, in books and in atlas production, in which the political and cultural landscapes were traced as the background and explanatory context to certain moments in human history.[12]

The picture we are given is of a widely recognized and enduring subject, unified in its concern to describe yet diverse enough to warrant meaningful internal divisions over what and how things were described. At the same time, geography had close connections with other subjects—not least with history with which it was conventionally bracketed—and with mathematics and other "particulars" within physics. Geography shared its position with these subjects as part of the "Science of Nature," within "Philosophy" under the general heading of "Reason." The "fruits" of geographical learning shared their root stock with cosmography and with chronology—the first apparent in geography's close connections with astronomy, the second reflecting links to history, in combination all subjects with an intellectual genealogy traceable to Ptolemy and to Strabo. Closer assessment of the place of geography reveals its engagement both with the physical and the moral bases of "Human Characteristics"—part, then, of the Enlightenment "Science of Man"—and a consistent category of definition alongside terrestrial and celestial characteristics. The measurable quantities of the physical globe dominated in one respect. But in addition to its human and physical concerns, geography was held also to embrace matters of practical utility resulting from comparative assessment of the features on the earth's surface—questions of navigation in relation to commerce, for example—as well as detailed topographical inquiry (figs. 24 and 25).

The principal authors of the *Encyclopédie* entries on geography were the geologist and agricultural writer Nicolas Desmarest, who wrote on physical geography, and two mapmakers, Jean-Baptiste Bourguignon d'Anville, geographer to the king, and Didier Robert de Vaugondy, author in 1755 of *Essai sur l'histoire de la géographie,* which stressed geography's place in the emergence of France and of French science in the early Enlightenment. The latter's father, Gilles Robert, had been geographer-in-ordinary to the king of France. Their articles were of high quality, with Diderot praising d'Anville's in particular. D'Anville was a distinguished mapmaker, member of the Académie des Sciences, and author of numerous papers on geography. Robert de Vaugondy provided maps for Montesquieu's *L'esprit des lois,* Brosse's *Histoire des navigations aux Terres Australes,* and Buffon's *Histoire naturelle,* among others. But their articles were few and brief. D'Anville's discussion "Etésiens," about the winds of the Mediterranean, was a mere fifty-five lines; Robert de Vaugondy's article "Globe" was an exact description, in four pages, of the manufacture of globes—but not how to use them—and his entry "Fuseau," a geometrical figure, was only twenty-five lines long.[13]

These men, together with others interested in geography—such as the hydrographer-mapmaker Jacques Nicolas Bellin and César François Cassini (Cassini de Thury), neither of whom wrote for the *Encyclopédie*—embraced *l'esprit encyclopédique.* They knew one another. They cited one another's work. They associated with one another. D'Anville was a member of the Académie des Inscriptions with other geographers such as Pascal-François-Joseph Gosselin, Nicolas Fréret, and Jean-Pierre de Bougainville. But they did not together constitute *un corps géographique.* Their connections were never strong—between the mapmakers Didier Robert de Vaugondy and d'Anville they were persistently acrimonious—and using terms like

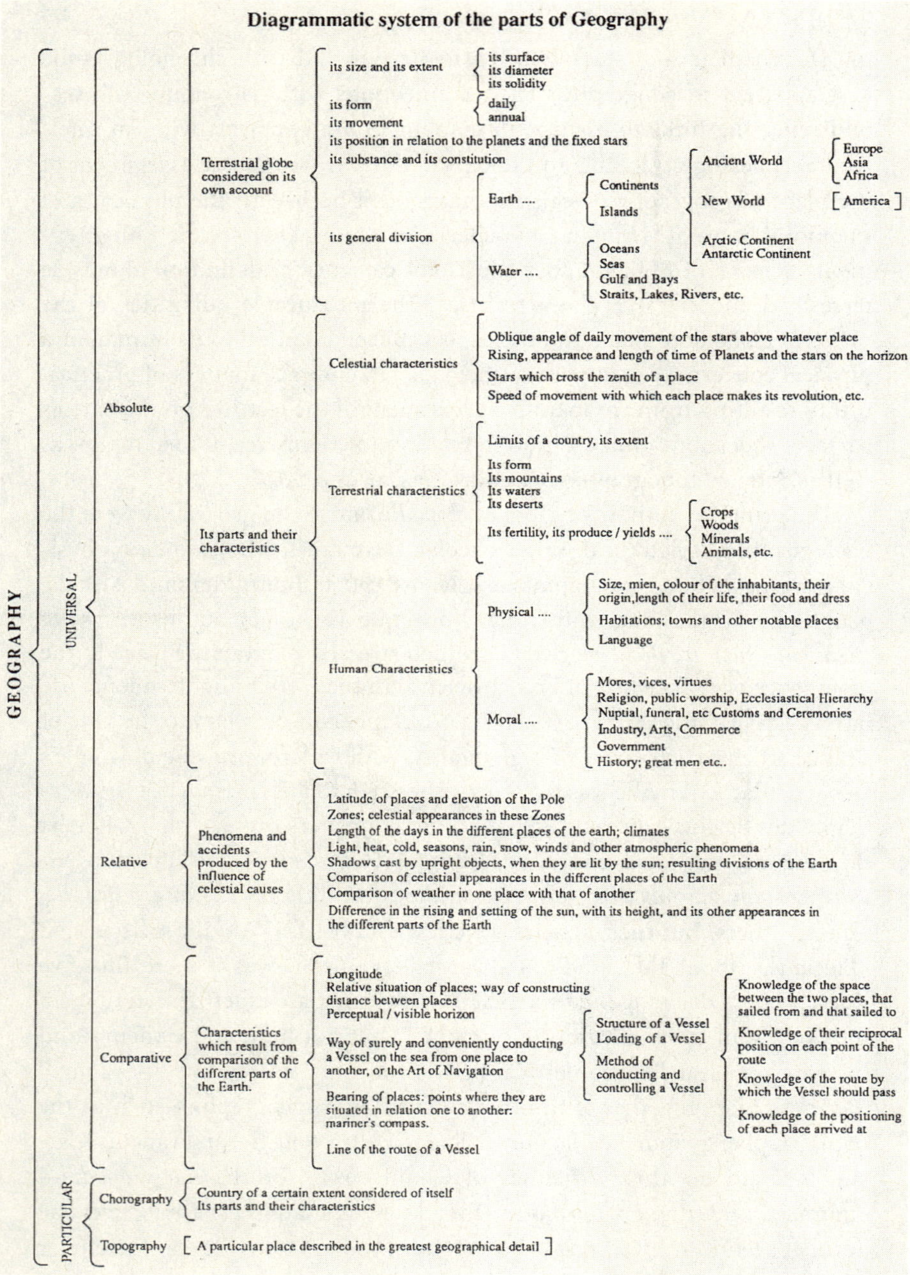

FIGURE 24

Geography charted. The diagrammatic systems of the parts of geography, from Denis Diderot and Jean d'Alembert, *Encyclopédie, ou dictionnaire raisonné des sciences, des arts, et des métiers* (Paris, 1751–68), *Supplement* (Paris, 1780), vol. 2. This view of geography's constituent parts and the relationships between them expands upon similar schemes in earlier encyclopedias. By permission of the Trustees of the National Library of Scotland.

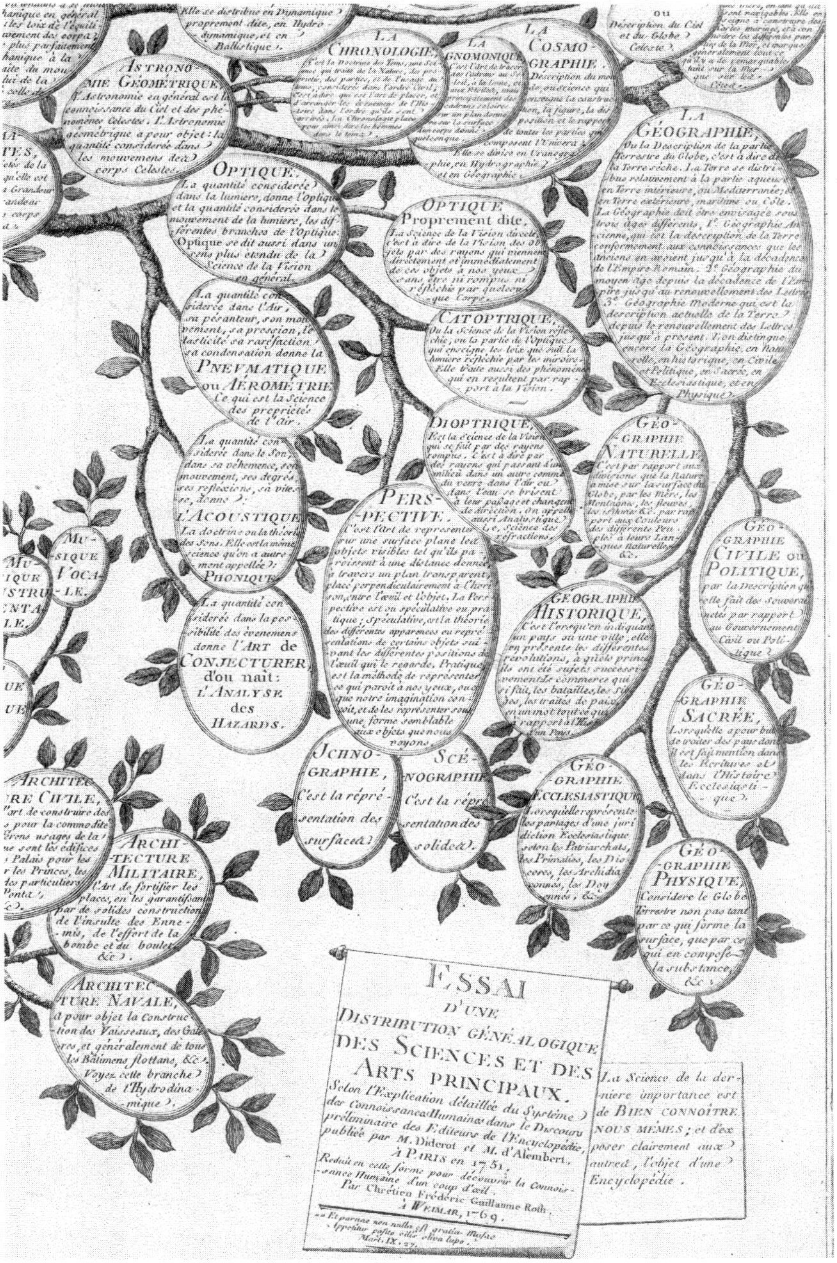

FIGURE 25

Geography in the Tree of Knowledge, from "Essai d'une distribution généalogique des sciences et des arts principeaux," in Denis Diderot and Jean d'Alembert, *Encyclopédie, ou dictionnaire raisonné des sciences, des arts, et des métiers* (Paris, 1751–68), *Supplement* (Paris, 1780), vol. 2. By permission of the Trustees of the National Library of Scotland.

"informal community" even may overplay the bonds between these men. By contrast, there appears to have been a greater sense of geographical *fraternité* among those mapmakers-cum-geographical engineers whose work was central to the work of the Dépôt de la Guerre.[14]

In the *Encyclopédie* then, we see both the classificatory place that geography held as part of the map of world learning and something of the term's comparative anatomy. Yet geography's epistemic space in the work did not have a clearly defined social or academic counterpart; its textual space was without institutional correlate. Geography, a means to descriptive world knowledge both human and physical, was placed within a map of knowledge that prioritized reason and signaled to its practical possibilities. But in *the* book of the Enlightenment, we see relatively little of geography's content and textual traditions and almost nothing of what it actually did. To understand these things of geography and its books in the Enlightenment, we must look elsewhere.

Geography's Books and Textual Traditions

"By Geography is understood a Description of the Surface of the natural terraqueous Globe, consisting of Earth and Water, which is represented by the artificial Globe." Most definitions of geography in eighteenth-century geography books and dictionaries echo Thomas Salmon's brief definition in his 1749 *New Geographical and Historical Grammar.* Behind such brevity and claims concerning geography's purpose as earth description lie some enduring constants as to what geography was, and variations in content, its genres, and in the authors and readers of its books.

Books of geography in the Enlightenment consistently emphasize geography as a coherent body of knowledge about a clearly defined object, namely, the situation of places on the earth and the content of those places in natural and human terms. Geography was understood to be formally separate from cosmography, the study of the position of the earth in the celestial system, and from topography and chorography. These distinctions in scale and substance were often blurred in practice, for cosmography and astronomy especially so, since many geographical works begin with a discussion of the earth's cosmological or astronomical place, and geography and astronomy were commonly taught together. Geography was closely associated with history—was, indeed, one of the "eyes" of history (by convention the left, chronology being the right)—in the sense that only by knowing geography (as place-names, physical distances, climates, and human cultures in place and across space) could historians know facts crucial to their own inquiries.

The association of geography and history in this way was a commonplace of post-Renaissance thought. Samuel Johnson insisted, for example, that knowledge of geography was necessary to read the Ancients, to understand the values different cultures placed upon universal categories such as hope and fear, and to travel with benefit. Edward Gibbon prefaces *Decline and Fall of the Roman Empire* with a chorographical-geographical study of Rome, and in the opening chapters of that work he cites d'Anville and other leading "ancient geographers" on the importance of "historical geography" in explaining the historical evolution of peoples. Gibbon's attention to "rational geography," as he put it, and particularly his debt to French geographical work—to Nicolas Fréret and notably to d'Anville, "the prince of geographers"—was essential to his political and historical writings. Eighteenth-century geographical texts also insisted on geography's scientific status. Geography was science as a form of knowledge—descriptive, locational, classificatory knowledge—not science as certainty grounded in demonstration. In sum, geography was defined and understood very tightly. "Geography was the verbal and mathematical description of the earth: it was a scientific subject, but not in our modern sense; it was restricted in its spatial scale to general discussions of the earth as a whole; it was distinguished from travel writing by its ordered and detached presentation of evidence; and it was created by textual collation rather than fieldwork."[15]

Who wrote geography's books? In Britain anyway, most of those persons who wrote books of geography were not geographers at all in the sense we now understand the term. Most authors who compiled information about the earth's surface at a world level were normally either historians or Grub Street journalists, the distinction often being opaque. "Grub Street," a term associated with cheap publication by hack writers, in effect "authors to let," was also a topographical reality in eighteenth-century London in that it was a street in the city's Cripplegate area. These people were compilers rather than fieldworkers. Geography was a career for them of sorts, but they were not geographers. This is not to say such works were not useful. Because geography was so closely linked to history, many texts established their utility by providing a review of the history of nations, offering what contemporaries took to be a "political geography" in which a nation's revenue or population was enumerated.[16]

Who read geography's books? Broadly, the intended readership had two constituencies: that community of humanist scholars—historians, classicists, and theologians—for whom geography had long been part of historical inquiry, and those persons needing a practical education of one sort or another. This second group may be further divided—between children and adults, and between adults with an interest in commerce or in geography's mathematical precepts or in history.

The first special geography written for children—George Meriton's *A Geographical Description of the World*—appeared in London in 1671. Dozens more followed in the eighteenth century. John Locke among others saw geography as necessary for a child's education. In catering to a particular audience, children's geography books used narratives and dialogues to make learning easier, were read alongside "instructive" geographical games, or presented fictional travel accounts in order to instill geographical principles. Newtonian principles—including summary versions of the debate over the shape of the earth—were incorporated, almost as performance, in the mathematical geography elements of the influential and widely read "Tom Telescope's" *Newtonian System of Philosophy* (1761): "We often play at *sham orations, comical Disputes, measuring of Land and Houses, taking the Heights of Mountains and Steeples, solving Problems and Paradoxes, on Globes and Maps,* and sometimes at *Natural Philosophy.*" In France, for example, the children's geography published in 1736 by Abbé Pierre Nicolas Lenglet du Fresnoy (an *encyclopédiste* but not of geographical entries) was popular and much reprinted. An English version first appeared in 1737 and by 1816 was in its twenty-sixth English edition. What made it useful was its style: "intelligible and short questions." At least one later edition had a companion map volume designed for children. So too Jean-Marie Bruyset's 1790 *Atlas des enfans,* for instance, which incorporated maps to pose questions on Europe's political topography: "What are the different forms of government in the states of Europe? We can distinguish five sorts of government: despotism, monarchies, rule by aristocracy, democracy, and mixed forms." Geography was not diluted in its definition and significance when written with children and pedagogy in mind: simply, particular markets demanded different textual strategies.[17]

In adult geographies, commerce and history were more evident—and commonly related—features. In George Bickham's 1743 *The British Monarchy,* for example, which depicted in prose, maps, and pictures "all the dominions subject to the King of Great Britain," enlightenment and knowledge of empire depended on geography. Geography was for Bickham "among the first and most necessary" of the "Liberal Arts" and was so because without it "all our Ideas of remote Transactions, Events, productions and whatever falls not under the Cognizance of our Senses, must be ever confused and imperfect." In also emphasizing writing in geography and in commerce, the art of penmanship in map work and as a polite accomplishment, Bickham used geography to connect the worlds of sociability and politeness with the worlds of commerce, empirical encounter, and imperial authority. So too Thomas Salmon, whose several geographical works were of a piece with his monumental *Modern History: Or the Present State of All Nations* (published in thirty-one volumes between 1725 and 1738), the first ten volumes of which focused on special geography in order to place the historical accounts that followed. Geography was history in the present—"the present state of the several kingdoms of the world" as well their "history . . . revolutions and memorable events."[18]

How, then, are we to understand as a whole geography's textual character in the Enlightenment? One leading commentator puts it thus:

> Looking from the three angles of the textual construction of geography, the readers and their sites of reading, and the writers and their sites of book production, a twin character of geography in the eighteenth century is apparent. On the one hand, geography ties in to a commercial and a practical milieu. The texts themselves emphasised their utility to statesmen and that merchants were a segment of their intended readership. The actual readership of geography books also showed that the public lectures and commercial academies encouraged aspirant merchants to read geography. Finally, the production of geography books was doubly commercial: the conditions of book production were those of cut-throat commercial competition, encouraging plagiarism: and moreover the authors of geography books wrote them in order to eke out a living. On the other hand, geography books were also linked to the tradition of late-humanist education, which emphasized Christian and classical scholarship. The books produced proclaimed their utility to those reading about classical civilisations and scripture.[19]

The truth of these remarks and the claims made should not deny another truth: that they very much constitute a view of geography's books and textual traditions in and from England. As Mayhew notes, "for all Salmon's appeal to Enlightenment values, his *Grammar* enforced the position of a Tory who defended orthodox faith in the Anglican Church, not that of a sceptical citizen of the world." Like others working with the rhetoric of enlightened cosmopolitanism, Salmon disclosed a readiness to stereotype other nations and cultures, placing them in relation to England in terms both of politics and of climate—"a good Medium between the Dutch and the French," with a constitution balanced between extremes. "Every *British* Gentleman is sensible, that he lives in a Country where Life, Liberty and Property are better secured that in any Kingdom in *Europe*." Salmon's geography is a work of the English Enlightenment—clerical and conservative, defensive of Anglicanism, critical of Whigs—even as it demonstrates two more general constants: those essential ideological continuities through which Europeans represented the nations of the globe and the textual strategies they employed to do so.[20]

In thinking about the different geographical dimensions of geography's books, we should of course not take literally all the claims of men like Pinkerton's anonymous reviewer. He, while lauding that work, dismissed books of geography in Enlightenment Germany as "dull and tasteless, . . . executed too much in the manner of the country in which they appeared, to render the study of geography easy, useful or interesting." He likewise dismissed the French. Aside from d'Anville's writings, that country's works were too brief, "by no means adequate to convey that portion of geographical knowledge which will rescue that nation from the charge of comparative ignorance." But beyond England and the limited views of this reviewer, geography's books were written and used in different ways, and they merit our attention accordingly.

In Enlightenment France as in England, books of geographical description—special geographies, geographical dictionaries and grammars, works for children, and so on—were common. So too, perhaps more so than elsewhere, were works offering textual explanations of the decisions behind the construction of given maps, so-called *mémoires,* and guidebooks to cartographic data collection and representation. Bellin's *Recueil des mémoires* (1751) is notable in this respect on hydrographic mapping. This relative attention to "mathematical geography" and to geographical description as an accompaniment to mapmaking was not unique to France. But its presence as a distinctive element in French geographical and scientific inquiry helps explain why men like Bouguer, Condamine, and Maupertuis were called on to help solve questions over the shape of the world. And it accounts for the importance in both ancien régime and post-Revolutionary France of the *ingénieur-géographes,* military geographer-engineers, whose work with map and survey helped shape France—helped shape the idea of France—as a geographical entity (see also chap. 4).[21]

Some have noted a relative "failure" in works of textual geographical description in France in that descriptive attempts to order and classify the world did not keep abreast of changing world knowledge and had less "rigor" than mathematical geography. This, together with the political significance of mathematical inquiry, led to a "crisis" in French geography by the late Enlightenment, an epistemic failure there to be "modern."[22] But reading backward from later developments does little justice to the complexities of geography's place in Enlightenment France and fails especially to appreciate geography's several textual strategies there and elsewhere. Textual classification was about description. D'Anville's work—on the mapping of Africa and in accounting for the different measurements prevalent in France, to name but two elements—was modern to his contemporaries. So too Phillipe Buache on the physical geography of the oceans, Bellin on hydrography, and Robert de Vaugondy's mapping. Textual and descriptive geography did encompass the new facts of exploration. Geographical knowledge was central to thinking about the human place in nature, most evidently in Buffon's work.

The geographical author and politician Edme Mentelle wrote different sorts of geography books to use in different institutional settings—and did so in order to stay alive during the Revolution. His *Élémens de géographie* (1758), for example, exhibits all the features of a "standard" special geography—basic facts about the continents, more detailed facts about France than other countries, an emphasis on "particulars." That and its anticlerical vision—geography was then a strong component in Jesuit schools and colleges—were enough to secure him a post, from 1760, teaching geography and history in a unified course at the École Royale Militaire, the officer-training institution. There he wrote extensively, including a three-volume contribution, *Géographie ancienne,* for Pancoucke's *Encyclopédie méthodique,* and became geography tutor to the royal household. After the Revolution, however, Mentelle wrote new and explicitly Republican geographies, works to mold new citizens: his *Méthode courte et facile pour apprendre aisément et retenir sans peine la nouvelle géographie de la France* appeared early in 1791. Similar works followed. From 1794, this "professeur public de géographie" and former geography teacher to the monarchy was lecturing in different institutions, using his *Méthode* and other texts such as his *Tableau élémentaire de géographie de la république française* (1792) to serve the needs of the new republic. Far from being divorced from the modern, Mentelle's case suggests that geography's books were integral to the very idea of modernity then shaping France (see fig. 3). As I discuss in more detail in chapter 10, making further sense of what geography's texts were and how they were used, in France and elsewhere, demands that we more closely connect what was written to the where of its usage.[23]

German geographers looked mainly to French and to British works rather more than to their own geography books, which were anyway few. Anton Friedrich Büsching's *Neue Erdbeschreibung,* published in eleven volumes between 1754 and 1792, was based on his lectures in geography at Göttingen, where he was professor of philosophy. Although Büsching's use of the German term *Erdbeschreibung,* "earth description," rather than "geography," perhaps reflects a rising German national consciousness in geographical scholarship, his use of population statistics certainly mirrored wider concerns in Germany with political-statistical enumeration. In other respects, Büsching's work followed the models of textual description outlined earlier. Immanuel Kant likewise worked within established traditions in not distinguishing between geography and history in his 1757 *Outline and Prospectus for a Course of Lectures in Physical Geography,* which was based on his own lectures at Königsberg. Nor did he see a strict separation between human and physical geography, despite his book's title. Moral geography, the study of human customs, was included within the "sensible world," that is, the world derived from sensory experience, itself part of the world of nature.[24]

English-language books of geography did not always privilege England. This is clear from the case of the Scots-born William Guthrie's *A New Geographical, Historical and Commercial Grammar,* first published in 1770 and much reprinted thereafter with maps by Thomas Kitchin and the astronomical sections by James Ferguson. Guthrie's work showed a "half-modern" world: late enough to incorporate findings proving the separateness of New Guinea from Australia but too early to outline for certain the shape of the Australian continent or of North America (see fig. 26, and cf. figs. 1 and 2 and plates 1 and 2). Guthrie, who had in the 1730s been a hack writer (alongside Samuel Johnson), echoed Salmon in also writing a history book prior to his influential geography. Unlike Salmon's Anglican text, Guthrie's *History of Scotland* (1767–68) was a work of moderate Presbyterianism—that, and not the agnosticism of David Hume, was the dominant tone of the Enlightenment in Scotland. So too was his *Grammar.* Conventional in all other respects, Guthrie's *Grammar* is notable for its content and style—political arguments are made through headed paragraphs on sensitive subjects—and because it provides a radical disjunction in political viewpoint from English authors.

More than that, Guthrie connects his geography to the Scottish Enlightenment through attention to stadial theories of social development—the conceptual model that, as we have seen, argued for societies developing in stages from agriculturalism to pastoralism to commercialism. In looking at the progress of societies, at the origins of nations, their laws, and government, and at the origins and progress of religion, Guthrie drew on Ferguson's *Essay on the History of Civil Society* (1767), Robertson's "View of the Progress of Society in Europe" in his *History of the Reign of Charles V* (1769), and the work of David Hume. Guthrie's synthesizing model of social development was then used to "read" the map of Europe, explaining

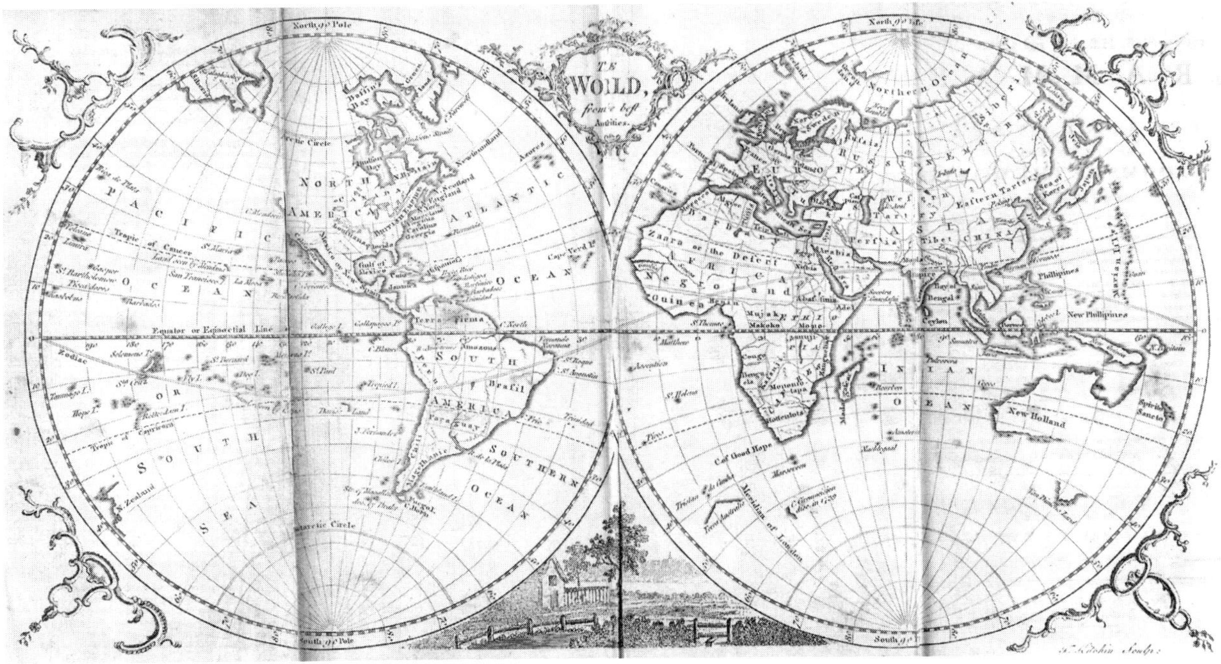

FIGURE 26

The world in 1770, from William Guthrie, *A New Geographical, Historical and Commercial Grammar* (London, 1770). The map shows New Guinea separated from Australia and California as a peninsula (cf. Moll's 1709 map of the world, plate 1) but does not incorporate the results of Cook's or others' voyages, hence the incomplete coasts of North America and New Holland (Australia). By permission of Edinburgh Library, Special Collections.

both the historical and commercial state of individual nations and the superiority of Europe in naturalistic terms. "Geography discovers to us two circumstances with regard to Europe, which perhaps have had a considerable tendency in giving it the superiority over the rest of the world. First, the happy temperature of its climate, no part of it lying within the torrid zone; and secondly, the great variety of its surface." If climate helped explain comparative moral geographies, the continent's providential physical geography was no less a factor in Europe's commerce, and thus in its overall progress and in the state of given nations. Referring to "the immense number of mountains, rivers, seas, &c which divide the different countries of Europe from one another," Guthrie reasoned that "these natural boundaries check the progress of conquest or despotism . . . [and] the seas and rivers facilitate the intercourse and commerce between different nations." The basis of his reasoning derives from Hume, Ferguson, and others. But in employing their work in his geography, Guthrie was innovative. In writing this leading work of Enlightenment geography, he distanced himself from the styles of English texts and brought together Scottish Enlightenment thought and its language of moderate Presbyterianism with established textual traditions in geography in order to understand geographical difference, in and beyond Europe.[25]

Elsewhere, others combined Enlightenment ideas and geography's textual strategies to explain different geographies and to argue for a new nation altogether. The Italian geographer and mapmaker Vincenzo Coronelli used his 1701 encyclopedia to promote geographical and linguistic unity for Italy. In Göttingen, the historian Gatterer used his sixty-six-map *Atlas complet des révolutions* in teaching comparative history to his students. In Greece, Meletios, archbishop of Athens, sketched a relative geography of European civility in his *Geography, Old and New* (1728), praising the French for their culture, the Swiss for their struggles for political freedom, and the English for being the "most civil and tame" of Europeans. His views on national identity being evident in geographical characteristics were echoed in Gregorios Phatseas's 1760 Greek adaptation of an Italian edition of Gordon's *Geography Anatomiz'd* and, in 1774, by Nikiphoros Theotokis in his *Elements of Geography.* Theotokis, who used his book as a teaching manual in the Princely Academy of Jassy, provides a comparative geography of the Enlightenment as liberty and freedom: Holland is identified for its religious tolerance, the Swiss and the Swedes for their liberty, the English for their seriousness of character and scientific disposition. The theologian-philosopher Iosipos Moisiodax used his *Theory of Geography* (1781)—partly written in Vienna and dedicated to the rulers of the Danubian principalities of Wallachia and Moldavia—to bring mathematical geography and explorers' accounts into the Greek Enlightenment, and as noted a means of making the cosmopolitan humanism of the wider Enlightenment world work locally. So too with the *Novel Geography* (1791) of Daniel Philippides and Gregorios Konstantas: Greece is positioned—reclaimed even—within a Europe defined by the dynamics of political change from old and corrupt monarchical regimes to republican communities (and, in Greece's case, from former glories to European nation).[26] But the clearest picture of geography books making Enlightenment work comes from America.

Enlightenment Geography and National Identity: Jedidiah Morse, American Geography, and the New Republic

Booksellers' adverts in eighteenth-century America tell similar stories. Where books of geography are for sale, they are works of special geography or geographical dictionaries and grammars written by Europeans—Gordon's *Geography Anatomiz'd,* Salmon, Guthrie, editions of Varenius with Newtonian additions. Public interest in geography in America was high before and after the American Revolution. There were, for example, over twenty-two public classes in Philadelphia and other cities and towns in Pennsylvania alone between 1728 and 1800. Benjamin Franklin deemed the position of geographer essential in advancing proposals in May 1743 for the establishment in Philadelphia of the American Philosophical Society. The national office of geographer general was established in 1777. Mapmakers were everywhere active. Geography was taught in schools and in universities. Political figures stressed the subject's importance—it was "the common concern of all polite Nations."[27]

Yet until the 1780s, Americans' interest in geography had to depend on books from anywhere but America. Worse still, much of what was written about America was not founded on truth. As Jedidiah Morse, Congregationalist minister in Charlestown put it bluntly, "Europeans have been the sole writers of American Geography, and have too often suffered fancy to supply the place of facts." For Jedidiah Morse, writing of the new nation of America so recently "risen into Empire," anything other than an American book of geography written by an American with Americans in mind was a reproach upon America. Morse's response was his *The American Geography, or A Present Situation of the United States of America* (1789). The work was hugely popular, much reprinted, and appeared in abridged and amended versions. With this book, Morse the minister became Jedidiah geographer to America—"America's leading geographer," even "the father of American geography." Although later views have sometimes misread its descriptive emphasis—Morse, conventionally, was a compiler not a fieldworker—*American Geography* is the first major book of geography in Enlightenment America to help shape America's national consciousness. Morse's *American Geography* was, in short, a book of Enlightenment geography with narrative authority and geographical "power." To understand why, we have to look more closely at its content, at its predecessors—notably, Morse's *Geography Made Easy* (1784) and Thomas Jefferson's *Notes on the State of Virginia* (1787)—and at the intellectual context in which these and other books worked "to write America."[28]

Textual renditions were only part of the geographical fashioning of Enlightenment America, as colonial territory and as the nascent nation of the early republic. What was the geographical extent of the new nation? How should administrators deal with the space in question? Geography on the American scale presented new challenges to politicians, to commercial and intellectual networks, and to the rules of law and God. Enduring myths of "howling wilderness" and penitential environments underpinned attitudes to nature and nature's reduction to habitable places. Political and religious diversity among the settlers, native inhabitants with unfamiliar cultures, foreign possessions to the north and south, to the west unknown lands, commercial opportunities but receding frontiers: each presented differences and distances, both moral and physical, which Americans needed to delimit.[29]

That is why the "print discourse of geography" in Enlightenment America matters. The "geographic culture of letters," even what has been termed "the geographic revolution in early America," was a vital means for Americans to become "geoliterate"—indeed, to become American: "it was the widespread diffusion of a geographic literature and geo-literacy that connected the Vermont lawyer and the Connecticut farmer, the New England statesman and the southern schoolgirl."[30] Two figures key to promoting this culture were Noah Webster, who emphasized the importance of knowing one's country in his 1788 *On the Education of Youth*—in which he also observed the "want of proper books" to do so—and Jedidiah Morse. In the various parts of his *The Grammatical Institute of the English Language* (1783–93), Webster placed increasing emphasis on thinking geographically about language and literacy in America: unifying the regional diversity of European dialects in use, Americanizing place-names. Readers were instructed to speak out loud the names of towns and states and the facts of their geography. Where in Britain "Tom Telescope" encouraged the performance of Newtonian principles and mathematical geography, Noah Webster's spellers encouraged

American readers to articulate geography as a means to literary reform and national identity. For Morse, advancing geography in America depended on inverting established textual conventions. Where Salmon privileged England and Guthrie Scotland, Morse described the United States and the Americas before Europe and the rest of the World. *Geography Made Easy* is the first book in Western print history to do so. By including a map of the new nation sited upon the "Meridian of Philadelphia," this geography book leaves no doubt that it is written from a particular geographical perspective, as a reaction against European convention and a symbolic starting point for America's view of itself.

Morse's *American Geography* built upon these ideas. It also extends and departs from Thomas Jefferson's *Notes on the State of Virginia.* Jefferson's work is a regional inventory of the state as a real place and as an "imagined" space in terms of its possible future geographies. It is also a counterblast to Buffon's views about the limiting effect of America's climate and soils and about the "inferiority" of America's natural species in terms of their smaller size relative to their European counterparts and the "degeneracy" in America of those animals that had been domesticated in both continents. Such views were to Jefferson "unfounded and degrading to one third of the globe." He sought also to reject the views of the Abbé Raynal, who depicted a stunted state for America's human cultures. Jefferson was no geographer in any formal sense, but where Webster helped establish geographical identity through national geoliteracy, Jefferson provided a geographical apologia for the moral integrity of American nature, physical and human.[31]

Morse's *American Geography* applauds Jefferson's anti-Buffonianism but does not so readily mirror Jefferson's views on the political landscape of the new America. Textually, Morse again rejects tradition. His description of America comes first and bulks large: in a book of 536 pages, Europe and the rest of the world are covered in only 45. After reviewing America's discovery, Morse follows convention within his "inverted" order: discussions of astronomical geography, descriptions of physical features, the situation and extent of places, and so on, before the detailed discussions ordered by state. Yet Morse does more than privilege America in terms of textual tradition. He offers a moral view of geographical difference *within* America, in ways that subscribe to Enlightenment ideas about climate and capacity. At the same moment, he privileges the moral topography of New England as the basis of American identity. Considering New Englanders to be characterized by "industry and frugality" and by what he called "that happy mediocrity . . . which, by inducing oeconomy and industry, removes from them temptations to luxury, and forms them to habits of sobriety and temperance," Morse saw them to be, constitutionally, the model for all America. Furthermore, Connecticut—Morse's own state—was, in matters of religion, "the best in the world, perhaps, for a republican government." In intellectual matters Morse was no less unequivocal: "in no part of the world is the education of all ranks of people more attended to than in Connecticut." The state was indeed the "Athens of America." In Virginia, good character was less common, with too many people there showing a "disregard for oeconomy." In North Carolina, there was "as little taste for the sciences as for religion." There, "political enquiries, and philosophical dispositions" were generally "too laborious for the indolent minds of the people at large," and temperance and industry were not to be reckoned among North Carolinians' virtues. In Georgia, climatic conditions—an "explanation" underlying Morse's view overall of America's geography—acted to induce a love of drink ("immense quantities of spirituous liquors"), a claim with clear echoes to Hume (see chap. 7). As for religion, politics, and literature there, "this state is yet in its infancy." Southern reviewers were predictably offended: Morse took their reactions as evidence that he had written an important book—a view echoed by commentators in Britain.[32]

Morse's *American Geography* is thus an American Enlightenment geography in contradistinction to European views. It is a regional and descriptive portrayal of the nation as it then was. It is also a particular view of what America might in the future be unless it held fast to established values and stood firm against those matters of "luxury" and "ostentation" that had undermined other, ancient, republics. In revising later editions and in other works, Morse continued to construct the new republic through books of geography. The book was bought and read widely—not just by town dwellers but by "hardscrabble families" on the rural and social margins.[33] In his *Elements of Geography* (1795), he again inverted established traditions—or, perhaps more properly, reaffirmed the invented American tradition of putting the Americas first—and notes that he did so to counter Guthrie's view of the world. Toward the end of *Elements,* readers were even made to swear aloud fealty to America and to its geography as part of their own enlightenment and that country's emerging national identity. "I am truly delighted, Sir, with the account you have given of my country, and I am sure I shall love it more than I ever did before. I hope I shall always be disposed to respect and obey my rulers."[34]

Not all books of geography were so explicit in their politics. As these examples show, the political dimension of books of geography in the Enlightenment was a constant, but it was not always utilized to the same ends. Salmon's geography was a work of the English Enlightenment. Guthrie's connected with the views of Scottish Enlightenment theorists concerning the nature of human social development. For Mentelle, different sorts of geography books could be mobilized to suit changing political and institutional circumstances. Morse intended a new moral geography for a new nation. Examining the textual traditions and strategies of geography's books allows insight into what "modern geography" in the Enlightenment was taken to be as well as revealing geography's connections with other subjects and practices such as history, astronomy, and mathematics and, to some extent, how the subject was used. If, generally, it is the case that books of geography in the Enlightenment were read for "voluntary self-enlightenment," it is also true that geography's books were read by different communities—children, adults, merchants, politicians—and to different ends.[35] American and French readers of Morse and Mentelle read their books from respective interests in nation building. In New England anyway, Morse's geographical works were certainly more commonly owned and read than the next commonest geography works—those of William Guthrie.[36] In Britain, others are more likely to have turned to Guthrie, to Salmon, even to Bickham to know the current extent of nations or the commercial potential of foreign markets.

Geography was not just a subject for geography books. What the term "geography" embraced in the Enlightenment and its place relative to broader concerns about the nature and structure of knowledge may be discerned in dictionaries and encyclopedias. The example of the *Encyclopédie* demonstrates—perhaps better than for any other Enlightenment book—that it is possible to trace the geographies of books: their sites of production and the provenance of the authors, books' movement over space, and their reception and readership. Books of geography have the same potential. Consider, for instance, that work with which I began this chapter—Pinkerton's *Modern Geography* (1802). Pinkerton intended his work to be an up-to-date scientific geography of the world as it had settled after the French Revolution. If we look at how he did this, however, particular geographies of scholarly attribution and of intellectual tradition emerge. For his European sections, Pinkerton looked to France and to Germany most readily in his sources, to d'Anville and to Büsching. But for his American and Asian sections, Pinkerton looked elsewhere—to Spanish, Dutch, and American authors. In short, Pinkerton was not only sensitive to varying European "spheres of interest" in the making of his geography, but his book exemplifies through the geography of its scholarly acknowledgments the Republic of Letters upon which it depended.[37] As Morse likewise recognized in appealing through his 1787 broadside "To the Friends of Science" as he undertook the further correction of others' geographies, producing new accounts of new nations demanded regulating the questions asked in "correspondence with several respectable characters in different states." Employing similar methodologies for other books in the Enlightenment may not, of course, be possible. But to do so for, say, Guthrie's *Grammar* or Morse's *American Geography* might further illuminate the ways in which books as a whole—and books of geography in particular—made the Enlightenment work in textual, social, and geographical space.

9

Geography in Practice

Looked at through its books, geography in the Enlightenment was a matter of collation aimed at summary descriptions of the whole world. Variations were apparent of course. Even in its literal sense of "earth writing," geography in its textual traditions embraced mathematical problems, incorporated Newtonian natural philosophy, and reflected the needs of different audiences. But as a matter of definition and of intellectual endeavor, geography was in a textual sense "fixed." A form of inquiry whose rationale was established by Renaissance humanism, it aimed at illuminating the Christian and the classical worlds as a basis of their historical examination. Geography's textual genres were also fixed: the geographical gazetteer arranged its information alphabetically, the geographical grammar by continent and nation. For Mayhew, "in this context, then, it is worth noting that the Enlightenment, however that vexed term is placed chronologically, does not pick out a distinct era in the practice of geography. However long your definition of the 'long' eighteenth-century, geography remained a practice defined by late-humanist theory and practices." Further, "it [geography] did not involve fieldwork or the geographer gaining accreditation by doubling as explorer or colonial administrator."[1] Understood thus, geography as an intellectual practice or "subject" was little more than synthesis, an insignificant element within that "intellectual revolution of the Enlightenment" that aimed at a Science of Man.[2] It was not "modern" at all as others have seen it, born in empirical encounter with the peoples and places of the Pacific Ocean and apparent in realistic description, systematic classification, and comparative explanation.

Such sensu stricto definitions and views of geography's fixedness are both true and false. They are true, but only if one takes a view based on textual traditions and self-definitions—what geography was said in its books to be. They are false if we want to understand geography's purposes and practices other than in its books—what geography did and how contemporaries thought and acted using geographical insights and language. Doing geography in the Enlightenment involved more than the writing and reading of books of geography written to one or other established convention. Indeed, even in geography's books, there was as we have seen considerable movement around geography's "fixed" nature. Geography's content matter reflected contemporary concerns with system in the emergent natural and the human sciences, incorporated the new facts of natural history and of ethnography, and included the results of discoveries then changing the shape of the world as contemporaries knew it. For these reasons, geography in the Enlightenment was continually being "modernized." Further, the facts of geography—understood in the form of climate, for example, or interpreted as differences over space—influenced ideas on human origin and diversity. And different authors—many of whom as Mayhew anyway notes were not geographers in any disciplinary, investigative, or administrative sense—took different views about the purpose and the practice of geography. For Büsching, geography was a means to the revelation of God's purpose. As we have seen, where Guthrie considered Europe's physical geography as one explanation of differences in nations' political progress, Morse saw geography as a means to educate Americans and create a new republic on earth, Turgot to chart "mappemondes politiques" of the world, past and present, and Mentelle to serve different political regimes.

All this suggests that Enlightenment thinkers and writers turned to geography in a variety of ways to understand and to represent their world. The seeming contradiction between textual particularity and matters of practice is, in fact, more one between the literal definition and the textual nature of geography on the one hand and, on the other, geographical knowledge, the ways in which people undertook to know their world as a geographical object. Such distinctions are never absolute. Suggestions of fixedness in the first are denied by variations within geography's books, by typological distinctions within that term—as apparent in the *Encyclopédie,* for example—and by the fact that geography was not always practiced by geographers. And, as the following chapter suggests, they are denied yet further if we look in more detail at where geography was being used, at the different local social spaces and sites in which geography in the Enlightenment was at work. In respect of geographical knowledge too, natural philosophers and others acted through geography—that is, they thought geographically—in understanding the distribution of things over space and time as part of their overall reasoning: Charpentier and Guettard in mineralogical mapping; Linnaeus, Soulavie, and others in botanical classification; and many others on the Science of Man as a Geography of Man.

Thinking geographically *about* the Enlightenment, about geography in the Enlightenment as more than a strictly intellectual and textual definition, must then incorporate geography in practice *in* the Enlightenment as well as considering the Enlightenment's geographical constitution and variations. Let me take three themes, each touched on in earlier chapters, to illustrate what I mean. What connects them is mapping: as a practice of measuring space, as a metaphor of classification, and as a way in which we now can understand the Enlightenment geographically. The first theme explores not the strict definitions of what contemporaries knew as geography and as astronomy but rather the overlap in practice between them, the notion of "mathematical cosmography." That mapping and mathematical cosmography were forms of geographical measurement is clear if we look at Enlightenment concerns with population as a category of statistical enumeration and of political economy. Such attention to statistical accounting, to "geographical accountancy" even, had one expression in "fiscal-military" geography. Simply, measurement through such geographical accountancy had at its heart the desire and capacity of states to know themselves. The second theme looks at the map world in Paris, at county mapping in England, and at the networks binding London- and Paris-based mapmakers. The final theme returns to the connections among environment, morals, and meteorology to discuss medical geography in the late Enlightenment.

Mapping and Measuring: Mathematical Cosmography, Military Geography, and the Capacity of the State

Textual definitions do not always work in practice. Consider the examples of astronomy—properly, cosmography—and geography. Where the first studied celestial bodies, including the earth's position relative to the stars and to the other planets, the second concerned itself with the study of the whole earth. In practice, the two subjects were closely related and commonly taught together. Many books of geography opened with a cosmographical discussion, often to help in the teaching of mathematical geography. For many Enlightenment figures, the two subjects were practically indistinguishable. In his "Idée général de la géographie," presented to the Académie des Sciences in Paris in 1752, for example, the physical geographer and encyclopedist Philippe Buache laid out three "charts" of geography: historical, physical, and mathematical or astronomical. Buache's third chart, "Géographie mathématique ou astronomique," was repeated identically by the German astronomer Johann Gabriel Doppelmaier in his 1752 *Atlas Coelestis* and is echoed in others' work. Such shared views of subjects reflected their conceptual and practical overlap—what has been termed "mathematical cosmography." Astronomy embraced terrestrial astronomy—including the earth's shape, its climates, and meridians—and celestial astronomy (the motions of the stars and planets, planets and their satellites, and so on). Geography was understood as theoretical and practical: scales of measurement and trigonometry as well as narratives of discovery in the first sense, and the making of celestial and terrestrial globes, sea charts, and land maps in the second. Whether in the teaching of mathematical principles, in navigation at sea, in astronomical or geodetic expeditions, or in the manufacture of instruments, neither subject, however strictly defined, made practical sense without the other.

Geography's practical significance in such ways was of course what shaped debates in Paris and in London about the shape of the earth, and their testing through fieldwork in Peru and in Lapland. And not just there. In Nuremberg, for example, a Cosmographical Society was active between 1746 and 1754. Founded by Johann Michael Franz, it included mapmaker-astronomers such as Georg Moritz Lowitz (who ran the Homann map- and globe-making firm in the town—of which Franz was a director—before succeeding Doppelmaier as professor of mathematics and director of the Eimmert Observatory there) and Tobias Mayer (Lowitz's brother-in-law). The society had two classes, a mathematical one, which dealt with geodesy, map projections, astronomical observations and calculations, and the design and construction of geographical instruments, and a historical class responsible for determining boundaries, coordinating the geographical data to be mapped, collecting maps, and writing school books. Other plans—never fully realized—aimed at establishing a geographical bureau in all states to collect geographical information, a Cassini-type survey of the whole of Germany, and a Corresponding Class through which interested persons elsewhere would report on the geography, economy, and natural history of their home area. Networks of correspondence, organized classes, the workings of an observatory, publishing ventures from the Homann firm, Franz's promotion of "the practices of geographical science," his idea of a *Staatsgeographus* or "state geographer," Mayer's unpublished "Collectanea Geographica et Mathematica": in Nuremburg, these things came together in "a programme of mathematical, geographical, astronomical and physical research feeding on the interplay of the one subject on the other, involving the simultaneous search for new principles and for new empirical data with which to discover and test them, and encouraging the construction of better instrumentation with which to put theory into practice."[3]

This example illustrates one local setting for concerns with terrestrial and celestial measurement. It is thus valuable in highlighting geography's practical and social constitution. Geography as it was put to work in Nuremburg—and in the Enlightenment more generally—did not reduce to a single essential definition. Outside its books, it was not precisely definable, everywhere the same. Geography was differently made in place, involved the distribution and measurement of places, and in combination with other subjects, helped give mathematical, astronomical, and textual order to those places. One common outcome—the map—was the result of geography working in combination with other subjects, not of "fixed" textual traditions linking geography only to historical explanation.

In England and in Bengal, James Rennell would certainly have understood this. His public life was spent in encountering and producing the geography of India through detailed maps and memoirs of that country. Educated Britons and others "back home" took his Indian work to be both the locus of Britain's imperial ambitions and an expression of geography's capacity to delimit the world. In his active retirement, Rennell turned to texts on ancient geography. His *The Geographical System of Herodotus examined and explained by a comparison with those of other authors and with Modern Geography* (1800), the only published part of a larger scheme that aimed to outline the historical geography of western Asia, helped establish new methods within geography by combining scrutiny of ancient and modern textual accounts with firsthand experience of the area under study. Rennell thus worked in an "old world" context and with the worlds of "new" geography in more ways than one. Yet Rennell should not be considered the father of modern English geography, as one nineteenth-century commentator has it, anymore than he should be seen as the father of Indian geography. Nor is it proper to separate Rennell's humanism from his militarism, Rennell the geography book writer and methodological "pioneer" from Rennell the mapmaker and practical geographer. In each of these several guises, geography in the Enlightenment was what Rennell did.[4]

In North America too, geography was practically at work among those mapmaker-geographers plotting the westward frontier of the United States and using maps to sell space as "safe" or "empty" to intending colonists. What others have seen as the "deepening" of knowledge about America in the Enlightenment was apparent in maps and related accounts before Morse's textual endeavors: in the work of Lewis Evans, for instance, or of "John Green"—pseudonym for the Irishman Bradock Mead, who, noted Thomas Jefferys, "I employed in many of my Geographical performances & in particular yr Chart of North and South America"—and in Jefferys' own 1776 *American Atlas*.[5] And we can most clearly see how geography was put to work—indeed, can see geography in the Enlightenment being "defined" through practice—in France.

For France's military geographers, the *ingénieur-géographes,* their particular form of geography was enshrined both by what they did in the field and how fieldwork was managed, and in the various *mémoires* of the Dépôt de la Guerre from its establishment in 1743. True, the Dépôt Général de la Guerre et de la Géographie, to give it its full title, was most active as a post-Revolutionary body between 1791 and 1830, but even before then it directed the strategic needs of military geography in times of war, and in peacetime it assigned the *ingénieur-géographes* to other state mapping projects. France's military geographers were taught mathematics and foreign languages and instructed to combine field and archival research with assessment of the answers to questions directed to appropriately chosen locals. Because its maps were combined with written descriptions, military geography provided an effective form of geographical synthesis. Especially useful in war, military geography was a handmaiden to state politics, an effective form of colonial power, and a distinctive form of geographical endeavor of use "to commerce, industry, and the advancement of general statistical knowledge, science and the larger civil society."[6]

To the proponent of this view, Étienne-Nicolas de Calon, onetime head of the Dépôt de la Guerre, geography was a form of Enlightenment "civic engineering." It was so because geography's own language helped provide a language for the Enlightenment's geographies, a means to territorial representation, a precision—or a seeming precision—based on a synthesis of mathematical accuracy and descriptive enumeration. Of course, the language used by and within geography differed in relation to context: evidence from travelers' tales or untutored locals' maps could not straightforwardly be incorporated without appropriate corroboration. Yet geography's language of scale and grids, of delineation and latitude, of empirical regularity and standardization (if more in theory than in practice), as well as its rhetoric of discovery and order helped determine a clear place for geography and a wider role for geographical thinking.[7]

At the same time, matters of geography were central to Enlightenment interests in nations' "vital accounts," registers of their population, resources, and economic and political capacity. What was known as "political arithmetic" was in 1698 defined by Charles Davenant as "the art of reasoning by figures, upon things relating to government." Political arithmetic has its origins in the work of men like John Graunt, whose *Natural and Political Observations Made upon the Bills of Mortality* (1662) indicates the reliance of this form of inquiry on numerical registration. As a form of knowledge, "political arithmetic comprised the virtues of quantification and the ideals of the Enlightenment."[8]

Why? In the Enlightenment, numbers of people fascinated—in registers of births, marriages, and deaths, by age, by type, by location. They did so because of the novelty of new "types" of humans and new ways of classifying humans in society—the use of "class" and "rank" in relation to humans was, like "species," part of this rapidly expanding analytic vocabulary—and because of the numerical basis to effective administration. As one contemporary put it, registers of numbers "must evidently be of the most important consequence to the politician, the philosopher, and the physician, in their several endeavours to relieve the miseries, and promote the happiness of human nature." For another, "there can be no well-ordered political machine, nor enlightenment administration in a country where the state of population is unknown."[9]

What politicians recognized as the utility value inherent in numerical registration, philosophers read rather differently. Physicians read it differently too, a fact apparent in their use of the ideas of "medical arithmetic" and "moral topography" in Enlightenment medical geography. For many philosophers and social theorists in the early Enlightenment, the significance of population centered on the virtues of a large population and on differences in population levels, and thus on the comparative economy and "happiness" of nations between modern and classical Europe. Beliefs that the Enlightenment world was less populous than the ancient world were a central feature of debates within conjectural history and stadial theory about the comparative state of "ancient" and "modern" societies. By the later Enlightenment, strands of "demographic pessimism" were apparent in the view of some writers that the populousness of past nations undercut claims to current modernity and thus weakened theories of progress. More optimistic views were held of course—notably by Hume, Turgot, Condorcet, and Godwin—but the writings of Robert Wallace and Thomas Malthus effectively provided a demographic objection to theories of social progress in emphasizing the high fertility potential of human populations and their tendency to outstrip the resources available.[10]

Why this matters here is because in debating the nature and consequences of political arithmetic, most writers took population to be the number of inhabitants in a given area. Geography, understood either as a particular place or location or an area—often, the parish—became the basis for the statistical enumeration and comparison of places and nations. More than that, those forms of mathematical accounting that underlay the emergence of the fiscal-military state in the Enlightenment depended on its constituent scales and spaces—the place, the parish, the nation, the overseas colony, the empire—fitting together. This idea, effectively the recording of a state's "geographical capacity," informed practice in several countries in the early Enlightenment. In France, plans for a national survey were based on one undertaken for the province of Vézelay. In England, Davenant used political arithmetic to render national space politically visible in the form of "excise geographies."[11] Because works of geography included trade statistics, indeed precisely because they enumerated what other countries contained as tradable commodities, it is possible to speak of geography then having a fiscal-military function in practical and in textual terms in addition to its realization in military mapping.[12]

Perhaps the best expression of such connections between statistical philosophy and geography as national political accountancy is to be found in late Enlightenment Scotland. There, in May 1790, Sir John Sinclair, MP for Caithness and a lay member of the Church of Scotland, laid before that body a proposal for a detailed survey of Scotland using parish ministers working to a common scheme. From parochial matters of fact, collected much as natural historians sought to understand "nature's economy" and physicians the working of the human body from its constituent parts, a picture of the nation as a whole would result. From that, better government would follow: "It is by pursuing the same method, in regard to political disquisitions, by analysing the real state of mankind, and examining, with anatomical accuracy and minuteness, *the internal structure of society,* that the science of government can alone be brought to the same height of perfections." A set of 160 questions was asked in four sections: questions 1–40 asked for information on the geography and topography of the parish, climate, natural resources, and natural history; 41–100 focused on population and related matters; 101–16 on "agricultural and industrial productions of the parish"; and the final section reserved for miscellaneous questions—on language, wages, house types, and so on. Despite some difficulties in compliance—Sinclair had to resort to sending "Statistical Missionaries to different parts of the country"—the project was complete by June 1799.

What Sinclair saw as "the completest survey of a kingdom, of which we have any knowledge" we should see, as did he, as a work of geographical inquiry exactly in sympathy with Enlightenment sensibilities about human happiness, good governance, and future progress. Sinclair represented his national survey both as "the Pyramid of Statistical Enquiry"—one national statistical account founded on 938 parochial accounts, with thirty-three provincial districts in between—and as "an inquiry into the state of a country, for the purpose of ascertaining the quantum of happiness enjoyed by its inhabitants, and the means of its future improvement." His method was far from unique, of course, and Sinclair acknowledged others' earlier efforts at national statistical survey, including those of the German Eberhardt

Zimmerman, whose zoogeography we have noted and whose *A Political Survey of the Present State of Europe* (1787) was a particular prompt. At the same time, Sinclair's work, while molded in the specific context of the Enlightenment in Scotland, was an outline for the wider and comparative utility of geographical accountancy as a basis of Enlightenment perfectibility: "if the same system of minute and careful investigation were adopted in other countries . . . the result of such extensive enquiries, contrasted together, would bring the science of government, and the happiness of the human race, in every civilized country, to greater perfection, than any other method that has hitherto been devised." [13]

Geographies of the Enlightenment Map World

We have seen how, in outlining his *Encyclopédie,* d'Alembert considered the map a metaphor for the organization and layout of all Enlightenment knowledge. Others had a more literal view of the map as something geographers did, the practical consequence of acquiring and displaying information about the nature of places. Then as now, however, there were differences in how a map was put together and in the social worlds of those involved in mapmaking.

Take Enlightenment Paris for instance. The French capital was distinguished by a strikingly local geography to the making of geography. Those *géographes*—a term embracing geographer and mapmaker in modern parlance—who were members of the Académie des Sciences and other academic institutions in the city were not artisans or part of any *communauté.* But they were closely dependent on networks of printers, engravers, and publishers (*libraire-imprimeurs*)—and patrons and clients—without whom the facts of geography would have remained unpublished and unread. Because of the advantages afforded by proximity to one another and from laws that anyway restricted their activities and locations, geographers and booksellers concentrated on the quai d'Horloge, the quai des Augustins, and along the rue Saint Jacques south of the Ile de la Cité. Here the local mapping fraternity likely would have figured within that "map of café talk" taking place in the many coffeehouses of the area (see chap. 4). Jean-Baptiste Bourgignon d'Anville, Paris and France's leading geographer and Johnson's "prince among geographers," was unusual in living and working north of the river on the Galerie du Louvre. Engravers were more dispersed within the city—they worked on numerous maps at any one time, often in association with different *géographes,* and so had less need to live close by any one client or near to one another. By the 1790s, however, this concentrated geography of geographers and their associates within Paris was much changed: court-appointed residences in the Galerie du Louvre disappeared (as did the post of *géographe du roi*); the quai d'Horloge, damaged by fire in 1776, ceased to be much used; and as restrictions were lifted on the locations and activities of the *libraire-imprimeurs,* so they and others moved out across the city.[14]

This geography of the Paris map world was a dynamic social one, a geography of local social spaces sustained by different patron–client relationships and the unequal movement of its material and its human constituents: engravers moved most often; engraved plates, part-finished maps, and atlases only when necessary; publishers and geographers hardly at all. For all that his geographical reputation was international, his sources likewise, D'Anville never left Paris—the epitome of the *géographes de cabinet.*[15] In other social and institutional spaces where geographers produced maps—such as the Dépôt de la Guerre or the Dépôt de la Marine—the social geography of the Enlightenment map world was different still. In the drawing room of the Tower of London, for example, military mapping developed significantly only after 1752, once that specified mapping space and its personnel had been recognized as a formal department of the Military Ordnance. Drawing room draughtsmen occupied a designated social and physical space, with rank determining workplace, and upward mobility in the first necessitating a new location for the second.[16]

Doing geography in the Enlightenment depended then on certain sites and social settings more than others. The mapping of Enlightenment England was prompted not just by advances in military mapping but by civic recognition of deficiencies in the country's geography. "I would submit to you as a friend," wrote the antiquarian William Borlase to a friend in 1755, "whether the state of British Geography be not very low, and at present wholly destitute of any public encouragement. Our Maps of England and its counties are extremely defective." Such entreaties, together with the perception that the French had stolen a march through the Cassinis, prompted the Royal Society of Arts to establish premiums for new and improved county maps of England. Numerous improved county maps, each emphasizing the utility of enhanced geographical knowledge to the betterment of transport and communication and so on, were produced in the half century from 1759.

Many were the work not of established professional mapmakers, however, but of amateurs in the field who offered cheaper rates. Thomas Jefferys, the leading map publisher, engraver, geographer to the king, and who produced numerous maps of Enlightenment America, did not receive a Society premium (he was told that he had let his Society membership lapse and had not followed proper application procedures in undertaking what maps he did produce—and so was never paid). His successor, William Faden, was by contrast highly successful. Where Jefferys sought to commission the surveying himself, Faden acted as a center of cartographic computation: "the resources of his engraving shop and publishing house were contracted out to the surveyor in a small way of business; as regular bidder at the auction rooms he concentrated the copper-plates of many county surveys in his own hands to publish new, but often ill-revised editions; but in other cases, if a county lacked a modern survey, he would, like Jefferys, commission a map."[17]

For Jefferys and Faden alike, as for their Parisian counterparts, being a geographer-mapmaker in the Enlightenment depended not on any fixed definition as to what geography was but on social connections of varying strength and duration sustained over and above other local affiliations and beyond national borders. Jefferys relied on Paris geographers for part of his stock, importing the work of d'Anville and Philippe Buache among others, and he also acquired items from the Homann firm in Nuremberg. To judge by his 1774 sales catalogue, William Faden bought over 175 items from d'Anville. Such trade was based in part on personal contact. In 1768, Jefferys visited Paris in 1768 on a map- and print-selling trip. Their venture ended shamefully: Jefferys and his companion, Robert Sayer, an otherwise reputable London map and print dealer, were caught selling pornographic prints.

Between 1777 and 1786, the Paris–London trade in maps was reversed, with William Faden dispatching several hundreds of maps to Jean Nicolas Buache de La Neuville in Paris, nephew of Philippe, and a *géographe du roi* and assistant keeper of the department of maps and plans in the Dépôt de la Marine. Even before his 1783 appointment as geographer to the king, Faden had been supplying members of the British government with maps, many of them marine surveys, some of which he obtained directly via Buache de La Neuville and the French Navy's mapmaking bureau. The Faden–Buache de La Neuville exchanges continued even when the two nations were at war—evidence of an eagerness for new data and information on the latest explorations and of geographer-mapmakers' place in the Enlightenment Republic's Map of Letters.

Here, albeit sketchy in outline, is one map of how and where geography was at work. The "map geographies" or "thread maps" connecting these two Enlightenment capitals and their geographers show an import rate into London greater than the export rate, a transnational network of cooperation between map sellers, exchanges between government offices facilitated by private brokers who were themselves geographers and mapmakers, and systems of production and of labor organization that demanded capital sums, often quite large ones, in advance of the product's sale, in order to cover design and engraving costs.[18]

Given such evidence, understanding how geography was put to work in the Enlightenment may demand that, in this context anyway, we need to look at the relationships among epistemic space, social space, and political intent. It is noteworthy, for instance, that English county atlases before 1776 used ornamentation as icons of political authority over the American colonies, and that the 1776 *American Military Pocket Atlas*—known as the "holster Atlas"—was printed to a format usable in the field and to fit in officers' coats: "calculated in its bulk and price [its preface noted] to suit the Pockets of Officers of all Ranks."[19] In France and for France, men like d'Anville, Buache de La Neuville, Jefferys, and Faden were geographers (though they did other things besides), and in their map work they made and represented new geographies. Their map world—sustained by correspondence and in networks of exchange, sites of specialized production, affiliations of expertise, reliance on the "latest authorities" (and a readiness to publish even without it)—was one of many such "maps" of intellectual and material exchange shaping the Enlightenment.[20]

Diseases, Quadrupeds, and Moral Topography: The Environment for Medical Geography

Medical science in the Enlightenment was distinguished by several key themes: diverse theories as to the nature and cause of illness and the philosophical and practical consequences of treating the ill body and mind; differing emphases in teaching and in treatment depending on person and place; changed understandings of diseases in terms of their classification and etiology; advances in pathology and in therapeutics; new surgical techniques; developments in the social status and the institutional management of the ill; and shifts in the meaning of terms like "madness," "diagnosis," "physician" even. Even when medical writers shared an understanding of the nature of medical advances and their potential contribution to mankind's well-being, different views were held concerning how it was to be done.[21]

It is such diversity, of course, that makes reference to *the* medical Enlightenment problematic—and, as we have seen, attention to its different geographical and social expression important. At the same time, because Enlightenment thinkers understood there to be significant connections between environment and "regimen," between morals and meteorology, it is erroneous to separate medical science from the earth or environmental sciences. Their concerns were in general terms apparent in an emergent "environmentalism" and in debates over "hygiene" and the "public good." They were evident, too, in particular questions: How did external conditions affect living things—in terms of disease and in terms of climate's influence on culture? How did soil or air affect well-being? How might the new sciences of life and the environment best be managed? These and other queries found expression in a range of Enlightenment medical writing—much of it by men who were not medically qualified—in which certain themes such as empiricism, utility, and the political management of health were central to the synthesis of the earth and the medical sciences, notably in the late Enlightenment.[22]

If there is a parallel with geography's books and geographical authors in the Enlightenment, there is also an important distinction. For where geography's books—many written by nongeographers—worked with textual conventions in which matters of geography were commonly offered as description, medical writers on the environmental basis of disease and the management of personal and public hygiene turned to the facts of geography for an explanation. They paid attention, that is, not just to the physical geography of places—to the physiological effect of the environment in relation to certain diseases—but to the moral and social correlates of environmental medicine. In discussing hygiene, for example, the French medical writer Jean-Noël Hallé was aware of the connections between what he termed "mankind's medical geography"—in which variations in climate and environment were central—and its medical history, in which matters of individual and social habit, temperament, and the political management of health over time were key issues. So too for the French Army physician François-Marie-Claude Richard de Hautesierck, whose plans from 1763 for recording details about the army's health and disease depended on networks of correspondents informing him about "Topographical and Medical Memoirs" and "Meteorological and Clinical Observations." Under the guidance of Fèlix Vicq d'Azyr in particular, the Société Royale de Médecine in Paris was, as we have observed, the center of a national correspondence network in which meteorological observations were allied to medical ones.[23]

Such things were, as also noted, the expression in part of longer intellectual traditions that consciously drew on Hippocrates' *Airs, Waters and Places,* for example. But they are also to be explained as part of Enlightenment practical interests in place and population numbers, in knowing how health might be regulated, and in providing for human betterment. This was an Enlightenment "biopolitics" as others have termed it, an Enlightenment "nosopolitics" or "noso-geography" even, in which the recording and classification of disease type—nosology—in relation to environment, was part of nation-states' geographical accountancy. This was just the view that Johann Peter Frank outlined from 1779 in his *System einer vollstandigen medizinischen polizey,* for instance, in urging in Germany—as Turgot, Vicq d'Azyr, and others were doing for France—that physicians should explore the condition and physical constitution of the inhabitants of even the smallest village.[24]

As commentators tried to account for the weather and for disease, their attempts to discern the explanatory significance of place and of geographical difference took different forms. Nosological systems reflected the classificatory programs of the botanists, whether of Carl Linnaeus, Bernard de Jussieu, or others. As vegetative zonation and classification offered one basis for a geographical nosology, some wondered if individual plants could be geomedical instruments. In his *Flora Scotica,* a regional geography of Scotland's plants on Linnaean principles, John Lightfoot observed that one species of trefoil acted as a natural hygrometer, its leaves "relaxed and flaccid in dry weather, but erect in moist or rainy." Others such as Joseph Priestley and Alessandro Volta used more conventional instruments in measuring not the plants but the air itself for moistness, seeing atmospheric experimentation as a means to testing the "virtue" of airs. In such ways, the atmosphere itself became a "scientific site" in which the principles of health and disease were produced and advances in pneumatic chemistry proposed. In its use in medical arithmetic and in meteorology, the barometer was both instrument and ornament—its use, like that of the eudiometer, at once conversation piece and calibrating device. For Nicolas Desmarest, even smell mattered. He considered the stunted stature of the inhabitants in one Swiss village to result from the powerful odor of their livestock. Yet others turned to the study of waters, their mineral content and medicinal virtues.[25]

For the travel writer and anthropologist Volney, Constantin-François de Chasseboeuf, matters of health and habitability resulted, in the United States anyway, from "the kind and quality of [the] currents of air" and from the soil, specifically soil temperature—the "natural and permanent warmth of the earth" itself. A geographically comparative eudiometry would be valuable—and would demand a different instrument: "it would be an interesting investigation, to pursue the inquiry in very opposite climates, and to compare the cold dry air of Siberia with a hot and damp air, as that of the West Indies, or with a hot and dry air, as that of Egypt and Arabia; and likewise to compare the strata of air near the earth with those of the middle and upper regions; for which purpose balloons may be of use." For

Volney, from the combination of facts on winds, soil, and temperature two corollaries arose that "throw great light on physical geography. One, that the temperature or climate of a country is determined by its habitual currents of air, or its *winds*. The other, that the topography of the land has commonly a decisive influence on the direction of these currents, and thus becomes an effective cause of the climate, and a constituent part of it" (see fig. 16).[26]

It is in this rich intellectual substrate that we find medical geography emerging as a distinctive form of geography in practice. One feature of Enlightenment medical geography was the concern to map disease. There was by the late Enlightenment a definable "medical cartography" in which disease was mapped and in which differences in the quality of places were understood to be vital to explanation of the intensity and the spread of disease. One expression of this type of geographical inquiry is to be found in the work of the German physician Leonhard Ludwig Finke. In his 1789 book on medical therapeutics, Finke advertised a forthcoming project: "I am thinking . . . about publishing a work with the following title: Geographical-Historical Essay on the General Indigenous Medication of Different Peoples of the World, including a nosological Map of the World." In that form at least, his plans were not realized. But in 1792, Finke published his two-volume *Versuch einer allgemeinen medicinisch-praktischen Geographie* (Attempt at a general medical-practical geography; the third volume, an addendum, appeared in 1795) (fig. 27). In the preface to volume 1, Finke explained that although the "endemic map" he was working on was not there published, "I do have it ready but have not sent it to be printed and I think it will not be printed soon." (It never was.)

Finke's medical-practical geography and his unpublished "Nosological Map of the World" have been seen as foundational moments in the emergence of "modern" medical geography.[27] I want to read them differently and not just because of the complex beginnings to that subject area.[28] Rather than see Finke's work as the beginning of something new, it is I suggest better considered as one product of that longer-run geographically informed synthesis of the earth and medical sciences noted here. Further, Finke's concern to provide a nosological map can only be made sense of in relation to his and others' efforts at thinking geographically—that is, by his using geography's classificatory methods—about the significance of place. In examining Finke's 1792 book for his attention to the languages of geography and in tracing the intellectual stimulus he drew from his fellow Germans such as Johann Peter Frank and Eberhardt Zimmermann, we can see how

Leonhard Ludwig Finke,
der Arzeney-Gelahrtheit Doctor und Profeſſor zu Lingen,

Verſuch
einer allgemeinen mediciniſch-praktiſchen

Geographie,

worin
der hiſtoriſche Theil der einheimiſchen
Völker-und Staaten-Arzeneykunde
vorgetragen wird.

Erſter Band,
welcher die Länder enthält, die ſich vom 45ten Grade, ſo wohl nördlicher als ſüdlicher Breite, bis zur Linie erſtrecken.

Leipzig,
in der Weidmannſchen Buchhandlung, 1792.

FIGURE 27
An Enlightenment medical geography, from Leopold Finke, *Medinisch-Praktischen Geographie* (Leipzig, 1792), title page to vol. 1. By permission of Edinburgh University Library, Special Collections.

Enlightenment thinkers, here medical practitioners of a certain sort, used geography's languages, methods, and textual definitions to make geography work in practice.

Finke was certainly not the first to "think with maps" in terms either of "noso-geography" or in regard to the political administration of medical space in the Enlightenment. In 1700, the Bolognese geographer and naturalist Luigi Ferdinando Marsigli prepared a plague map of the eastern Adriatic coast, outlining in his "Mappa Geographica" the precautionary measures to be taken in the wake of recent Turkish incursions: mountains appear as the basis for points of isolation, cordons sanitaires are depicted, the scale is given in hours not linear distance—a reflection, perhaps, of different treatment times. Marsigli may have borrowed from a similar map by Filippo de Arrieta, an administrator for the kingdom of Naples, who produced a map of plague controls including marine quarantine for the province of Bari in 1694.[29] Where Marsigli and de Arrieta drew lines to delimit medical space, Valentine Seaman, a surgeon in New York's hospital in the 1790s, used dot maps to make sense of the sites and social geography of the yellow fever then raging in the city, consulting as he did so the epidemiological tracts of Noah Webster.[30] And although he never drew maps to accompany his works, James Lind, the Scottish surgeon whose 1753 *Treatise on the Scurvy* had a major impact on naval health, was certainly thinking geographically throughout his *Essay on Diseases Incidental to Europeans in Hot Climates* (1768). The first part of Lind's book offers a global medical geography—the world is divided into four major regions, which are in turn subdivided—but his appeal to geography was deeper than to matters of scale alone. In arguing for the production and spread of diseases, Lind was "fundamentally struck by the significance of location and the character of place in trying to understand the nature of disease," his conviction being that "if the geography of place A is different than the geography of place B then the disease profile will be different."[31]

Finke—who knew Lind's work—reasoned similarly. Graduating from Halle in 1772, Finke practiced as a physician, first in Lengerich, later in Kassel, before taking up the post of rural medical officer for the district of Tecklenburg. By visiting the towns and villages of their district, medical officers prepared medical topographies, syntheses of meteorological, hydrological, botanical, and demographic data geared toward more efficient political administration: "In this way," as Frank earlier put it, "they would prepare for each district a kind of special geography." Finke's work in this respect first appears in his book of 1780 on bilious fever in Tecklenburg, a work that is, effectively, a "noso-geography," a medical and political arithmetic of the outbreak and its consequences.[32]

In bringing personal experience to his 1792 work, Finke also brought an understanding of geographical method to what, in terms of medical geography, he knew still needed to be clarified as a precise form of knowledge. "A medical geography:—What is it?—What should it contain?—How should it be set up?—In what sequence should countries follow one another?—What are the sources to be drawn upon?—Where do you find it discussed?—What are the benefits to be expected?—For whom is it intended?—How can it be made more perfect?" Answers to these questions, argued Finke, depended on understanding the relationship between chorography, topography, and geography—that is, in recognizing the relationship of scale between the chorographical (the local) and the geographical (the global) as they were expressed topographically (in regional terms). "Now since nobody has any objections to a medical description of all of the features of a locality of a country as being called a *medical topography* or *chorography,* there cannot be any objections to the simple title 'a *medical geography*' either, assuming that it covers all of the inhabited countries of the world." His emphasis was geographical rather than medical in that comparative questions of location and of scale were crucial to an understanding of medical and moral circumstance. His end-in-view was political in using "general medical-practical geography" as a tool for political enlightenment: "Thus medical geography is a light with the aid of which some idea of the strengths and weaknesses of nations can be obtained and from which examples to be followed or avoided can be drawn."[33]

There is a further sense in which Finke's geographical rationale and method engages us—namely, how he connected with other geographical work. In thinking about his geographical description of diseases at a global scale, Finke divided countries according to their specific latitude, grouping them by divisions of 10 degrees of latitude. He turned for his projected nosological map to the regional topographies employed by the statistician-zoogeographer Eberhardt Zimmermann: "The map will be in agreement with the Zimmermann zoological one. However, instead of listing the names of the animals, as in the case of the latter, I will replace them with the names of indigenous diseases." Zimmerman's "Tabula Mundi Geographico Zoologica" (World geographical-zoological map) was in turn based on a world map drawn by D'Anville, "improved"—as Zimmermann noted—"with the mountain formations by Buache" (fig. 28). Whether or not Zimmermann and Finke drew directly on Buache's work—who also influenced Hallé's discussion of Africa's medical geography—is unclear,

FIGURE 28

An Enlightenment animal geography, from Eberhard Zimmermann, *Specimen Zoologicae Geographicae* (Berlin, 1777). In this inset of the world map of quadrupeds, the false topography of African mountain chains has the effect of "explaining" the absence of species in the Saharan region. By permission of Edinburgh University Library, Special Collections.

for Buache's work was also taken up in the geography lectures of Johann Christoph Gatterer, professor of history at Göttingen and a perhaps more likely influence on his fellow Germans. Whatever the actual route of this exchange of geographical information, Buache shaped its intellectual genealogy by introducing the concept of natural divisions of the earth using major river basins divided by continuous mountain chains. That Buache was in several respects wrong in his delimitation of these mountain chains and drainage basins is, arguably, less important than what his efforts represent as an attempt at geographical classification. For what Zimmermann saw in Buache's topography and Finke in Zimmermann's zoogeography was a means to order the world through geography.[34] What we see is a shared understanding that differences in geography had practical medical correlates and political consequences.

Considering geography in practice in the Enlightenment—instances of geography at work—has here revealed geographer-mapmakers in Paris, London, and in Bengal, medical officers of health and state administrators in Germany and in Italy, and unfinished proposals for global medical geographies whose form drew directly from others' geographical work and for which the intended end-in-view was the improvement of the human condition through the application of geographical principles. Knowing that geography in the Enlightenment was put to work in such ways would suggest that we see it not as something "fixed" or closed, bound only by its textual definitions and untouched by other concerns such as astronomy or mathematics, but as something actively made by different social interactions in place. If this is so, two things at least follow. The first is that identifying the ways in which geography was at work *in* the Enlightenment is to highlight further the multiple geographies *of* the Enlightenment. The second is that if we want to know better what geography in the Enlightenment was, and how it was practiced, we need to consider more closely the "where" of its making and use.

10

Spaces and Forms of Geographical Sociability

Caught by the Parisian authorities in 1768 for selling pornography with their geography, the mapmakers and sellers Robert Sayer and Thomas Jefferys were not alone in seeing connections between the two. In the erotic book *A Voyage to Lethe* (1741), for example, is the illustration "A Map or Chart of the Road of Love, and Harbour of Marriage." The map, which is based on the Mediterranean world, depicts a cautionary cartography of the dangers of courtship and marriage, a geographical allegory of the journey from the Sea of Common Life to Felicity Harbour by the eponymous sea captain Samuel Cock (fig. 29). Sayer would certainly have known the map and would have tried to sell it—he was its author.

Whether this map was the reason behind Sayer's first arrest for peddling indecent material, in 1749, is unclear, but it was certainly part of a longer-run history of "geo-pornography" in which allegorical maps and textual descriptions used erotic images and geography's language. Such cartographical allegory worked straightforwardly enough: a map's words and pictures were replaced with the commonplaces of practical reasoning in life—such as marriage—and often with a geographical account of the female body. In the salons of Enlightenment France in particular, the mapping of such relationships was a form of what we might term *géographie galante* whereby geography's formal language was transferred to the world of the everyday and used in courtly society to symbolize social relationships and to secure social prestige. In one 1684 *Carte du chemin d'amour*, for instance, social circumstance and social standing are mapped as way markers: only in achieving (and going beyond) Fidelité, Constance, and Discretion and so on could Amour be really found. Other "Maps of the Island of Marriage" were produced, often as illustrations to satires on marriage, such as Matthew Seutter's 1730 map of the "Assault of Love" on the island fortress of "Manhood," or the 1734 "Carte de l'Isle du Mariage," which depicted marriage as a journey through a landscape of strife, or the map by "Simon Single" of the "Province of Cuckoldshire," principal city "Hornborough."

None, however, was as explicit as that textual expression of geo-pornography, *A New Description of Merryland* (1741), which described the female body in the geographer's language. Written by Thomas Stretser

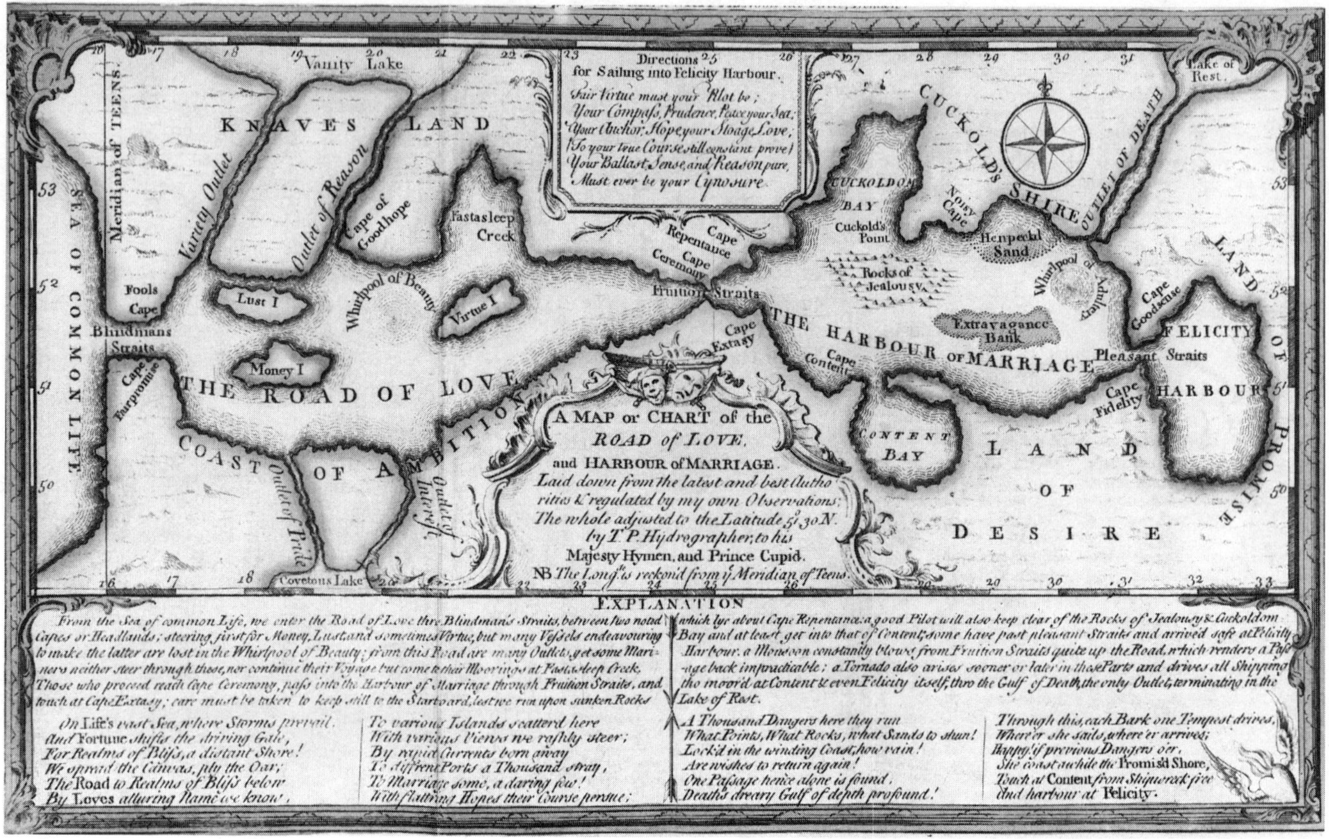

FIGURE 29

Tracing life's path. "A Map or Chart of the Road of Love, and Harbour of Marriage," from *A Voyage to Lethe* (London, 1741). By permission of the British Library (Cup.1001.c.4).

(under the pseudonym Roger Phequewell), the work ostensibly provided a corrective to the work of "modern Geographers" and to their instruments, atlases, and globes, which neglected to describe the country in question. "Merryland," the female form, is topologically enumerated in medical terms—LBA (labia), CLTRS (clitoris), HMN (hymen), and so on—and, for the want of an accompanying map, which Stretser knew "would considerably enhance the price of the book," he directed readers to the anatomical atlas of the French physician François Mauriceau, noting that the reader may see there "all the noted places & divisions laid down exactly as they are situated." In a following work, *Merryland Displayed,* Stretser satirized his own work as a dismissal of geographers' claim to truth: "The first Conception [i.e., *A New Description of Merryland*] was owing to our Author's accidentally reading in Gordon's Geographical Grammer these words, which Mr. Gordon uses in speaking of Holland, viz. 'the Country lying very low, it's Soil is naturally wet and fenny.' Ha! Said he, the same may be said of a * * * * as well as of Holland; this Whim having once entered his Noddle, . . . his wise Head fancied here was a fine Scope to ridicule the Geographers, so he sets to scribbling."[1]

Such evidence further illustrates how the languages of geography permeated the different literary genres of the Enlightenment, including pornography from which context it has been overlooked, and that they did so through the rhetoric of exploration and mapping as well as of medical topography.[2] In this chapter, my concern is not with the various genres of geography per se or with the metaphorical utility of geography's language. It is with geography's different social spaces and the types of geography conducted and practiced in them. Whether in the form of didactic-satirical cartography or as textual geo-pornography, the *géographie galante* of the salon was certainly not the same as the military geography of the French *ingénieur-géographes.* Neither was it the same as the formal instruction in geography in Enlightenment universities. And each, we may suppose, was different from the geography undertaken in Enlightenment academies, and from those geography lessons using geographical games, books, and globes of one sort or another given in the privacy of one's home.

Drawing on what we know of geography's textual genres and different practices discussed earlier—geography's epistemic "what"—the central focus in what follows is with geography's social "where" and in linking geography's spaces and practices: teaching spaces and instructional discourse; private lessons in the home or salon and conversation; reading and experimentation. Four sets of examples are discussed: geography in Enlightenment academies, in Enlightenment universities, in the public sphere, and in the home. These examples do not presume strict differences between "private" space and an Enlightenment "public sphere" for geography. Questions of audience and differences of intent behind the use of geography in different social spaces make any neat dualism between public and private space problematic. Strict correlations between, for example, geography's books, their authors, readers, and the ends sought in doing geography at home, in a university class, or through attendance at public lectures in the subject are difficult to draw. In exploring the local sites and social spaces of geography's use, my aim is to reveal something of where, how, and why geography was used as a means to Enlightenment sociability.

Knowing Places: Geography in the Learned Academies

It is difficult to know in detail which Enlightenment academies and learned societies paid formal attention to geography: the geography of Enlightenment academies—the "general cartography" of scientific production, as Roche has it, of regional academies in France—is itself no clear guide to the place of geography in these social and institutional spaces (see fig. 6). Yet as we have seen for a few such bodies—in Paris, Berlin, and Nuremberg to name just three—geography in the form of mathematical cosmography was undertaken there, and people understood as geographers by the books they wrote and the classes they gave were members or fellows, not least because their contemporaries so described them. In several cases, the presence of a geographer was part of more general institutional concerns of national or local self-knowledge.

Reckoning in 1743 that "the first Drudgery of Settling new Colonies, which confines the Attention of People to mere Necessaries, is now pretty well over: and there are many in every Province in Circumstances that set them at Ease, and afford Leisure to cultivate the finer Arts, and improve the common stock of Knowledge," Benjamin Franklin was insistent that the society of "ingenious Men" outlined in his "Proposal for Promoting Useful Knowledge"—what became in Philadelphia in 1769 the American Philosophical Society—should have a geographer from its foundation (alongside a physician, botanist, chemist, mathematician, a "Mechanician," and a general natural philosopher). Geography there then was part of the shared concern for useful knowledge as a means to improvement. In this regard, the Philadelphia society was little different from its many counterparts in mainland Europe, in the later eighteenth century especially so, where as we have seen geography featured as part of those practices of *landeskunde*, regional economic description aimed at advancing national well-being, as well as in medical topographies and natural history surveys. We must allow for differences too: between like and unlike bodies in the same country as well as for the fact that some members of such bodies did geographical work but under the label "natural history," "mineralogy," or even "forestry." Members of the Physical Society of Zurich, for example, undertook surveys of "the economic condition of every village" in the vicinity of the town (but did not describe what they did as geography). The Bohemian Academy of Sciences, the Economic Society in Saxony, and the Economic Society in Leipzig each undertook large-scale geographical work, predicated on the assumption that knowing geography was essential to economic improvement. Such rationalist economic impulses were quite different from Maupertuis's strategic vision for the foundational place that geography as cultural exploration should have in a revived Berlin Academy. His schema for developing the academy, laid out in part in a 1752 essay "Sur les progrès des sciences," begins with geography and anthropology, topics that encompassed the opportunities for new knowledge presented by the encounter with the Pacific world and that also embraced new languages and practices of terrestrial measurement.[3]

We can know something of the place of geography in Enlightenment academies and societies by assessing the content of societies' journals and the specialist interests of members. Neither measure is without its problems. The first is not simply an expression of local members' interests; the second obscures problems of definition, in terms of the subject and the sites of its undertaking. In his study of western Europe's scientific community between 1660 and 1760, for example, Gascoigne notes that of the 614 enumerated "scientists," only about 4 percent were geographers, including geodesists.[4] But since other fields—notably astronomy (9 percent of the total), natural history (19 percent), and mathematics (12 percent)—were strongly geographical in content and method, we might reasonably claim of the Enlightenment academies that there was more geography being done in and through them than there were geographers doing it. Evidence on the role of mathematical cosmography in the Nuremberg Cosmographical Society hints at this. What we know of geography in such bodies in Paris, Edinburgh, London, and Saint Petersburg supports the claim more strongly.

In terms of occupation, there were only seven *géographes du roi* as members of the Académie des Sciences between 1699 and 1793 (about 1 percent of the known occupations listed), but seventeen persons listed geography as their principal disciplinary interest. We know that among the 133 botanists and natural historians and the 108 mathematicians were people like Maupertuis and La Condamine with geographical interests and capacities—who advised the Académie on geographers' papers as they came forward for publication—and we may presume that to be the case for the handfuls of members who recorded cartography, hydrography, and meteorology as their disciplinary interests.

Much the same is true of the geographical activities of the Society for Improving Arts and Sciences and particularly Natural Knowledge, commonly known as the Philosophical Society of Edinburgh, between its foundation in 1731 and its demise in 1783. The society divided its intellectual responsibilities into two "provinces," with "some of the general Parts of Geography" falling within geometry, astronomy, mechanics, and optics. Its mathematician members reported on astronomical observations, notably Colin Maclaurin, professor of mathematics at the University of Edinburgh, who actively engaged as a Newtonian in debates about the shape of the earth. Meteorological data were discussed and shared with other societies in the city, and the society promoted the mapping of Scotland's northern coastline and of Orkney and Shetland. Such national self-description was rooted in a discourse of civic improvement. Maclaurin understood the significance of the sought-for Northwest Passage to the Asian markets: "I wish the Northern passage to the E. Indies could succeed. It has long been my Castle in the Air (for every body has some) for the benefit of poor Scotland, considering its northern situation. I should be glad to hear what they think from Petersburgh [Saint Petersburg]." The Lisbon earthquake exercised the society's members considerably and prompted correspondents across Scotland to write on the local effects of tremors.[5] Yet analysis of the topics discussed in the society shows that "geography" did not feature formally at all. Its practice was promoted as a basis to national and international natural knowledge. Cognate concerns—astronomy, meteorology, mapping, botany—were regularly addressed in print and word. But the subject was never the basis of formal papers delivered or of debate among the members.

In London's Royal Society, although nobody held the formal position of geographer, the subject featured strongly in the society's *Philosophical Transactions* between 1720 and 1779, within the categories of "natural history" and "mixed mathematics" especially. In mixed mathematics—astronomy, mapmaking, surveying, and navigation, as well as geography (in short, "mathematical cosmography")—264 papers published in the *Philosophical Transactions* between 1720 and 1779 were in astronomy (43 percent of the total), a further 31 percent in geography (191 papers). Yet to reiterate an earlier point: strict distinctions are misleading. Much of the astronomical work was concerned with fixing latitude and longitude and so belonged to geography. In fact, in the Royal Society, "in this very practical sense, astronomy was almost entirely subordinate to geography . . . observatories were explicitly founded and funded to advance navigation and surveying, not to test Cartesian or Newtonian theories of the cosmos." Mixed mathematics was crucially important to Britain's interests as a mercantile and imperial nation. We can see in the Royal Society's journal one expression of geography's significance in parallel with the commercial rhetoric of men like George Bickham and the fiscal-military tones of others' special geography books.[6]

In posing questions about geography, exploration, and commerce, Colin Maclaurin had good reason to look eastward: geography's most evident institutional space in Enlightenment academies was Russian—the Geographical Department of the Saint Petersburg Academy of Sciences and Arts. Although several geographical enterprises had been initiated by Peter the Great—notably the German Daniel Messerschmidt's expedition to Siberia 1720–27 and the Imperial Senate's 1724 questionnaire about the geography of the Russias which became the basis of Kirilov's *Flourishing Condition of the All-Russian State* (1727)—the Geographical Department in Saint Petersburg properly began in 1739, as part of the academy's observatory under the direction of Joseph-Nicholas Delisle. The Geographical Department flourished under the direction of Mikhail Lomonsov from 1757 and under Stephan Rumovski from 1765. Together with Leonhard Euler from 1769, Rumovski initiated mapping, natural history, and surveying projects, including work that culminated in the "Great Map of Russia" in 1776. Being an auxiliary department of an academy meant little, however, when personnel changed: the Geographical Department declined after Euler's death in 1783.[7]

In Washington and in Richmond, the view held of geography's purpose in knowing one's place was much the same as that from Saint Petersburg. In Washington, the Geographers' Department established in July 1777 by the Continental Congress—the first official authorization by the federal government of a geographic agency—specifically arose from the need for maps for warfare. After the Treaty of Peace in 1783, the emphasis shifted to surveying and mapping. Thomas Hutchins, New Jersey–born and geographer to Washington's army but formerly a British officer, was appointed to the post of geographer to the United States to administer the General Land Office. Whereas Hutchins worked for the nation, William Tatham was appointed in 1789 by Virginia's state governor to organize and supervise a geographical department from the capitol building in Richmond. Institutionalizing geography was essential for a nation only just coming to terms with itself.[8]

Teaching Spaces: Geography in Enlightenment Universities

Colin Maclaurin might not have heard much geography among his Edinburgh peers or read it in the publications of the Philosophical Society, yet he certainly encountered it with his students: Maclaurin taught geography in the city's university as part of his final mathematics class, in 1741 at least. Maclaurin's attention to mathematical geography shared teaching space with "Surveying, Fortification and other practical parts." In the same year, John Ker, the university's professor of humanity, was teaching the essentials of "ancient geography" within his history lectures. Here, in one leading site of Enlightenment learning, two different conceptions of geography were being taught—the one mathematical, the other humanist. Both were practically useful in different ways. Both men made reference in their teaching to others' texts, and both produced geography as an instructional discourse, something heard and written down because dictated, not read silently from books. Both Ker and Maclaurin had counterparts elsewhere in Enlightenment Scotland and were part of a longer-run practice of geographical teaching. Briefly charting geography's university place in Scotland reinforces my argument about local variations within national contexts and reiterates the importance of considering the different sites and social spaces in which the Enlightenment—and geography—was at work.

From at least the 1670s, the teaching of geography in Scotland's universities mirrored more general shifts away from Aristotelian and classical thinking toward Cartesianism and, latterly, Newtonian natural philosophy. In Marischal College, Aberdeen, for example, geography was included in the teaching of Newtonian mathematicians and natural philosophers. John Stewart, professor of mathematics there, taught geography to his first-year class from 1748—and may have done so from 1727. The "mathematics texts" he used included that prompt to Stretser's 1741 satire, Patrick Gordon's *Geography Anatomiz'd,* and a 1733 edition of Varenius's *Compleat System of General Geography, . . . since Improved and Illustrated by Sir Isaac Newton and Dr Jurin.* In the second year at Marischal, geography was taught alongside "the most usefull parts of Natural History and the Elements of Civil History." In Marischal in the 1750s, Thomas Blackwell, professor of Greek, gave lectures on geography and ancient history, as did Thomas Gordon in King's, whose classics lectures studied "the Geography of the Antients" in order that his students should understand the where of the ancient world. Also in

King's College, the Common Sense natural philosopher Thomas Reid gave a course in 1752 titled Elements of Geography. His course was in four parts: a general regional geography of the globe; the main lands and seas named; "a very few of the Most remarkable Things belonging to each Country or City"; and accounts of the trade winds. This last may appear odd now, yet it was then quite in keeping with one view of geography's utility, namely, making commerce efficient. As Reid put it, geography was an essential part of that philosophy "which may qualify Men for the more useful and important Offices of Society," rather than making them simply "subtle Disputants." That, Reid noted, was "a Profession justly of less value in the present age."

At the University of Glasgow, geography was part of the mathematics class, with prizes being awarded for the "the best exercise in geography" from the 1784–85 academic year onward. Although no such prizes were awarded after 1792, medals were given for essays of a geographical-mathematical nature—such as that given in 1797 to David Warden, an Irish student, for "the best historical and philosophical account of the application of the barometer to the mensuration of heights." James Millar, professor of mathematics at Glasgow, delivered a separate lecture course in geography to his students every year from 1802 until 1821. In his 1765–66 course at St. Andrews, Nicholas Vilant, professor of mathematics, taught navigation, fortification, architecture, spherical trigonometry, projections of the sphere, use of the globes, fluxions, and astronomy, much of which material became the basis to his *Elements of Mathematical Analysis* (1786).[9]

It is one thing, of course, to locate geography on the maps of Enlightenment knowledge, one here constituted by the geography in and of Scotland's universities. It is quite another to know its meaning and importance for contemporaries there and elsewhere, or to know if and how the cognitive content of what was held to be geography connected with the other intellectual concerns of the Enlightenment. Doing geography through practical propositions, dictates, and reading exposed university students to geography's close affiliations with astronomy and geometry on the one hand and to civil history on the other. But did it mean that other geographical concerns were omitted? Did discussion of fluxions and spherical projections mean that the geography of the world as a matter of practical discovery, human diversity, and its physical makeup was a closed book? There are partial glimpses to suggest that it was not. In Glasgow, Millar's geography lectures exposed students to the competing theories of Hutton and Werner; in Edinburgh, students encountered competing cosmographical theories in learning Newton's work, had access to globes, and were directed toward history books in which geography figured as part of the explanation for the rise and fall of societies. This is not to ask inappropriate questions predicated on the false supposition that geography was—or should be—something more "modern" than this engagement with mathematical cosmography, civil history, and practical utility would suggest. It is to consider universities as one sort of Enlightenment site and social space and to inquire what relationships geography's presence there had with other themes of importance here, namely, the geographies of the Enlightenment and the Enlightenment world as an object of geographical knowledge.

The fact that our answers to date are limited to only a few places does not diminish the importance of understanding the Enlightenment geographically. Far from it, for in different sites across Scotland, we can note the presence of geography as a form of Enlightenment knowledge at work. So too for France. In Paris at the École Normale, for example, geography was central to an understanding of human cultures in comparative context, with Edme Mentelle emphasizing not just a geographical and cartographical methodology based on observation—"geography is a science which can only be learned properly by using one's eyes"—but also a conception of geography that was materialist and utilitarian. Political geography (the third part of the course) underwrote all the human sciences. Physical geography (the second part) provided the link between mankind and nature. Mathematical geography (the first part of the course) provided a unifying basis in measurement and in language. In Paris, questions of fundamental importance attributed to geography within the Enlightenment Science of Man were most apparent in the Class of Moral and Political Sciences at the National Institute from 1795. Buache de La Neuville and Mentelle provided a direct continuity between the École Normale and the Institute and helped shape the geography teaching in both. Even recognizing that the Institute's geographers achieved only modest results, and that the class was shut down by Napoleon from January 1803, we can see in its activities something of what geography was held to be and how it connected with other concerns.

For Condorcet, the geography section of the National Institute represented an opportunity—never fully realized—to explore on a grand scale the effect of the environment on the body and the mind. Volney, although not formally in the geography section, was particularly keen that the Institute promote the study of the influences of climate and terrain on the moral qualities of peoples. America, its native inhabitants and its colonizers, was his laboratory, even if he never completed the intended "political" section to his physical geography of that continent. Together with Bougainville, Volney petitioned the new legislature for a vast experimental plot on which comparative analytic tests on plants and animals could be run in order to improve stock and quality, domesticating the far away and naturalizing the exotic with a view to moral advance and improved practical understanding. As these two sketched grand schemes to connect geographical work with that of the hygienists in order to "ground" material advance, Buache de La Neuville was drawing up instructions for the ill-fated La Pérouse. Mentelle, unsuccessfully, advocated the promotion of a centralized "statistical geography" using France's political ambassadors: a network of credible sources whose foreign returns would sustain a domestic archive of others' fiscal-military capacity. Arguably, the most successful research coordinated by the Institute took place farthest from home: during the voyage of Nicolas Baudin to Australia in the ships *Géographe* and *Naturaliste* between October 1800 and March 1804.[10]

The fact that the geographers of the Class of Moral and Political Sciences produced work of limited success reflected political circumstance more than their own shortcomings or any failure to position geography in relation to wider intellectual concerns. What is also revealed within Paris is a differing "institutional topography," a different typology even, for geography's sites and spaces. Where the École Normale provided a setting, notably under Mentelle, for the teaching of an essentially descriptive geography, the Institute was home to an ambitious and theoretical program for the study of human-environment relations in America and the human geography of Pacific peoples. As sites for Enlightenment geography, both differed from the Académie des Sciences with its practical emphasis on mathematics, geodesy, astronomy, and navigation.

Even if this is an incomplete picture, it highlights the significance of local difference within national context, here within one city's institutional spaces. Recognizing the fact that geography as taught or as proposed as a research program was not everywhere the same demands comparative understanding at different scales. In Enlightenment Germany, for instance, students at Königsberg under Immanuel Kant experienced a more strongly philosophical focus to their geography, there largely a matter of physical geography, than did their counterparts in Göttingen, who engaged with mathematical geography and the language and methods of mapmaking through the teaching of, variously, Tobias Mayer, Anton Büsching, and Johann Christoph Gatterer. Kant first lectured on geography at Königsberg in 1756. His *Outline and Prospectus for a Course of Lectures in Physical Geography* (1757)—which Kant undertook, given the lack of suitable texts (although Kant's lecture notes circulated for over thirty years in manuscript form before being published as *Physiche Geographie* in 1802)—is noteworthy. Kant made no distinction between geography and history and placed emphasis on physical geography as the description of nature, the world as the object of "outer sense." Geography for Kant was the descriptive study of simultaneous occurrences in the present. Thus understood, history was a sort of "continuous geography," events in time but also, crucially, in space. Physical geography so considered—as Kant put it, the "geography of Nature"—was thus a necessary basis to all further knowledge and understanding. Geography's place was as the basis to all learning: in Königsberg at least, the Science of Man rested on the Natural Geography of Man.[11]

In Göttingen, geography was less a foundation to all Enlightenment knowledge than an applied practice, a route to state understanding. Earlier teaching under Johan Koehler had considered ancient and medieval geography, but with Tobias Mayer from 1751, who occupied himself with mathematical geography, mapmaking, and what he called *Weltbeschreibungswissenschaft* (the science of world description), geography at Göttingen from the mid-eighteenth century was conspicuously practical. Mayer brought the Nuremberg mapmakers and his fellow cosmographers Johann Michael Franz and Georg Moritz Lowitz to Göttingen: the first became professor of geography; the second lectured on mathematical geography and gave courses on drawing maps. Göttingen geography was not just practical. Although Gatterer, professor of history at Göttingen, lectured on geography—seeing it, much like Ker in Edinburgh and Blackwell and Gordon in Aberdeen, as a basis to historical understanding—the main proponent for the subject as the descriptive textual study of the earth's surface was Büsching, professor of philosophy. Concerned more with accurate depiction of known features—Büsching excluded speculation about the polar regions because too little was known for certain—Büsching's lectures were descriptive more than mathematical, textual and teleological dictates even, not practical propositions.[12]

In Geneva, geography was different still. The lectures in physical geography given by Horace-Bénédict de Saussure, professor of philosophy, and recorded in 1775 by a theology student, Jacques-Louis Peschier, suggest that the subject would have been more familiar to student audiences in Königsberg, or perhaps Glasgow, than in Göttingen or Aberdeen. Any full and proper course of geography, Saussure argued, should begin with astronomical geography, but "since this requires a certain knowledge of mathematical principles which you have not yet attained," he began his course with the simpler yet fundamental principles of physical geography, understood as the earth in regard to its structure and materials.

The details of the course—which include Saussure discoursing on the geodetic expeditions of La Condamine and Maupertuis—need not detain us. Of interest is the fact that the Saussure-Peschier syllabus further reveals the geographical movement and reception of Enlightenment ideas—in this case, Saussure's engagement with Buffon's work. The encounter was, in part, personal: the two men had met, once, in Paris in 1768. Saussure disliked Buffon, thinking him a generalizer, his *Théorie de la terre* without proof. Buffon considered Saussure a provincial pedant. But the encounter was also intellectual. That Saussure should use Buffon's work in his teaching illustrates the role played by geography lectures in airing fundamental differences in terms of how the earth was understood in the Enlightenment and over what physical geography was. For Saussure, Buffon's preconceived theory of the earth lacked empirical proof—Saussure's knowledge of mineralogy and stratigraphy was certainly more up-to-date—and Buffon's substantive accounts, of mountain formation, the role and scale of marine currents, and so on, lacked logical order and did not adequately describe variations in the earth's features. As Saussure lectured and, later, as he reworked his manuscripts (never finishing their revision), so he variously mediated—accommodated, taught, and rejected—Buffon's ideas. In this *vignette* of speculative Parisian theory meeting pragmatic Genevan empiricism, we can site differences over the earth's geography and what it meant. In one university setting, we can see in miniature how differences in geography—understood in terms of location, as an intellectual program and as the movement of ideas, books read and words delivered or written in certain social spaces—acted to constitute the Enlightenment as a whole.[13]

Much the same is true for geography in schools during the Enlightenment. Knowledge of geography was commonly recognized as part of a gentleman's education, and increasingly, in the Enlightenment, for women too. This is reflected in the growth in the number of children's geography books and in different forms of geographical instruction such as geographical games. Yet relatively few people received any formal education at all. And our knowledge of geography in schools in the Enlightenment is still very uneven. Given this, what follows pays more attention to spaces of educational sociability other than the schoolroom, notably to geography's place as a civic discourse in public space and to geography's place and instructional role in the home. But recording the role of a few schools as sites of geographical learning helps illustrate more general points about the Enlightenment understood geographically.

In Warrington Academy, for example, where he helped establish geography in the curriculum, Joseph Priestley accorded geography its traditional place in underpinning history. He may have used Gordon's *Geography Anatomiz'd* and Guthrie's *Geographical Grammar* in doing so: his peers in other dissenting academies and many grammar schools certainly did. Priestley's purpose was less traditional, however, for he saw history teaching as emancipatory, helping "to free the mind from many foolish prejudices." In contrast, in most grammar schools in England, geography was an adjunct to Latin and Greek, a conservative part of a humanist tradition that emphasized geography as an "ocular science" both because it was one of the "eyes" of history and because it was taught through looking at maps and reading texts in order to put historical information in the right place. Where map reading was regularly included, map drawing and other mathematical geography were as often ignored, not because they were not part of geography but because they were not part of a grammar school education.

By contrast, the geographical curriculum at Perth Academy under John Mair began by placing geography within a global natural history, a positioning Mair considered "well calculated to fix the attention and awaken the curiosity of young people," not least because he illustrated his teaching with experiments before engaging with practical problems in astronomy. His second-year course treated geography as an introduction to civil history before addressing geography's commercial utility. Mair had no time for Gordon or for Guthrie but wrote his own teaching text, *Brief Survey of the Terraqueous Globe* (1762). His successor, Robert Hamilton, used Mair's book alongside Guthrie's but took teaching through demonstration even further, persuading the town's authorities to purchase globes and air pumps for the academy. Training his pupils, Hamilton advanced himself: leaving Perth Academy in 1779, he took up the position of professor of natural philosophy at Marischal College, Aberdeen, where he included geography in his university classes and in his public lectures.[14]

Geography's Public Places

Moving from Perth to Aberdeen, from schoolteacher to university teacher and public lecturer, Robert Hamilton at least merited the title "professor." Others, like John MackGregory, itinerant lecturer and self-styled "Professor of Universal History and Geography," did not. He nevertheless used that title to promote himself and his public lectures in geography and universal history in Edinburgh and London between 1707 and 1715, advertising his classes through printed flyers and news sheets distributed around the cities' coffeehouses. Later in the century, among the several public lecturers in geography in Enlightenment Edinburgh was James Dinwiddie. In 1777, Dinwiddie had advertised lectures on geography and astronomy in the small Scottish market town of Dumfries. By 1779, he was in Edinburgh, advertising public classes in French, mathematics, and experimental philosophy as well as in geography. By 1792, however, Dinwiddie was part of what we have seen was Enlightenment Europe's first significant encounter with China, Lord MacArtney's diplomatic mission—officially as an astronomer—from which position he then moved to India, to teach science and geography in the College of Fort William in Bengal, before returning to Britain in 1806.[15]

In Edinburgh, a further twenty-two men gave geography lessons in the city between 1708 and 1800, and more in the decades following. Some, like Robert Darling in 1776 and again in 1793 and 1994, variously termed "private teacher of geography" and "teacher of mathematics and geography," lent geography commercial utility by teaching writing, bookkeeping, mathematics and geography, "and Gentlemen to Measure and Plan their own estates." Supplementing his ministerial stipend, the Rev. William Smart instructed in "the most useful problems of geographie" by going to his pupils' houses or having them come to him in his lodgings, "or at any other convenient place as the party shall direct." One set of classes given by George Paterson in the city ran for over twenty years. In Glasgow, Robert Lothian, chaplain to the city's Trades House in the 1780s, taught classes—with separate sessions for women—in geography, military mathematics, and astronomy by using self-made "improved machinery," including a planetarium, an orrery, and an armillary sphere. Lothian was one of eight men teaching geography in the city between 1755 and 1800: two, Robert Dobson and James Stirling, even ran separate classes at the same address, "Old Coffee House Land," between 1755 and 1764. Even in some small fishing villages, geography classes were held, often there with an emphasis on navigation.[16]

In Philadelphia, Benjamin Franklin was providing for the town's geographical public even as he campaigned for a geographer in founding the Philosophical Society. Among geography books sold by Franklin and his colleague David Hall in the 1740s—before Jedidiah Morse's was available—the most regularly purchased were Gordon's *Geography Anatomiz'd,* his namesake's *Introduction to Geography, Astronomy, and Dialling* (1726), and editions of Varenius. Classes in geography were held in the city from at least 1738, perhaps as many as twenty-two separate classes in different locations at one time or another between then and 1794. Several of the teachers advertising geography offered mathematics too—promising to teach "Newton's Mathematical Philosophy and Planetary System," as did Charles Fortescue, schoolmaster, or a whole course in "Experimental Philosophy," as did Mr. Baron, the syllabus of which could be had at the town's post office, with the subscription payments to be given to David Dove, master of the grammar school. Some men like Gabriel Nesman, a Swedish minister, taught geography in day classes in January 1751 alongside French, Latin, High German, and Dutch. Others such as Alexander Power between 1766 and 1771 ran an "Evening School" for geography, "the use of the globes and maps; and how to make maps." Not unlike Hamilton in Aberdeen, John Heffernan, latterly of the University of Pennsylvania, opened a private "English and Mathematical Academy" on Philadelphia's Cypress Street at which he taught geography and the use of globes. Local newspapers carried notices of Büsching's *Geography* just in, of "New Setts of Geographical Cards" for use in domestic learning, and, in 1787, of Noah Webster's speller and how it might be used "to instruct them [the public] in the Geography, History and Politics of the United States." Geography classes for young women were a feature of Mr. Poor's Young Ladies Academy on Arch Street, Philadelphia. There, in December 1789, Eliza Dolby was given a money prize for "excelling in geography"—in fact for reading and discoursing to her fellow students upon the recently published *American Geography* of Jedidiah Morse. Later that century, geography classes are recorded in several smaller Pennsylvania towns—as also in Virginia, New York, and in South Carolina.[17]

The sociable worlds of public geography classes in Enlightenment Philadelphia were different in intent, in their practitioners, and in their transient nature from the large-scale mapping projects then being coordinated by Hutchins and Tatham in Washington, D.C., and in Richmond. Because this is so, the idea of "America's geographers" in the Enlightenment ought not to focus just on the producers of geography or uncritically on "founding fathers" such as Morse and Hutchins.[18] In looking at sites of geography's reception and at its audiences, we might distinguish between, on the one hand, an emergent national geography—of which Morse's book, schemes for mapping national boundaries, and the presence of geography at Princeton and at the Academy of Philadelphia (what became the University of Pennsylvania) were different expressions—and, on the other, different local publics engaging in geography by reading Morse to their fellow students or by learning to draw maps. Even so, such distinctions should not be too firmly drawn, for people moved across and between these different social spaces. Take Andrew Porter, for example, who offered geography within his English and Mathematical School on Union Street in Philadelphia between 1767 and 1776. Joining the revolutionary army in 1776, Porter fought with distinction, was commended by Washington, declined the professorship of mathematics in Philadelphia, but, by virtue of his self-taught mathematical-military geography and astronomical skills, became a boundary commissioner for Pennsylvania and worked on the Mason-Dixon line.[19]

These glimpses of lives and audiences raise more general matters concerning geography's place in the Enlightenment's public worlds. Here were people, not geographers by any professional definition of the term, teaching the subject in different ways to different audiences who, we may likewise presume, did not in any formal sense see themselves as students of geography. Yet the presence of men like MackGregory, Dinwiddie, Porter, and others shows that geography was on the map of Enlightenment public knowledge. This and other evidence—public classes in the subject, booksellers dealing in geography's books, audiences and tutors on the move, women taking geography lessons, students drawing maps—point to a more detailed geography of geography in the public sphere than has been supposed, a geography of networks, movement, overlapping social spaces, and epistemic practices. For men like Dinwiddie and Porter, geographical mobility was also social mobility, their social and occupational mobility a prompt to engage differently with geography in meeting different needs. Elsewhere, others were doing the same: Manuel do Cenáculo in Lisbon, for example, who for much of 1772 spent his mornings preparing the king's grandson for rulership by teaching him geography as an art of statecraft, and his afternoons reforming learned institutions and planning for military schools in which Portuguese students would be taught military geography, civil history, and foreign languages like their counterparts in Paris and Madrid.[20] But of most public lecturers, we still know little. It is not always easy to distinguish among men like Hamilton the legitimate professor, educated tutors seeking only to make an honest living through geography, and men like MackGregory, the Grub Street jobbing hacks of Enlightenment geography. We cannot easily know the audiences and how they were taught, or discern in detail differences in emphasis placed on geography in the several "commercial academies," private classes, or public lectures.

Nevertheless, men and women came together as Enlightenment publics for geography. If this is a geography of geography in the Enlightenment, however partial, it is also something more. It is an illustration of how and where those forms of knowledge that made up the Enlightenment—mathematics, botany, civil history, and so on—were engaged with as public discourse and how they might thus be thought about and located, individually and as complex networks of association. Thinking geographically means that we might approach the question of sociability not just as a historical sociology or through a temporal framework. For as these illustrations show, Enlightenment sociability did not exist "above" geography: it was made in and between places, through certain practices in given sites and spaces as well as being made through geography itself. Thinking about the different sites and spaces of and for sociability and to the ways things happened there—whether as books or more specialist journals read or as maps followed in popular magazines such as the *Gentleman's Magazine*[21]—means that matters of scale, location, and the forms taken to make and move knowledge are crucial to what we take the Enlightenment to be. Let me illustrate this by looking for a moment at geography in the home.

Domestic Geographies and Practical Instruction

Recalling her Burntisland childhood in the 1790s, Mary Somerville, physical scientist and author of the influential *Physical Geography* (1848), noted how Mr. Reed, the village schoolmaster, came to the house on winter evenings to teach her the use of the globes. He also taught Latin and navigation, "but these were out of the question for me." Homework involved more than direct instruction in the parlor: "my bedroom had a window . . . I spent many hours studying the stars by the aid of the celestial globe."

For Émilie du Châtelet, in contrast, knowledge of astronomical principles was shaped by salon conversation and private writing removed from her peers. Sociability in and about natural knowledge involved mannered talk with people at the forefront of the Enlightenment's scientific and philosophical controversies, with Newtonian physics and English philosophy providing particular "grist for the conversational mill."[22] For Somerville it came from instructional dialogue and untutored observation at home as cosmographical knowledge locally made. Somerville's experience of home learning rather than salon politeness was widely shared and often involved instruments other than words and books. For women especially, the home was a site of conversation, instrumental instruction, and dialogue, a space for the mutual production of geography and of sociability. Instruction aimed at instilling appropriate learning, particularly at producing "politeness" as a moral quality, commonly involved geography and astronomy in the home and use of instruments like the orrery and the armillary sphere.[23] These, together with the telescope and terrestrial and celestial globes, were geography's conversation pieces in a wider battery of demonstration devices.

Instruments of one sort or another were common features of science in the Enlightenment home, "demonstration" embracing a sense of disclosing or revealing natural philosophical principles through artifice and a degree of showmanship necessary to both amuse and instruct one's audience. Machinery and instruments, shop bought or homemade, were employed in various places—public houses, coffeehouses, university common rooms, as well as parlors—and they did more than simply transmit fixed notions of natural philosophy. Rather, they helped shape it in certain ways according both to the interests of given audiences (in ways we have seen for Volta's battery) and to how, in certain places, the operators' conduct differed. If in general, then, "the material culture of natural philosophy, its instruments and models, was a vital part of its doctrinal authority," we can note of geography that its production and reception were not just a matter of texts and words but of other devices used in various ways.[24]

Geographical games reflected a growing interest in travel and in travel narratives, often from a particular perspective. The Geographical Pastime series of the London-based publisher John Wallis, for example, included a *Tour of Europe,* the game ending with the first person to traverse Europe successfully and return to London, "the first city in Europe." Others used cards similarly to "instruct" in geography. John Holmes's *Grammarian's Geography and Astronomy* (1751) was used in combination with playing cards, each suit signifying one of the continents, each number intended to trigger a fact of geography in the student's memory. The stereotypes were common: the king of diamonds stood for the Empire of China, and the Chinese were typified by "Fawning, Hypocrisy and Over reaching." In the Abbé Gaultier's *Complete Course of Geography, by Means of Instructive Games* (1795), geography was taught using "common" (that is, annotated) maps and "plain maps" and "a set of counters, having the name of kingdoms, provinces, islands, seas, and rivers, &c marked on them," which, from memory of the common maps, were placed on the plain maps.[25] A further popular form of instructional geographical game was the jigsaw, or as John Spilsbury, the leading maker in England termed them, "dissected maps." Spilsbury's own geography and biography connected him to others' worlds. Between 1753 and 1760, Spilsbury, "Engraver and Map Dissector in Wood, in order to Facilitate the Teaching of Geography" as he styled himself, had been apprenticed to Thomas Jefferys in London. But he made and marketed his work only after leaving Jefferys' employment—products that Jefferys and others further developed after Spilsbury's death in 1769 at age thirty. Most "dissected maps" and other geographical games were used in the home to instruct through play and as a pastime—to show the location of places in relation to one another and thus how the world fitted together (see plate 10).[26]

The teaching of geography in the home was commonly associated with astronomy and thus, as Somerville's case illustrates, with the use of the terrestrial and celestial globes. It is just this sense that Girodet-Trioson brings to his painting *The Geography Lesson,* which illustrates the cover of this book. Indeed, the combination of astronomy and geography and their instrumental instruction within the home perfectly suited the agenda of politeness in the Enlightenment: the subjects were part of established traditions of education, were useful in themselves, and, for geography, could be allied to historical understanding and to commercial concerns. For Benjamin Martin, instrument retailer, itinerant lecturer, and author of *The Young Gentleman and Lady's Philosophy* (in which geography's teaching in the home figured for all the just-stated reasons), "the necessity of the Knowledge of the Use of the *Globes, Sphere,* and *Orrery,* for an easy Conception and due Understanding of the first Principles of *Geography, Astronomy, Dialling, Navigation, Chronology,* and other liberal Sciences, is known to every one; and is of the first Consideration among those Qualities requisite for forming the SCHOLAR and the GENTLEMAN."[27] In Greece too, Moisiodax understood that his text would have greater value if used with other instruments: all of the thirty-three geographical problems listed in his 1781 book were solvable through the use of globes.[28]

But exactly which globes were being described here? And with what use in mind? Just as forms of allegorical mapping in the Enlightenment carried meanings about the voyage of life, so the conjunction of domesticity, private space and attention to the globes could afford an opportunity for encounters with heavenly bodies that was anything but polite—a far from *galante* geography. This is the central message in Pietro Longhi's painting *The Geography Lesson* (1750–52) (plate 12). Longhi's work satirizes contemporary pretensions—a daughter is here being educated in the polite refinements. Posed beside a celestial globe, the young woman holds a pair of compasses and is thus caught in the act of global measurement, an open atlas at her feet. This moment of geographical enlightenment—the light cast across the painting strikes her, the globe, and the atlas but not the books in the background—is also one of sexual frisson. The painting plays on the word "globe," on bodies of knowledge and knowledge of the body. The globes on display are not instructional-conversational pieces but the swelling breasts of the young woman herself, as the standing instructor's gaze reveals. Here, partially glimpsed, lies a forbidden "merryland." The globe at once material and symbolic is an instrument of secular learning, of Newtonian principles, and an object of private desire.[29]

Symbolically and substantively, geography in the Enlightenment worked differently in different places. Charting geography's public presence or its role in university teaching illuminates our understanding of the Enlightenment—and of geography—because it illuminates matters of local meaning and the dynamics of local practice. This is not to particularize the local, and it is certainly not to claim that geography in the Enlightenment had significance above other forms of knowledge. It is to note that in addressing the ways in which the Enlightenment was sustained and what it meant to people, we must keep its geography in mind and recognize how and where the subject of geography was part of Enlightenment inquiry.

Recovering geography's place in the Enlightenment is to reveal a world of books, maps, institutional endeavor and individual meaning, mobile audiences, peddled geo-pornography, and domestic politeness. Only by seeing how and where these and other features of geography worked can we appreciate better the complex nature of Enlightenment geography and of the Enlightenment itself. What we have seen in these chapters concerning Patrick Gordon's *Geography Anatomiz'd* may be taken to illustrate my more general point. Written by a Scot in 1693, published in London, intended for the children of gentry, the book was satirized by geographical hacks, bought by Philadelphians, translated by Italians and by Greeks, and was in its twentieth edition by 1760, having by then been "supplemented" and "enlarged" by other geographers.

Yet books of geography do not tell the full story. They did not even all carry the same meaning, nor were they written and used in the same way. Even the *Encyclopédie, the* book of the Enlightenment, was encountered differently on account of its geographical content and depending by whom and where it was bought and read. Similarly, maps differently represented and constituted the Enlightenment world. In their making, we can trace how mapmaking depended on accommodating selected facts, networks of affiliated individuals, and claims to a language of exactitude. Maps gave the statistical philosophy of the Enlightenment spatial expression. Regulating space depended on making connections among topography, climate, and the constitution of national space. In medicine and in morals, the biopolitics of the Enlightenment was also a geopolitics.

Looked at locally, the map of geography in the Enlightenment reveals features of difference because of geography's settings and forms—between universities and domestic space, between books read during a nation's building and mathematical and philosophical propositions taught to instill politeness, knowledge of trade winds, or a working comprehension of astronomy and navigation. What geography meant depended on where it was taught, by whom, and how.

11

Conclusion
The Enlightenment—Questions of Geography

In thinking about the Enlightenment geographically, I have proposed here an argument that the "where" of Enlightenment knowledge should be taken as seriously as its "what," "who," "how," and "why." To questions such as "The Enlightenment: what and who?" and "What is Enlightenment?" we should now be able to add "The Enlightenment: where and why?" and "Where was Enlightenment?" and be confident of reasonable answers. Further, the practices of geographical knowledge and the textual definitions and different meanings of geography within the Enlightenment ought now to be cited more readily within the range of what is held to be Enlightenment studies, and to be more commonly sited on the "map" of what contemporaries took to be Enlightenment. In one regard, to consider the presence of geography in the Enlightenment, to document the geographical practices by which the world then was revealed as an object of geography, and to outline the metaphorical and rhetorical power of thinking geographically is to do no more than was done by numerous contemporaries, albeit in different ways and places—David Hume, Diderot, and d'Alembert, for example, or Jedidiah Morse, Voltaire, Buffon, Eberhardt Zimmermann, James Rennell, William Robertson, and William Guthrie, to name just a few. But what I hope also to have done is to have extended the range of modern thinking on how now the spatial dimensions of the Enlightenment can be made sense of by emphasizing the importance of space and of place, of local differences, the mutability of taken-for-granted categories such as "nation"—and geography—and the importance of travel.

Part 1 presented evidence about geographical scale, the language of geography, and the importance of place in proposing the effective dismissal of *the* Enlightenment as a unitary and philosophical phenomenon reducible only to the level of the nation. My concerns lay with the geographical modification of essentialist accounts, in stressing the difference that thinking geographically about transnational and local scales makes to the picture of the Enlightenment as a whole. In part 2, I showed how the earth was opened up as a subject of geographical inquiry in the Enlightenment, and more, how the new geographies of the world fundamentally shaped Enlightenment knowledge about nature's diversity and human difference. Here too the language and practices of geography pervaded what were held to be the botanical and the geological sciences. And the facts of geography and of geographical difference over space in terms of culture were integral to those social and historical theories concerning human society over time that marked the emergent human sciences. As part 3 revealed, geography was itself a form of Enlightenment knowledge, a means to polite sociability and of public and institutional interest. Yet its books were read differently, its use in instructional discourse varied—between the physical earth theories of Saussure in Geneva, for example, and the humanities curriculum of professors in Aberdeen and in Edinburgh. Because this is so, we can of course place geography in the Enlightenment yet at the same time recognize the difference that place makes to what geography was held to be and to how it was put to work.

These ideas represent some of the main strands of my argument. Let me close, however, by offering a series of questions and by examining two further and related issues pertaining to the Enlightenment understood geographically. The first is to return to the theme of what we may call the Enlightenment's "future geographies." This is the view in the Enlightenment that progress meant betterment, that ideas of progress and reform were not just moral and intellectual issues but were materially "earthed" in the sense that political economists, philosophers, natural historians, and others could build a better physical environment, by "seconding Nature" as Buffon had it, by using the "Science of Man" and models of historical change to lay out new prospects. In short, how did Enlightenment writers imagine the future geographically?

The second theme explores what we might think of as an antithesis to the first: namely, how, now, is the Enlightenment our geographical contemporary? Here, I consider modern moments and spaces of the Enlightenment's representation, its geographical "reenactment" even, in relation to the claims of those whose interests focus on the Enlightenment's continuing legacy.

The Enlightenment—Geographically

From consideration of what I have taken here to be "the geographies of the Enlightenment," we now ought to be more sensitive than has been the case to the importance of geographical difference in thinking about the Enlightenment, in terms of its production, mobility, and reception. It is likely, of course, that the Enlightenment will continue to be understood in relation to different national perspectives, perspectives that may themselves differ in the objective of doing "national" Enlightenment studies—to highlight an originating intellectual moment, thinker, or scientific discovery as part of one's national history, to see in the Enlightenment political and intellectual circumstances not realized today, or, simply, because there is always more to know about the period for a given national context.

My focus on national, cosmopolitan, and local dimensions, most notably in part 1, and on questions of social and epistemological space is not to prescribe one geographical scale or sense of space above another. It is to call for the geographical modification and qualification of *the* Enlightenment—as have others, of course, in speaking of an Atlantic Enlightenment or the Baltic Enlightenment[1]—to suggest that, in future, our studies should be more attentive than they have commonly been to questions of location, meaning, and translation as geographical questions, and so aware of the implications of the uncritical use in Enlightenment studies of terms like "core" and "margin." It is also to move away from conceptions of the Enlightenment's geographies as being about static maps of national political boundaries and toward a greater sense of mobility, of dynamism, of process. Rather than see a single Enlightenment moving as just a European phenomenon, it is to visualize maps of networks, lines of movement connecting points in space. The lines, of varying strength and duration depending upon the nature of the connections wrought in and over space—through regular correspondence, the long-distance exchange of artifacts, daily conversation in a café, the movement of texts in original and in translation, the passage of a navigator's vessel—connect people not just in different geographical locations but in different social settings, in other social spaces, where the reception and rearticulation of Enlightenment ideas gave them new meaning.

"Pluralizing Enlightenment" or "decomposing the Enlightenment" is to call for a typological widening of what the Enlightenment was, what it did, and why it mattered. Thinking geographically about the Enlightenment in the ways I have is consistent with such concerns. But thinking geographically is also to recognize the importance of geographical particularity. Notions of "the Enlightenment project" weaken in the face of evidence about geographical difference: of the different scales used on maps, of the use of different base meridians (in Paris, London, or Philadelphia) as start points for national depiction; of the different places used to propose notions of a supposedly ordered globe (Lapland, Quito, Paris, London, Berlin, Saint Petersburg), or of the simply faltering moves to accommodate geographical diversity—on the shore, on ships' decks, and on ruled lines on a map—and to represent it in secure maps of the world.

If these views regarding the geographical conceptualization of the Enlightenment have any wider purchase, two implications at least follow. The first has to with the possibility of using this conception of the Enlightenment as a heuristic in explanation of other Enlightenment themes. The figurative "maps" of sites of production and of reception signaled to here, the social spaces in which ideas were talked about, the practices by which the world was imagined and represented, and the textual strategies by which certain forms of knowledge moved have here been illustrated through geography, most clearly in part 3. Yet there is every reason to consider that this conceptual schema, of Enlightenment as a certain epistemological space, could be used in explanation of the making, movement, and reception of, say, natural history, practical chemistry, medical discourse, moral philosophy, religious moderatism, or political economy. Indeed, since many of the illustrations I have used have come from just such subject areas, it seems legitimate to suppose that further thinking about these and other discourses and their attendant practices and audiences in the ways identified here might produce a yet richer understanding of Enlightenment.

The second is the extent to which thinking geographically about the Enlightenment has the effect—or, to put it even more strongly, has the end in view—of displacing the privileged geographies of metropolitan Europe. I have here shown that, in terms of the geographical sources of that raw material that helped fashion Enlightenment theories about social development and human difference, the Americas represented the Enlightenment's first but partial "ethnographic laboratory" until the later 1760s, at which point the Pacific world became the Enlightenment's principal geographical test bed. Particular locations—geodetic measuring sites in Quito and Lapland, associational and computational sites in Saint Petersburg, Nuremburg, and Göttingen, a botanic garden in Uppsala, Fort Venus on Matavai in Hawaii where Cook observed the Transit of Venus from a tent—have had particular significance at one time or another. The geographical settings for Enlightenment views about the Science of Man may have varied, but in several senses, the "margin" was the "core," the movement of the "far away" back to Europe a key feature in how the world was geographically encountered and represented. But noting the centrality and the displacement of geographical information in this way is not the same as challenging the Eurocentrism, indeed the strongly northern European emphasis, of the Enlightenment's geography.

Perhaps in this respect the question should be "Whose Enlightenment was it anyway?" Just such a question has been asked in relation to the

Enlightenment in Spanish America. As Cañizares-Esguerra has noted, with particular reference to that wider conception of the Enlightenment favored as we have seen by Stephen Shapin, Dorinda Outram, Sanki Muthu, Laurent Dubois, Peter Hanns Reill, and others, the broadening of notions of "the public" beyond Britain, France, and Germany is welcome. But it is also "limiting and only partially right," since the questions and controversies of the world with which it worked were those that characterized northern Europe. "But the language and rules of the northern European discourses and controversies were not passively transferred to the New World, Spain, or the Papal States." The themes that were significant locally to the public in Spanish America had less to do with building new religious and political languages "than with constructing alternative, critical epistemologies," indeed with "explicitly attempting to develop a critique of Eurocentric epistemologies." And they centered not on theories about reason or in the mobility of texts and artifacts, but on the interpretations that could be placed upon Aztec monuments and their inscriptions, upon the dynamics of local Mexican scholarship in a colonial context, and upon the activities of the Creole intelligentsia who acted to ensure that the stones and other finds never traveled to Madrid. Read critically, the monumental stones established a longevity to Amerindian culture that disproved Buffon—the "New" World was older than "Old" Europe. This, together with local geographical investigations into Mexican flora and fauna that challenged Linnaean schemata, undermined efforts to "naturalize" a foreign, that is, European Enlightenment in native soil.[2]

This Spanish American example confirms the importance of "where" questions because, like those other "peripheral" cases of Greece and Spain and Portugal discussed here, it demands that we take seriously not just the ideas inherent in the Enlightenment and their variant social meanings but the place in which they took shape and in what form there, how they traveled (or did not), and how they were received elsewhere. Recognizing the importance of the Enlightenment's geographies challenges decisively conceptions of the Enlightenment either as a uniform thing, embodying a single set of principles, the everywhere-the-same view, or as only a philosophical and limited European thing, the just-in-Paris view.

The Enlightenment's Future Geographies

Enlightenment contemporaries modeled the "Natural History of Man" with reference to the "Geography of Man." As part 2 in particular has shown, in looking to explain human society in relation to physical determinants, notably climate, writers advanced theories of social progress and of universal and of conjectural history that sought to explain the present state of human culture by comparative examination of the human past. Whether such theories were progressive, in looking forward to a new "stage" of human development, or were retrospective, in seeking to explain the past in the present or to chart the decline of ancient cultures to their condition of present "decay," they offered different interpretative frameworks for considering the relationship between the human and the natural worlds. To some, the physical environment limited human society—Enlightenment "happiness" was geographically constrained. For others, the physical earth afforded no obstacles to human society—Enlightenment perfectibility was not geographically determined. Even when geographical conditions presented insurmountable barriers to human betterment, and despite political reform, human geographies could never overcome Nature's geographies.[3] Where, for example, Robert Wallace concluded in *The Various Prospects of Mankind, Nature, and Providence* (1761) that the physical geography of the earth would always restrict human endeavor, Condorcet's vision in his *Sketch for a Historical Picture of the Progress of the Human Mind* (1793) was of human society overcoming its physical limits: as he put it, "nature has fixed no limits to our hopes." Yet even for Condorcet, the world's future perfectibility would be measured by certain geographical standards:

> Will not every nation one day arrive at the state of civilisation attained by those people who are most enlightened, most free, most exempt from prejudices, as the French, for instance, and the Anglo-Americans? Will not the slavery of countries subjected to kings, the barbarity of African tribes, and the ignorance of savages gradually vanish? Is there upon the face of the globe a single spot the inhabitants of which are condemned by nature never to enjoy liberty, never to exercise their reason?[4]

Such speculative musing on future geographies, social condition, and physical environment formed part of a wider genre of Enlightenment utopian writing, of which geographical utopias were a popular part. A few such "robinsonades," so named because of the influence of Defoe's *Robinson Crusoe,* were set on islands, others in America.[5] The idea of utopian geographies also encompassed the reaction of *voyageurs naturalistes* to the paradisal geographies of the Pacific islands in particular. For Bougainville, reflecting on New Cythera (Tahiti), "one hopes for the sake of these people that Nature has refused to produce here the objects of Europe's cupidity; they need only the fruits which the land provides in abundance, and the rest, by drawing us Europeans here, would only bring to them the evils of the iron age. . . . I will always remember you with delight, and as long as I live I will celebrate the happy island of Cythera: it is the true Utopia."[6]

The imagined utopias of Bougainville, based as they were on real places, are different from the imaginative geographies of literary figures such as Restif de la Bretonne's 1781 *La découverte australe par une homme volant* (Southern explorations by a flying man), in which the Pacific world is revealed, island by utopic island, before the lead character arrives in the settlement of "Megapatagonia." Both types of work are different from Turgot's forward-looking "mappemonde politique" outlined as part of his theorizations on universal history and geography, and different still from the concerns of Volney in his *The Ruins, or a Survey of the Revolutions of Empires* (1791). There, in gazing upon the lost cities of Middle Eastern antiquity, the author reflects upon the rise and decline of different peoples and regions of the world, the nature of human society and the nature of the future.[7] None is as remarkable as Louis-Sebastian Mercier's vision of the future, *L'an 2440* (published in Amsterdam in 1770). Waking from sleep, the author encounters a world, a France, a Paris transformed—"horrible little houses" that had cluttered the capital have been cleared, the Académie Française (where once "donkeys munched on thistles") has become a temple to learning, where salons, open to the public, everywhere "raise the souls of the citizens."[8]

All this is to observe that geographical thinking—as part of historical methodology, in imaginative literature, in concerns to do with utility—featured strongly in the Enlightenment within discourses of the future, with what sort of world Enlightenment would or ought to produce, with what Condorcet called "the real improvement of Man."[9] For some, the question was "What would the Enlightenment become?" As others looked to past evidence in explaining their present, the Enlightenment sought answers to the question "Why are our geographies different?" People then were aware of what we may think of as the transformative power of geography. And this is one reason, of course, why the Enlightenment continues to fascinate. For those concerned with the Enlightenment's legacy, critical reason has not improved the human condition everywhere and for everyone. The map of the "geography of hope" remains to be completed on a world scale.

The Enlightenment—Our Geographical Contemporary

In several ways, the Enlightenment is with us still. It continues as an object of modern historical and geographical inquiry because of its political implications and for the ideas and languages it employed. Despite the fact that its meaning was being debated even as it was being advanced, that its thinkers varied in what they said and, as I have shown here, in where they said it, why, and about which parts of the world, many modern scholars concerned to consider "what's left of it," often with reference to "the Enlightenment project," have found fault with the Enlightenment for failing to deliver. The Enlightenment is our contemporary, cautions Outram, because it has been both intensively studied and readily seen by some as the basis of our own period.[10]

Considering here the Enlightenment's "where," less its "what," and hardly at all its "so what" has disclosed a variety and a dimension to the Enlightenment that, for me, denies the views of those we might term guardians and the views of those we might term critics. By the first I mean those whose view of the Enlightenment is of an intellectual movement in a defined historical period. This is a relatively easy claim to defend, for as is clear from my remarks in the introduction, this book sits within a longer critical tradition in Enlightenment studies, one that has reshaped the Enlightenment from French intellectual elitism to the social history of ideas to the historical geography of ideas and practices. By the second term I mean those whose view of the Enlightenment is of a failed connection with ourselves, with modernity, and with its inequalities. The Enlightenment is with us still, they claim—but only because it has failed. This is a less easy critical position to oppose, not least because, in several ways, we moderns continue to locate ourselves in the Enlightenment.

That is to say, we still commonly represent, and even reenact, the Enlightenment. Consider, for example, the recent retracing by Robert Whitaker of the expeditionary route of La Condamine, Godin, and others on the South American geodetic expedition of the 1730s. Evaluating the Enlightenment's geographies means re-creating them—experience and direct observation, then as now, being the warrant of the truthful.[11] Consider too the much larger re-created geographical enterprise that is the modern-day *Endeavour,* a replica of Cook's first expeditionary ship, conceived in 1987 and launched in December 1993, and the Pacific voyaging in 1976 of *Hokule'a,* a Hawaiian double canoe similar to those pictured by Bougainville two hundred years before (see fig. 19). Both are re-created Enlightenment vessels of a geographically complex modernity. The first helps commemorate, for Europeans and white Australians anyway, the significance and, to an extent, the fragility of the Enlightenment's geographical encounters. The second has established a historical legitimacy for Pacific peoples, has borne witness to a geographical capacity denied by the Enlightenment's narratives.[12]

Elsewhere, Enlightenment is displayed, not performed. In London's British Museum, visitors to a new permanent gallery, Enlightenment: Discovering the World in the Eighteenth Century, which opened in 2004, can see Tahitian ancestor figures, ancient Indian writings, scientific instruments, plant specimens from across the world, and much else besides without the need to circumnavigate the world's oceans or push through jungles in order to plot lines of triangulation or to take instrumental readings to compute the shape of the earth. Perhaps, then, our question should be "Where does the Enlightenment end?" for in different places and social spaces, in different ways and with different meanings, at sea, and in the differing interpretations we place upon its artifacts and legacy, the Enlightenment is with us still.

Notes

Chapter One

1. Contemporaries' views of Enlightenment, including Kant's and others' response to Mendelssohn, are discussed in Schmidt, *What Is Enlightenment?* Schmidt's introductory essay to that volume (1–44) summarizes how the Enlightenment has been dealt with up to the mid-1990s. A useful essay emphasizing the Enlightenment as our contemporary is Outram, "Enlightenment Our Contemporary," 32–42.

2. The commentator in 1789 was Karl Friedrich Bahrdt, as quoted in Umbach, *Federalism and Enlightenment in Germany,* 25. A more recent review of how the Enlightenment has been considered by modern commentators is Hunt, with Jacob, "Enlightenment Studies," in Kors, *Encyclopedia of the Enlightenment,* 1:418–30. This *Encyclopedia* is the most up-to-date guide to the Enlightenment as a whole. Readers interested in tracing the evolving historiography of Enlightenment studies might usefully chart the differences in content and organizing structure between this work and other Enlightenment encyclopedias: for example, Reill, *Encyclopedia of the Enlightenment;* and Delon, *Encyclopedia of the Enlightenment.* Other general guides include Hyland, Gomez, and Greensides, *The Enlightenment;* Jacob, *The Enlightenment;* and Porter, *The Enlightenment* and *Enlightenment: Britain and the Creation of the Modern World.* It is noteworthy in terms of my argument here that part of the content of the 2003 *Encyclopedia* edited by Kors is organized around the themes of political geography of the Enlightenment and agencies and spaces of the Enlightenment.

3. Modern interpretations of the Enlightenment as a definitive period marked by the philosophical enquiries of notable thinkers begin with Cassirer, *Philosophy of the Enlightenment* (published first in German in 1932). English (1951) and French (1966) translations stimulated further discussion about *the* Enlightenment. Other accounts of the Enlightenment as a uniform intellectual endeavor include Berlin, *Age of Enlightenment;* Crocker, *Age of Enlightenment;* Gay, *The Enlightenment;* Hampson, *The Enlightenment;* Smith, *The Enlightenment.* The idea that the Enlightenment began earlier than the eighteenth century and was broader in its range than these studies suggest is first associated with Hazard, *Crise de la conscience européenne.* Hazard's view is echoed in Israel, *Radical Enlightenment.*

4. It is noteworthy in this respect that Darnton has recently emphasized this view in arguing against what he sees as the "multiplication" of Enlightenments (a trend to which,

in other work, Darnton was a leading contributor): Darnton, "George Washington's False Teeth," 34–38; Darnton, *George Washington's False Teeth,* 3–24.

5. On interpretations in the 1970s of the Enlightenment as complex and multiple, see Darnton, "In Search of the Enlightenment"; and Bender, "New History of the Enlightenment?" The Enlightenment as eighteenth-century social history is stressed, for Britain, by Roy Porter in *The Enlightenment,* and *Enlightenment: Britain and the Creation of the Modern World.* For one view of the contradictions offered by Darnton's views of the Enlightenment, see Popkin, "Robert Darnton's Alternative (to the) Enlightenment." For one view that emphasizes Porter's influential role in "socializing" the Enlightenment, see Schaffer, "Enlightenment Brought down to Earth."

6. The quote is from Shapin, "Social Uses of Science," 111–12; on the Enlightenment and book history, see Darnton, *Business of the Enlightenment.* On the Enlightenment and women, see Tomaselli, "Enlightenment Debate on Women"; and Hunt, *Women and the Enlightenment.* The following provide good introductions on medicine: Cunningham and French, *Medical Enlightenment of the Eighteenth Century;* and Porter, *Medicine in the Enlightenment.* The question of sexual attitudes is covered in Rousseau, *Perilous Enlightenment;* and in Rousseau and Porter, *Sexual Underworlds of the Enlightenment.* On race, Eze, *Race and the Enlightenment* is valuable. On religion, Vila, *Enlightenment and Religion,* and Young, *Religion and Enlightenment in Eighteenth-Century England* offer good guides to the literature. Works emphasizing the Enlightenment's diversity include Outram, *Enlightenment;* and Hulme and Jordanova, *Enlightenment and Its Shadows.* The view that we should talk of "enlightenments" rather than "the Enlightenment" is articulated by Pocock in *Barbarism and Religion.* For the view that, despite this recognition of plurality, Pocock's Enlightenment is too centered on England and religion, see Emerson, "How Not to Deal with Enlightenments." The point about pluralizing "the Enlightenment" for reasons of historical accuracy is also made in Muthu, *Enlightenment against Empire,* 264. Some commentators insist on a particular character for the Enlightenment. Crocker promotes it as a political affair, in "Interpreting the Enlightenment" and "Enlightenment: What and Who?" Williams, *Enlightenment,* incorporates contemporary and modern views on the politics of the Enlightenment. Im Hof, *Enlightenment,* associates the Enlightenment with a specific period but stresses its diversity. For the view that the Enlightenment's definitional diversity incorporates "the rise of individual autonomy over traditional community, the rise of secularizing reason over inherited authority, the disengagement of nature from a supernatural worldview, the rise of methodical and institutionalized criticism, the rise of science as both technique and worldview, the rise of historical consciousness and the practice of historical method, and the establishment of bourgeois institutions and a democratic ethos in public life," see Shea and Huff, introduction to *Knowledge and Belief in America,* 1–14, quote at 8. There is even a cartoon version of the Enlightenment: Spencer and Krauze, *Enlightenment for Beginners.*

7. Michel Foucault has been one of the most attentive modern critics of the Enlightenment, not least in emphasizing its incompleteness as a historical and contemporary political project: see Foucault, "What Is Enlightenment?" and his "Kant on Enlightenment and Revolution." For the view that Foucault's work on the late eighteenth-century natural and human sciences—*The Order of Things: An Archaeology of the Human Sciences*—misrepresents Enlightenment inquiry, see Rousseau, "Whose Enlightenment? Not Man's: The Case of Michel Foucault." Horkheimer and Adorno's *Dialectic of Enlightenment* (written in Los Angeles in 1944 and published first in German in 1947 and in English in 1972) is perhaps the best-known

work on the Enlightenment's "dark side." For a commentary on it, see Sherratt, "Adorno and Horkheimer's Concept of 'Enlightenment'"; and Outram, *Enlightenment*, 8–10.

8. The relationship between postmodernism and the Enlightenment is notable given the paradox that postmodern critics are evaluating the very period and ideas they have, by definition, distanced themselves from. On this point, see Todorov, *Deflection of the Enlightenment;* Racevskis, *Postmodernism and the Search for Enlightenment;* Harpham, "So . . . What *Is* Enlightenment?"; Gray, *Enlightenment's Wake;* and Cascardi, *Consequences of Enlightenment.* Edited collections on this theme are Liedman, *Postmodern Critique of the Project of Enlightenment;* Geras and Wokler, *Enlightenment and Modernity;* and Baker and Reill, *What's Left of Enlightenment?*

9. The Enlightenment in terms of its national expression is the subject of Porter and Teich, *Enlightenment in National Context.* Livingstone and Withers offer a discussion of this work as a "way marker" in terms of geographical studies of the Enlightenment; see their *Geography and Enlightenment,* 1–28. For studies of America, other countries outside the "Enlightenment hearth," and for recent work on the Enlightenment within it, the following are illustrative: Robertson and Timms, *Austrian Enlightenment and Its Aftermath;* Brewer, *Discourse of Enlightenment in Eighteenth-Century France;* Goodman, *Republic of Letters;* Faull, *Anthropology and the German Enlightenment;* Gargett and Sheridan, *Ireland and the French Enlightenment;* Schmidt, *Hearing Things;* Ferguson, *American Enlightenment;* Behrens, *Society, Government and the Enlightenment;* Bartlett and Hartley, *Russia in the Age of the Enlightenment;* Klein and La Vopa, *Enthusiasm and the Enlightenment in Europe;* Ferrone, *Intellectual Roots of the Italian Enlightenment.*

10. On the Enlightenment in Europe's periphery, the essays in Gavroglou, *Sciences in the European Periphery during the Enlightenment,* cover Portugal, Spain, Italy, Scandinavia, and the Greek-speaking regions (and see chap. 2 here). For one view that the attempts to discuss the connections between geography and Enlightenment and to "de-center" Europe in the essays in Livingstone and Withers, *Geography and Enlightenment,* were only partially successful given their failure to look at conceptions of "Enlightenment" elsewhere, see Raj, "Bounded Landscapes."

11. The point about the Enlightenment in Scotland as a label for a relatively recent line of inquiry I take from Wood, introduction to *Scottish Enlightenment,* 1. On the question of social differences within Glasgow and geographical differences between that town and others, see Emerson and Wood, "Science and Enlightenment in Glasgow." The example of the international nature of the Enlightenment in political economy is from Robertson, "Enlightenment above National Context." For a recent review that stresses the diversity of the Scottish Enlightenment, see the essays in Broadie, *Cambridge Companion.* For a local study outside Aberdeen, Edinburgh, and Glasgow, see Allan, "Scottish Enlightenment and the Politics of Provincial Culture."

12. Hulme and Jordanova, introduction to *Enlightenment and Its Shadows,* 7. The Enlightenment's engagement with the "New Worlds" of the Pacific and American worlds is the subject of attention in Outram, *Enlightenment;* and in Rousseau and Porter, *Exoticism and the Enlightenment.* "West–Rest" geographies of contact are the subject of MacLeod and Rehbock, *Nature in Its Greatest Extent;* Pagden, *European Encounters with the New World;* Clarke, *Oriental Enlightenment;* and Ching and Oxtoby, *Discovering China* (and see also chaps. 5 and 6 here).

13. The question of disciplinary "origins" in the Enlightenment is one element of Foucault's *The Order of Things.* On the "human" or "social" sciences, both Saiedi, *Birth of Social Theory,* and the essays in Fox, Porter, and Wokler, *Inventing Human Science,* are useful.

14. On geography's "birth" through these practices in the Pacific, see Stoddart, *On Geography,* 28–40.

15. The phrase "variegated geography" is from Clark, Golinski, and Schaffer, introduction to *Sciences in Enlightened Europe,* 20.

16. The quote is from Golinski, "Science *in* the Enlightenment," 419. The idea that "dot" and "thread" maps might plot Enlightenment sites and connections is indebted to Harris, "Long-Distance Corporations, Big Sciences, and the Geography of Knowledge." For an introduction to what science in the Enlightenment was, see Heilbron, "Natural Philosophy and Science"; Hankins, *Science and the Enlightenment;* and Golinski, "Science *in* the Enlightenment." Useful edited collections on science in the eighteenth century are Frängsmyr, Heilbron, and Rider, *Quantifying Spirit in the Eighteenth Century;* Rousseau and Porter, *Ferment of Knowledge;* and Clark, Golinski, and Schaffer, *Sciences in Enlightened Europe.* The fullest treatment of eighteenth-century/Enlightenment science is provided by the essays in Porter, *Cambridge History of Science.*

17. For reviews of the "geographies" of science, see Smith and Agar, *Making Space for Science;* Shapin, "Placing the View from Nowhere"; Shapin, *Social History of Truth;* Withers, *Geography, Science and National Identity,* 1–29; and, in introducing a theme set of papers on the issue, Naylor, "Introduction: Historical Geographies of Science." Other discussions of the movement of scientific ideas include Montgomery, *Science in Translation;* the essays in Bourguet, Licoppe, and Sibum, *Instruments, Travel and Science;* and those in Simões, Carneiro, and Diogo, *Travels of Learning.* These and other works cited here are very different from the uncritical reductionism favored by Dorn, *Geography of Science,* and Nisbett, *Geography of Thought.*

18. Livingstone, *Putting Science in Its Place,* 12, 164.

19. Said's discussion of "traveling theory" forms chapter 10 of his *The World, the Text and the Critic.* Quotations are from 226, 241–42.

20. Miller, "Joseph Banks, Empire, and 'Centers of Calculation' in Late Hanoverian London," 25. Latour's notions are spelled out in his *Science in Action,* 215–57.

21. Bourguet, Licoppe, and Sibum, introduction to *Instruments, Travel and Science.*

22. Harris, "Long-Distance Corporations, Big Sciences, and the Geography of Knowledge," 272–73.

23. Livingstone, *Geographical Tradition,* 102–38, provides a summary of geographical practices in the Enlightenment. On the political languages of Enlightenment geography in Britain (mainly for England), see Mayhew, *Enlightenment Geography.* For a Scottish perspective, see Withers, *Geography, Science and National Identity.* On France, see Broc, *Géographie des philosophes;* and Godlewska, *Geography Unbound.*

24. On the *Encyclopédie* as a map of knowledge and of the relative state of the peoples of the world as "the Great Map of Mankind," see Yeo, *Encyclopedia Visions,* and his "Classifying the Sciences"; Marshall and Williams, *Great Map of Mankind;* and Withers, "Geography in Its Time."

25. Voltaire's cartography of reason is noted in Chartier, "Man of Letters," 163–64. Moses Mendelssohn's view about culture and the Enlightenment is quoted in Schmidt, *What Is Enlightenment?* 54.

26. Dubois, "Enslaved Enlightenment."

27. Quoted in Beaglehole, *James Cook's Journals,* 2:175.

28. Darnton, *Business of the Enlightenment.*

29. Thomson, "North Africa and the Levant"; Maggs, "Asia."

30. Secord, "Knowledge in Transit," 668.

31. Brewer, "Spaces of Enlightenment," quotations at 173, 182. Similar uses of the term "space" inform Etlin's discussion of French Enlightenment architecture; for example, he uses notions of "the space of magnificence" (in reference to built form), "the space of hygiene," as an epistemological referent, and "Revolutionary Space" and "space of liberty" as social, taxonomic, and architectural descriptors; Etlin, *Symbolic Space,* passim.

32. On these general matters, see Heilbron, "Natural Philosophy and Science"; Porter, introduction to *Cambridge History of Science,* vol. 4; and Reill, "Legacy of the 'Scientific Revolution.'" On Europe's first geographical society beginning in Venice in 1680, see Cosgrove, "Global Illumination and Enlightenment in the Geographies of Vincenzo Coronelli and Athanasius Kircher," 37.

33. On this point, see Colten and DeLyser, "Louisiana Purchase Territory"; and Allen, "Thomas Jefferson and the Mountain of Salt."

34. Godlewska, "From Enlightenment Vision to Modern Science."

Chapter Two

1. Gay, *Enlightenment,* 1:3. The other quotes are from, respectively, Marsak, *Enlightenment,* 3; Crocker, introduction to *Blackwell Companion to the Enlightenment,* 1. On Paris as the Enlightenment's capital, Roche, *France in the Enlightenment,* 641. The Darnton quote is from his *George Washington's False Teeth,* 6.

2. The phrase "complex revisionisms" is from Porter, *Enlightenment: Britain and the Creation of the Modern World,* 3; it is from his typically pithy discussion there of Enlightenment historiography (1–23) that I cite Voltaire and Shaftesbury. For summaries of the revisionist interpretation of England in the Enlightenment, see Porter, "Enlightenment in England"; and Porter, "England."

3. This quote and much of the paragraph (with its original emphasis) is from Pocock, "Enlightenment and Revolution," 252.

4. The American example is May, *Enlightenment in America,* xvi–xvii. The Italian evidence is discussed in Chadwick, "Italian Enlightenment," quote at 90; and in Venturi, *Italy and the Enlightenment.* On the third variant, see, for example, Roche, *France in the Enlightenment;* and Bartlett and Hartley, *Russia in the Age of the Enlightenment.* My remarks on the two principal variants here are adopted from Withrington, "What Was Distinctive about the Scottish Enlightenment?" 9–10.

5. On music, Till, *Mozart and the Enlightenment;* Verba, *Music and the French Enlightenment;* Fubini, *Music and Culture in Eighteenth-Century Europe;* and Christensen, "Music." On social theory, Griswold, *Adam Smith and the Virtues of Enlightenment;* Tribe, "Economic Thought"; and Osborne, *Aspects of Enlightenment.* On the body, Haidt, *Embodying Enlightenment;* Stafford, *Body Criticism;* and Vila, *Enlightenment and Pathology.*

6. Smoliy, "Enlightenment and State-Formation in Eighteenth-Century Ukraine"; Shevchenko, "Contribution of Ukrainian Thinkers to the Political Culture of the Enlightenment"; Bajkó, "Ideas of Enlightenment in the Colleges of Hungary and of Transylvania"; Duțu, "National and European Consciousness in the Romanian Enlightenment." On the use made of Western ideas in reforming Hungary in the Enlightenment, see Pajkossy, "Western European Models and National Traditions." Regional differences within Hungary are

discussed in Kovács, "Hungary." On the role of national consciousness helping form the idea of Russia and the idea of Enlightenment in Russia, see Serman, "Russian National Consciousness and Its Development in the Eighteenth Century."

7. Jones, *Peasantry in the French Revolution,* 22–23; Ozouf, *Festivals and the French Revolution,* 126. On the ideas behind "Mastery over Space," see Roche, *France in the Enlightenment,* 41–74.

8. Withers, *Geography, Science and National Identity,* 25–29, 112–57.

9. The literature on the meaning of "nation," "national identity," and "nationalism" in relation to eighteenth-century Europe is huge. For guides, see Anderson, *Imagined Communities;* Gellner, *Nations and Nationalism;* Hobsbawm, *Nations and Nationalism;* Tilly, *Formation of National States in Western Europe;* and Smith, *Theories of Nationalism.* My remarks on the "civilizational geopolitics" of Enlightenment Europe and the changing geographies of France and of Europe are taken from Heffernan, "Changing Political Map," and his *Meaning of Europe,* 23–41. See also Paul, "Europe the Way It Was and Is"; Wangermann, "Conditions of National Consciousness in the Epoch of Enlightenment." On the place of eastern Europe in the Enlightenment, see Wolff, *Inventing Eastern Europe.*

10. One extreme expression of such thinking is Camic, who in his *Experience and Enlightenment* considered only five intellectuals to "comprise the Enlightenment" in Scotland: David Hume, Adam Ferguson, William Robertson, Adam Smith, and John Millar.

11. Porter and Teich, *Enlightenment in National Context,* vii.

12. Chadwick, "Italian Enlightenment."

13. Wangermann, "Reform Catholicism and Political Radicalism in the Austrian Enlightenment"; Frängsmyr, "Enlightenment in Sweden." The phrase "the disappearing tailcoat of a learned marquis" is from Withrington, "What Was Distinctive about the Scottish Enlightenment?" 10.

14. Porter, "Enlightenment in England," quotes at 4, 5, and 3, respectively.

15. Phillipson, "Scottish Enlightenment," 28.

16. Hampson, "Enlightenment in France," 43.

17. Schama, "Enlightenment in the Netherlands," 55.

18. Taylor, "Enlightenment in Switzerland," 88.

19. Whaley, "Protestant Enlightenment in Germany"; Blanning, "Enlightenment in Catholic Germany"; Teich, "Bohemia."

20. Dukes, "Russian Enlightenment," 187.

21. Pole, "Enlightenment and the Politics of American Nature." The idea of *the* (unified) Enlightenment in America was epitomized in Commager, *Empire of Reason,* from which the quote is taken (xi). The more complex picture was epitomized in May, *Enlightenment in America.*

22. May, *Enlightenment in America,* 133–49, 181–96.

23. In no order of importance and only as a selection, these would include, for America, Koch, "Contest of Democracy and Aristocracy in the American Enlightenment"; Koch, "Aftermath of the American Enlightenment"; Nybakken, "Enlightenment and Calvinism"; McDermott, "Enlightenment and the Mississippi Frontier"; Tucker, "Beyond Reason and Revelation"; for Latin America, Whitaker, *Latin America and the Enlightenment;* Weinberg, "Enlightenment and Some Aspects of Culture and Higher Education in Spanish America"; for Italy, Venturi, *Italy and the Enlightenment;* of England, Voitle, "Reason of the English Enlightenment"; for Sweden, Frängsmyr, "Swedish Science in the Eighteenth Century"; and for further illustration, Hall, "Development of Enlightenment Interest in Eighteenth-Century

Corsica"; Thomas, "Enlightenment and Wales in the Eighteenth Century"; Carrato, "Enlightenment in Portugal and the Educational Reforms of the Marquis de Pombal"; and Okamoto, "Enlightenment in Japan."

24. I refer here to Venturi, whose final chapter of *Utopia and Reform in the Enlightenment* is titled "The Chronology and Geography of the Enlightenment"; quotes at 126, 136.

25. In a huge literature and in addition to what is cited elsewhere here, see the following: on Austria, Bodi, "Austrian Enlightenment"; Cole, "Nation, Anti-Enlightenment, and Religious Revival in Austria"; and Robertson and Timms, *Austrian Enlightenment and Its Aftermath;* on America, Ferguson, *American Enlightenment;* and Jaffee, "Village Enlightenment in New England"; on Italy, Ferrone, *Intellectual Roots of the Italian Enlightenment;* on China and West–East relationships, Clarke, *Oriental Enlightenment;* Demel, "China's Changing Image during the Age of Enlightenment"; and Maggs, "China"; on Russia, Marassinova, "Russian Enlightenment and the Educated Nobility"; Mikeshin, "Enlightened Russians in the Eighteenth Century and Today"; Moracci, "Influence on the Russian Enlightenment of the Cultural Policy of Catherine II"; Schlafly, "Russian Travellers Discover Western Europe in the Age of the Enlightenment"; and De Madariaga, "Russia"; on Germany, Lowood, *Patriotism, Profit, and the Promotion of Science in the German Enlightenment;* Umbach, *Federation and Enlightenment in Germany;* on Portugal, Domingues, "Science and Nationalism"; Pereira, "Portuguese Enlightenment"; Simões, Carneiro, and Diogo, "Constructing Knowledge"; and Storrs, "Portugal"; on Spain, Goodman, "Science and the Clergy in the Spanish Enlightenment"; Pagden, "Reception of the 'New Philosophy' in Eighteenth-Century Spain"; and Storrs, "Spain"; on the Netherlands, Jacob and Mijnhardt, *Dutch Republic in the Eighteenth Century;* and Buisman, "Some Considerations on the Social Diffusion of Enlightenment Ideas in the Netherlands"; and on Greece, Dialetis, Gavroglu, and Patiniotis, "Sciences in the Greek-Speaking Regions"; Tabaki, "Greece"; and Vlahakis, "Greek Enlightenment in Science"; on England, Porter, "England"; on France, Ravel, "France"; on Scandinavia, Christensson, "Scandinavia"; on Hungary, Kovács, "Hungary."

26. Consider and compare, for example, the accounts of Venturi, *Italy and the Enlightenment;* Ferrone, *Intellectual Roots of the Italian Enlightenment;* Findlen, "Science as a Career in Enlightenment Italy"; Findlen, "Forgotten Newtonian"; and Cerruti, "Dante's Bones." Cerruti is particularly sensitive to the geographical differences informing the idea of "Italy" in his attention to the history and geography of Italian science.

27. Wood, *Scottish Enlightenment;* Withers, "Toward a Historical Geography of the Enlightenment in Scotland"; Withers and Wood, *Science and Medicine in the Scottish Enlightenment.*

28. Herman, *Scottish Enlightenment.* In this context, in a work of roughly similar date and summary intention—Enlightenment and the birth of modernity—Scotland is "cavalierly spliced" into Britain by Porter in *Enlightenment: Britain and the Creation of the Modern World,* thus denying both the premise on which Herman's account depends and Porter's own earlier declarations concerning the importance of geographical (national) difference.

29. McDermott, "Enlightenment on the Mississippi Frontier"; and Jaffee, "Village Enlightenment in New England."

30. I take these points from Kitromilides, *Enlightenment as Social Criticism,* passim, but esp. 185–91; and Kitromilides, "Europe and the Dilemmas of Greek Conscience," 3–5.

31. Tabaki, "Greece"; Vlahakis, "Greek Enlightenment in Science"; and Dialetis, Gavroglu, and Patiniotis, "Sciences in the Greek Speaking Regions," from which the argument about "appropriation" is taken (42); Kitromilides, *Enlightenment as Social Criticism.*

32. On Portugal, Pombal, and the Enlightenment, see Storrs, "Portugal"; Carrato, "Enlightenment in Portugal and the Educational Reforms of the Marquis of Pombal"; Pereira, "Portuguese Enlightenment"; Domingues, "Science and Nationalism"; Maxwell, *Pombal,* esp. 159–62; and Simões, Carneiro, and Diogo, "Constructing Knowledge." Storr offers less of an "antidote" view of Portugal's Enlightenment than does Maxwell. On the *estrangeirados* and the idea of communication networks in Enlightenment Portugal, see Carneiro, Simões, and Diogo, "Enlightenment Science in Portugal."

33. Goodman, "Science and the Clergy in the Spanish Enlightenment"; Pagden, "Reception of the 'New Philosophy' in Eighteenth-Century Spain"; Storrs, "Spain."

34. On Enlightenment in Latin America, see Goodman, "Science and the Clergy in the Spanish Enlightenment," 131–35; Weinberg, "Enlightenment and Some Aspects of Culture and Higher Education in Spanish America"; the essays in Whitaker, *Latin America and the Enlightenment;* and Pimentel, "Iberian Vision." The idea of a romance Enlightenment is briefly developed by Ricuperati, "Enlightenment in the Romance Countries." The quote on Brazil is from Marchant, "Aspects of the Enlightenment in Brazil," 115.

35. Cañizares-Esguerra, *How to Write the History of the New World,* 267. See also Lanning, "Reception of the Enlightenment in Latin America."

36. As quoted in Gould, "Lisbon 1755," 406.

37. Clarke, *Oriental Enlightenment,* 37–53.

38. On this constitutive idea of scale, see McMaster and Sheppard, introduction to *Scale and Geographic Inquiry.*

39. See, for example, Pyenson, "An End to National Science"; and Livingstone, *Putting Science in Its Place,* 13–14.

Chapter Three

1. Pyenson, "An End to National Science"; Robertson, "Enlightenment above National Context" and, at greater length, *Case for the Enlightenment.*

2. Venturi, *Italy and the Enlightenment;* and see also Robertson, "Franco Venturi's Enlightenment."

3. Pyenson, "An End to National Science," 273.

4. Goodman, *Republic of Letters,* passim.

5. On this issue (to which I return in chap. 5), see Shapin, *Social History of Truth,* passim; Shapin, "Rarely Pure and Never Simple"; Livingstone, *Putting Science in Its Place,* 140–53. The point on the "strength" of one's links in networks of correspondence is adapted from Lux and Cook, "Closed Circles or Open Networks?"

6. There is a large literature on such questions, prompted mainly by poststructural writings in literary theory, notably in Germany. As a summary guide, see Holub, *Reception Theory,* passim.

7. Whelan, "Republic of Letters," 437.

8. Daston, "Ideal and Reality of the Republic of Letters," 368, 369. The observation about the Republic of Letters as not a "free country" is from Goldgar, *Impolite Learning,* 150.

9. Gascoigne, *Science in the Service of Empire,* 147–65; Gascoigne, "Joseph Banks, Mapping and the Geographies of Natural Knowledge," 156–59.

10. Popkin, "Periodical Publication and the Nature of Knowledge in Eighteenth-Century Europe."

11. Daston, "Ideal and Reality of the Republic of Letters," 369.

12. These remarks, and the quote from Voltaire's entry "Gens de lettres" in the *Encyclopédie,* are based on Chartier, "Man of Letters," 145.

13. Chartier, "Man of Letters," 147–51.

14. Shapin, "Image of the Man of Science"; Schiebinger, "Philosopher's Beard."

15. Schiebinger, "Philosopher's Beard." These remarks on epistolary formulae and letter writing as a social practice are taken from Cook, *Epistolary Bodies;* Barton and Hall, *Letter Writing as a Social Practice;* Earle, *Epistolary Selves;* Gilroy and Verhoeven, *Epistolary Histories.* On women, science, and correspondence in the Enlightenment, see Fara, *Pandora's Breeches.*

16. Bazerman, "Letters and the Social Grounding of Differentiated Genres."

17. That is why there is value—and further work to be done—in prosopography as a research technique in the Enlightenment. On this, see Allen, "*Arcana ex Multitudine*"; Gascoigne, "Eighteenth-Century Scientific Community"; Clark, "Pursuit of the Prosopography of Science."

18. Boschung, "Göttingen, Hanover, and Europe." For a full examination of Haller's world of correspondence, see Boschung et al., *Albrecht von Haller,* passim.

19. Spary, *Utopia's Garden,* 61–78; Williams, *French Botany in the Enlightenment.*

20. I take this term from Goodman, *Republic of Letters,* esp. from chap. 4, "Into Writing: Epistolary Commerce in the Republic of Letters."

21. This example and the quotes concerning Volta's different maps are from Pancaldi, *Volta,* 174–75.

22. Widmalm, "Professor Celsius and Don Andrea."

23. Darnton, "What Is the History of Books?" 65.

24. Four works in particular are usually understood to have shaped the field. These are, in order of publication: Febvre and Martin, *L'apparition du livre* (translated into English in 1976 as *The Coming of the Book*); Davis, *Culture and Society in Early Modern France;* Darnton, *Business of the Enlightenment;* and Eisenstein, *Printing Press as an Agent of Change.* See also Johns, *Nature of the Book.*

25. Darnton, "What Is the History of Books?" This essay has been much reprinted. For one accessible discussion that places it in relation to later work in book history, see Finkelstein and McCleery, *Book History Reader,* 7–26.

26. Darnton, "George Washington's False Teeth," 34. On this point and the paradox of Darnton's treatment of the Enlightenment through its book history as a reading of the Enlightenment's "Other," see Popkin, "Robert Darnton's Alternative (to the) Enlightenment."

27. For example, Cavallo and Chartier, *History of Reading in the West;* Davidson, "Introduction"; Darnton, "History of Reading."

28. On this, see Rose, "History of Books"; and the other essays in Mason, *Darnton Debate.*

29. Febvre and Martin, *Coming of the Book,* 167–215. The view of this work as the bible of book history is taken from Davidson, "Introduction," 8.

30. For an example of the different geographies of reviewing Humboldt in the Enlightenment, see Rupke, "Geography of Enlightenment"; and Rupke, *Alexander von Humboldt.*

31. Secord, *Victorian Sensation,* pt. 2. For the point on the localist studies of reading in London, Liverpool, "the imaginary town of Oxbridge," and Edinburgh, and the methodological implications of Secord's work, see MacPherson, "Essay Review." For a work that draws

on Secord's approach, see Livingstone, "Science, Religion and the Geography of Reading." Secord's interest in "the study of communicative practices" more generally is discussed in his "Knowledge in Transit."

32. On Smith (and other political economists) in Lisbon, for example, see Cardoso, "Economic Thought in Late Eighteenth-Century Portugal."

33. These remarks are from Oz-Salzberger, "Translation."

34. Tribe, *Governing Economy,* 133–48.

35. Oz-Salzberger, *Translating the Enlightenment,* 60–62, passim. For other examples of the role of translation of Scottish Enlightenment figures in the German Enlightenment, see the introductions to the individual volumes edited by Klemme, *Reception of the Scottish Enlightenment in Germany.*

36. Kontler, "William Robertson's History of Manners in German"; Kontler, "William Robertson and His German Audience on European and Non-European Civilisations"; Renwick, "Reception of William Robertson's Historical Writings in Eighteenth-Century France." For a short commentary in similar vein on the reception of the British Enlightenment in central Europe, see Kontler, "Enlightenment, *Aufklärung* and Conservatism."

37. Maggs, "Voltaire and the Balkans"; see also Forrest, "Problem of Influence."

38. Quoted in Maggs, "Voltaire and the Balkans," 96.

39. Gargett, "Voltaire's Reception in Ireland"; Kennedy, "Readership in French"; Sheridan, "Irish Literary Review Magazines and Enlightenment France."

40. Ó Ciosáin, "Attitudes towards Ireland and the Irish in Enlightenment France"; Gargett, "Voltaire's View of the Irish."

41. For introductions, see Cunningham and French, *Medical Enlightenment of the Eighteenth Century;* Porter, "Medical Science and Human Science in the Enlightenment"; Porter, *Medicine in the Enlightenment;* Porter, *Greatest Benefit to Mankind,* 245–303; Porter, *Flesh in the Age of Reason.*

42. On examples of the different geographies of medical provision, see Cunningham, "Medicine to Calm the Mind"—who notes that "the founding of the famous Edinburgh medical school can justifiably be looked on as the first public act of the Scottish Enlightenment" (57); Geyer-Kordesch, "Georg Ernst Stahl's Radical Pietist Medicine and Its Influence on the German Enlightenment"; French, "Sickness and the Soul"; Martin, "Sauvage's Nosology"; Brock, "North America, a Western Outpost of European Medicine." On the background of Edinburgh's medical students as one further illustration of this point, see Rosner, *Medical Education in the Age of Improvement.*

43. The phrase is from Wilson, "Medicine," 3:47.

44. Perdiguero, "Popularization of Medicine during the Spanish Enlightenment"; Szlatky, "Tissot as Part of the Medical Enlightenment in Hungary"; Porter, "Spreading Medical Enlightenment"; Brock, "North America, a Western Outpost of European Medicine," 201–2, 215.

45. The terms "untransferability" and "misreception" I take from Oz-Salzberger, *Translating the Enlightenment.* I take "appropriation" in this sense from France, "Lost for Words." On these questions, see also Holub, *Reception Theory;* and Thompson, "Reception Theory and the Interpretation of Historical Meaning."

46. This last example—of the tree of Cracow, a chestnut tree that stood in the heart of Paris in the gardens of the Palais-Royal and acted as the focal point for certain news networks in mid-eighteenth-century Paris—is discussed in Darnton, "Early Information Society."

47. Vlahakis, "Note on the Penetration of Newtonian Physics in Greece."

48. Langer, "German Translations of 18th Century Italian Geoscience Writing."

49. As quoted in McDougall, "Charles Elliot's Medical Publications and the International Book Trade," 216. For a discussion of the book trade as a central feature of the Enlightenment in Scotland, see Sher, "Science and Medicine in the Scottish Enlightenment."

50. Hayward, "Emmanuel Mendes da Costa"; Withers, "Walker and the Practice of Natural History in Late Eighteenth-Century Scotland."

51. Miller, "Joseph Banks, Empire, and 'Centers of Calculation,'"; Mackay, "Agents of Empire"; Drayton, *Nature's Government;* Gascoigne, Science in the Service of Empire.

52. Banks's view that plants made good presents for visiting dignitaries is discussed in Drayton, *Nature's Government,* 46–48. The view of Banks as a "meer toad eater" is Sir James Edward Smith's (the founder of the Linnaean Society and a student of John Walker's), as quoted in Miller, "Joseph Banks, Empire, and 'Centers of Calculation,'" 22. The modern critique of Banks is Pyenson, "Over the Bounding Main," 409–10.

53. Omai has been the subject of considerable attention: these remarks are based on McCormick, *Omai Pacific Envoy;* and Alexander, Omai "Noble Savage."

54. The example is from Pancaldi, "Appropriating Invention"; and Pancaldi, Volta, esp. 254–56.

55. Franklin, "Maritime Observations," 314–15.

56. The phrase is from Van Damme, "Reason and Sentiment," 385. See also France, "Lost for Words."

Chapter Four

1. The example is from Daston, "Enlightenment Calculations," 182–83.

2. For example, Smith and Agar, *Making Space for Science;* Livingstone, *Putting Science in Its Place,* 17–86.

3. Ogborn, *Spaces of Modernity.*

4. Galison, *Einstein's Clocks, Poincaré's Maps.*

5. Gooday, "Premisses of Premises."

6. The term "idioms of discourse" is based on Pocock's reference to "idioms" or "languages" of political thought (Pocock, *Virtue, Commerce and History*). The other is from Wuthnow, *Communities of Discourse,* passim.

7. Jacob and Mijnhardt, introduction to *Dutch Republic in the Eighteenth Century,* 5.

8. Mijnand, "Dutch Enlightenment," 211–12.

9. Ibid., 212.

10. Roberts, "Going Dutch," 351.

11. Ibid., passim.

12. Snelders, "Professors, Amateurs, and Learned Societies." For a view about the anatomy theater as a site for the making of the "new science" in Holland, see Rupp, "New Science and the Public Sphere in the Premodern Era."

13. Vartanian, "Annales School and the Enlightenment," 246.

14. Golinski, "Science *in* the Enlightenment," 420.

15. Clark, Golinski, and Schaffer, introduction to *Sciences in Enlightened Europe,* 3.

16. In an extensive literature, see especially the essays in Calhoun, *Habermas and the Public Sphere;* Mah, "Phantasies of the Public Sphere"; Jacob, "Mental Landscape of the Public Sphere"; Goodman, *Republic of Letters;* Goodman, "Introduction: The Public and the Nation"; La Vopa, "Herder's *Publikum*"; Melton, *Rise of the Public in Enlightenment Europe,*

1–15. On its neglect of science, see Broman, "Habermasian Public Sphere and Science *in* the Enlightenment"; Wood, "Science, the Universities, and the Public Sphere in Eighteenth-Century Scotland"; Stewart, *Rise of Public Science;* Golinski, *Science as Public Culture.*

17. Jacob, "Mental Landscape of the Public Sphere," 96.

18. Unsurprisingly, the literature on this topic is vast. For useful introductions, see the work of James McClellan: *Science Reorganized,* "Scientific Institutions and the Organisation of Science," and "Learned Societies"; and, under various authors, the entry "Academies" in Kors, *Encyclopedia of the Enlightenment,* 1:4–19.

19. A note on this map is necessary. It is based on the outline map and the two appendices, "Official Scientific Societies, 1660–1793" and "Notable Private and Semi-Private Scientific Societies, 1660–1793," in McClellan, *Science Reorganized,* 6–7, 261–80, and 281–91, respectively. McClellan's material is, however, inconsistent. He lists a Collegium Curiosum sive Experimentale, a private "Renaissance Academy" in Altdorf, Germany, active between 1672 and 1695 (appendix 2, 281–82), but does not locate it on his map. Other bodies are listed—in Auxerre, Bratislava [Pressburg], Bremen, Cosenza, Cuneo, Jena, Mainz, Milan, Newcastle-upon-Tyne, Nuremberg, Oxford, Palma, Reggio d'Emilia, and Rome—but not located on his map. McClellan does not include, in either appendix or on his map, a variety of other private societies whose remit was scientific—for example, the Aberdeen Philosophical Society, or Wise Club as it was also known, begun in 1758, or the Glasgow Literary Society, active between 1752 and 1803, which, despite its name, was much involved in scientific matters. I have mapped them here. Many smaller societies throughout Europe that were engaged in practical scientific pursuits—agricultural societies, chemical societies, medical societies—are neither listed nor mapped by McClellan. McClellan's map and appendices—and the map derived from them of scientific societies in 1789 in Goodman and Russell, *Rise of Scientific Europe,* 245, must be used with caution. It is for this reasons that I term this figure an outline map.

20. As quoted in McClellan, *Science Reorganized,* 5, 8.

21. McLellan, "Scientific Institutions and the Organisation of Science," 93.

22. On France's provincial academies, see Roche, *Siècle des lumières en Province.*

23. Raeff, "Enlightenment in Russia and Russian Thought in the Enlightenment"; Schulze, "Russification of the St. Petersburg Academy of Sciences and Arts in the Eighteenth Century"; Gordin, "Importation of Being Earnest."

24. In a wide literature on the Académie, see Hahn, *Anatomy of a Scientific Institution;* McClellan, "Académie Royale des Sciences"; Sturdy, *Science and Social Status;* Briggs, "Académie Royale des Sciences and the Pursuit of Utility."

25. These points are from Terrall, "Culture of Science in Frederick the Great's Berlin," quote at 356. See also Brown, "Maupertuis *Philosophe.*"

26. On these points, see Ferrone, "Accadèmia Reale delle Scienze," 545–49. An illustration of what I mean in terms of the long-run assessment of Enlightenment "institutional space" and "social space" is provided by Emerson's work on the Philosophical Society of Edinburgh: Emerson, "Philosophical Society of Edinburgh, 1737–1747," "Philosophical Society of Edinburgh, 1748–1768," "Philosophical Society of Edinburgh, 1768–1783," and "Scottish Enlightenment and the End of the Philosophical Society of Edinburgh."

27. Clark, *British Clubs and Societies.*

28. Jones, "Living the Enlightenment and the French Revolution," quote at 157. On the Lunar Society as an Enlightenment body, see Uglow, *Lunar Men.*

29. Elliott, "Birth of Public Science in the English Provinces"; Flavell, "Enlightened Reader and the New Industrial Towns"; Porter, "Science, Provincial Culture and Public Opin-

ion in Enlightenment England"; Costa, "Marketing Mathematics in Early Enlightenment England"; Cooter and Pumfrey, "Separate Spheres and Public Places"; Millburn, "London Evening Courses of Benjamin Martin and James Ferguson"; Millburn, "James Ferguson's Lecture Tour of the English Midlands."

30. Stewart, "Meaning for Machines."

31. Allan, "Scottish Enlightenment and the Politics of Provincial Culture"; Money, "Science, Technology and Dissent in English Provincial Culture"; Gascoigne, *Cambridge in the Age of Enlightenment;* Jaffee, "Village Enlightenment in New England"; Jankovic, "Place of Nature and the Nature of Place"; Jankovic, "Arcadian Instincts."

32. Cochrane, *Tradition and Enlightenment in the Tuscan Academies,* 151–56.

33. Shafer, *Economic Societies in the Spanish World,* 29–30.

34. McClellan, *Science Reorganized,* 5.

35. Lowood, *Patriotism, Profit, and the Promotion of Science in the German Enlightenment.*

36. On these points, see Golinski, *Science as Public Culture;* Golinski, "Utility and Audience in Eighteenth-Century Chemistry"; Hufbauer, *Formation of the German Chemical Community;* Donovan, "Scottish Responses to the New Chemistry of Lavoisier"; and, more generally, Money, "From Leviathan's Air Pump to Britannia's Voltaic Pile."

37. This phrasing is from Golinski, *Making Natural Knowledge,* 84.

38. On coffeehouses in general, see Ellis, *Coffee House.* The quote is from Robinson, *Early History of Coffee Houses in London,* 150. On gender and "flash talk," see Cowan, "What Was Masculine about the Public Sphere?"; Berry, "Rethinking Politeness in Eighteenth-Century England."

39. Stewart, "Other Centres of Calculation, or, Where the Royal Society Didn't Count," 153. See also Stewart, "Public Lectures and Private Patronage in Newtonian England," and his "Public Culture of Radical Philosophers in Eighteenth-Century London." On performance as public spectacle in the Enlightenment, see also Schaffer, "Natural Philosophy and Public Spectacle in the Eighteenth Century"; Schaffer, "Machine Philosophy"; Secord, "Newton in the Nursery"; Sutton, *Science for a Polite Society;* Licoppe, *Formation de la pratique scientifique;* Terrall, "Experimenting with Enlightenment."

40. Levere, "Natural Philosophers in a Coffee House," 136, 143; Golinski, "Conversations about Chemistry."

41. Stewart and Weindling, "Philosophical Threads."

42. The points on medical lecturing and the Enlightenment are taken from Morton, "Lectures on Natural Philosophy in London"; and Porter, "Medical Lecturing in Georgian London." On the spaces for London's art world, see Dias, "World of Pictures"; Solkin, *Painting for Money;* and Solkin, *Art on the Line.* On the Royal Society, see Gascoigne, *Science in the Service of Empire,* 16–33; Gascoigne, "Royal Society and the Emergence of Science as an Instrument of State Policy"; Miller, "Into the Valley of Darkness"; and Miller, "Usefulness of Natural Philosophy."

43. Darnton, "Early Information Society."

44. Terrall, "Gendered Spaces, Gendered Audiences"; Fara, *Pandora's Breeches,* 88–105 (on du Châtelet), 167–85 (on Marie Paulze Lavoisier). On Bassi, see Findlen, "Science as a Career in Enlightenment Italy"; Logan, "Desire to Contribute." For overview pieces, see Koerner, "Women and Utility in Enlightenment Science"; Tomaselli, "Enlightenment Debate on Women"; Schiebinger, *Mind Has No Sex?;* Schiebinger, "Philosopher's Beard"; and Schiebinger, "European Women in Science."

45. Jacob and Sturkenboom, "Women's Scientific Society in the West."

46. The distinction between "High" and "Low" is one proposed in Darnton, *Business of the Enlightenment.*

Chapter Five

1. Bourguet, "Explorer"; Sörlin, "National and International Aspects of Cross-Boundary Science"; Iliffe, "Science and Voyages of Discovery"; Fulford with Bolton, *Travels, Explorations and Empires.*

2. Sorrenson, "Ship as Scientific Instrument in the Eighteenth Century." The phrase "vessels of modernity" is from Murray, "Notwithstanding Our Signs to the Contrary," 61. On the ship as, variously, a space of cross-cultural encounter, engine of radical proletarian consciousness, and heterotopia, see Wilson, *Island Race,* 173.

3. Forster, *Voyage round the World,* 610, 649.

4. On Enlightenment interest in the Pacific, see Dunmore, *French Explorers in the Pacific;* MacKay, *In the Wake of Cook;* Smith, *European Vision and the South Pacific;* Smith, *Imagining the Pacific;* MacLeod and Rehbock, *Nature in Its Greatest Extent;* Miller and Reill, *Visions of Empire;* Williams, "Pacific: Exploration and Exploitation."

5. The 1773 Phipps expedition is examined in Savours, "Very Interesting Point in Geography."

6. Williams, *Voyages of Delusion.*

7. Dampier, *New Voyage round the World,* 2:108. On Dampier generally, see Preston and Preston, *Pirate of Exquisite Mind.*

8. On the rhetoric of explorers' accounts, see Lamb, Smith, and Thomas, *Exploration and Exchange.* On Enlightenment commentators such as Bougainville and Diderot thinking differently about Europe because of what they thought about the Pacific, see Vogel, "Sceptical Enlightenment."

9. The enlargement of the "planetary consciousness" through exploration and travel is a central theme of Pratt, *Imperial Eyes.*

10. On which topic, see, for example, Colley, *Captives;* Muthu, *Enlightenment against Empire;* Wilson, *Island Race.*

11. Frost, "Pacific Ocean, 785; my reference to the ideas of theater and performance in the South Seas is from Dening, *Mr. Bligh's Bad Language,* and his *Presenting the Past.* On the actual theatrical performances around Omai, see Joppien, "Loutherbourg's Pantomime *Omai.*"

12. Pinkerton, *Modern Geography,* 1:3, 2:510.

13. African Association, *Proceedings of the Association for Promoting the Discovery of the Interior Parts of Africa,* 1:3–4.

14. Mauviel, "Volney, Constantin-François de Chasseboeuf"; Thomson, "Ottoman Empire."

15. Savage-Smith, "Islam."

16. Said, *Orientalism;* Said, *Culture and Imperialism;* Fulford with Bolton, *Travels, Explorations and Empires,* vol. 4.

17. Barley, "Africa"; Withers, "Mapping the Niger."

18. Cranmer-Byng and Levere, "Case Study in Cultural Collision."

19. Cribb, Harrison-Hall, and Clark, "Trade and Learning"; Dikötter, "China"; Kumar, "India"; Nakayama, "Japan."

20. Late seventeenth-century guides to travel are discussed in Sörlin, "National and International Aspects of Cross-Boundary Science," and it is from this I take the term "scientization of travel." See also Withers, "Geography, Natural History and the Eighteenth-Century Enlightenment."

21. For a fuller discussion of these points, see Shapin, *Social History of Truth.*

22. Frost, *Bering,* passim.

23. The importance of terminology—*voyageur naturaliste* and "explorer"—is stressed in Bourguet, "Explorer." The numbers of scientists are given in Sörlin, "National and International Aspects of Cross-Boundary Science."

24. On knowledge, trade, and empire, see Drayton, "Knowledge and Empire." Malaspina is discussed at length in Kendrick, *Alejandro Malaspina;* and David, *Malaspina Expedition.* The related "revival" of Enlightenment science in Spanish colonial institutions is examined in Pimentel, "Iberian Vision."

25. The examples of Humboldt and Pallas are discussed in Outram, "On Being Perseus."

26. Rennell, "On the Rate of Travelling, as Performed by Camels," 130–31. On Rennell's interest in precision more generally, see Bravo, "Precision and Curiosity in Scientific Travel."

27. Bourguet, Licoppe, and Sibum, introduction to *Instruments, Travel and Science.*

28. The question of longitude has a large literature, both in the eighteenth century and among modern scholars. The best single account is Quill, *John Harrison;* and there are valuable essays in Andrewes, *Quest for Longitude.* Popular accounts are Sobel, *Longitude;* and Sobel and Andrewes, *Illustrated Longitude.*

29. This point is made by Bennett, "Travels and Trials of Mr. Harrison's Timekeeper" (and Bennett there offers the view that Sobel, in *Longitude,* overplays Harrison's "lone genius" role).

30. The example of Vancouver testing instruments from a "local" meridian is from Davies, "Testing a New Technology." For fuller accounts, see Fisher and Johnston, *From Maps to Metaphors;* and Clayton, *Islands of Truth.*

31. Lapérouse's encounter on Sakhalin's shore I take from Bravo, "Ethnographic Navigation and the Geographical Gift."

32. Fleurieu, *Discoveries of the French,* vi.

33. James Cook's and other explorers' meeting with the Maori is discussed in Salmond, *Two Worlds,* and her *Between Worlds.*

34. For summary histories of mapmaking in the Enlightenment, see Edney, "Cartography"; and Withers, "Mapping."

35. On mathematical cosmography, see Edney, "Mathematical Cosmography and the Social Ideology of British Cartography"; and Forbes, "Mathematical Cosmography."

36. On France's Enlightenment mapping and the Cassinis, see Konvitz, *Cartography in France;* and Pelletier, *Carte de Cassini.*

37. This point is made in Konvitz, "Nation-State, Paris and Cartography in Eighteenth- and Nineteenth-Century France," 4.

38. The connections between "national" and "natural" mapping in Enlightenment France are explored in Sahlins, "Natural Frontiers Revisited"; and Nordman, *Frontières de France.*

39. The idea of the Military Survey's mapping of Scotland as one of political surveillance is from Widmalm, "Accuracy, Rhetoric and Triangulation." On the Military Survey, see also Withers, *Geography, Science and National Identity,* 149–53.

40. The complex origins of the Ordnance Survey in Britain are summarized in Delano-Smith and Kain, *English Maps,* 216–24.

41. William Roy, as quoted in Close, *Early Years of the Ordnance Survey,* 12.

42. Gascoigne, "Joseph Banks, Mapping and the Geographies of Natural Knowledge," 156–59.

43. Godlewska, "Napoleonic Survey of Egypt."

44. Russia's Enlightenment mapping is discussed in Goldenberg and Postnikov, "Development of Mapping Methods in Russia in the Eighteenth Century." Dolan, *Exploring European Frontiers,* discusses Russia's physical size and the role of "ethnic mapping," 73–112. The quotation is from Williams, *Rise, Progress, and Present State of the Northern Governments,* as quoted in Dolan, 79.

45. Roy, *Account of the Mode Proposed to Be Followed,* 38.

46. The Languedoc example is from Licoppe, "Project for a Map of Languedoc in Eighteenth-Century France."

47. The quote from Cassini de Thury appears in Pedley, "Mapping Brittany," 14.

48. These remarks on exploration, mapping, travel, and translation in North America follow Belyea, "Amerindian Maps," and her "Inland Journeys, Native Maps."

49. Raj, "Colonial Encounters and the Forging of New Knowledge and National Identities."

50. Banks, as quoted ibid., 131.

51. Raj, "Refashioning Civilities, Engineering Trust."

52. The map examples are taken from Salmond, *Two Worlds,* and her *Between Worlds.* On native–Western views of Cook, see Obeyesekere, *Apotheosis of Captain Cook;* Sahlins, *How "Natives" Think;* and King, "Some Thoughts on Native Hawaiian Attitudes towards Captain Cook."

53. The terms "islands of truth" (itself from Immanuel Kant), "cultural effacement," and "cultural engagement" I take from Clayton, *Islands of Truth.*

54. Christian, "Paul Sandby and the Military Survey of Scotland."

55. Smith, *European Vision and the South Pacific.* The idea of this "voyage into substance" as one of factual depiction is associated with Stafford, *Voyage into Substance.*

56. On picturing generally in Enlightenment science, see Shea, *Science and the Visual Image in the Enlightenment.* Connections among maps, graphs, and paintings as forms of graphical information are addressed in Headrick, *When Information Came of Age.* On topographic depictions and conventions in Alpine Europe and elsewhere, see Klonk, "Science, Art and the Representation of the Natural World."

57. On British Guiana, see Burnett, *Masters of All They Surveyed;* and for Brazil, Martins, "Art of Tropical Travel."

58. On travel, the Grand Tour, and geographical writing, see Chard, *Pleasure and Guilt on the Grand Tour;* Chard and Langdon, *Transports;* and the essays in Renwick, *Invitation au Voyage.*

59. Rousseau's quote on China appears in Davis, "China, the Confucian Ideal, and the European Age of Enlightenment," 16.

60. As quoted in Sambrook, *Task and Selected Other Poems,* 144–45.

61. These distinctions are made and discussed in Wolff, "Travel Literature."

62. See Obeyesekere, *Apotheosis of Captain Cook;* and Sahlins, *How "Natives" Think.* For useful reviews of the Obeyesekere–Sahlins debate, see Bravo, "Anti-Anthropology of Highlanders and Islanders"; and Stoddart, "Captain Cook and How We Understand Him."

63. Bourguet, "Explorer"; Bravo, "Ethnographic Navigation and the Geographical Gift."

64. The idea of books' "voyage into narration" is discussed in Bourguet, "Explorer," 296.

65. The library borrowing statistics are from Kaufmann, *Borrowings from the Bristol Library.* Hawkesworth's role as a compiler is examined by Pearson, "Hawkesworth's Alterations."

66. Lamb, Smith, and Thomas, *Exploration and Exchange,* 73–91.

67. Leask, *Curiosity and the Aesthetics of Travel Writing,* 54–101; Withers, "Towards a Geography of Trust."

68. On these points, see the essays in Fulford, Lee, and Kitson, *Literature, Science and Exploration in the Romantic Era;* and Ballantyne, *Science, Empire and the European Exploration of the Pacific.*

Chapter Six

1. As quoted in Kaye Lamb, *Voyage of George Vancouver,* 1:275–76.

2. For summary discussions of these issues, see Porter, "Terraqueous Globe"; Rousseau and Porter, *Ferment of Knowledge;* Frängsmyr, Heilbron, and Rider, *Quantifying Spirit;* Terrall, "Heroic Narratives of Quest and Discovery"; Heilbron, "Natural Philosophy and Science." The translated quote from Buffon is from Buffon et al., *Histoire naturelle* (1749), 1:97.

3. Ray, *Wisdom of God Manifested in the Works of the Creation,* 169–72.

4. Porter, "Terraqueous Globe," is useful on this point. For accessible recent summaries, see Heilbron, "Natural Philosophy and Science"; and Gascoigne, "Ideas of Nature." For particular disciplines, see, for example, Spary, "Botany"; Oldroyd, "Geology"; Farber, "Natural History"; Rappaport, "Earth Sciences."

5. This point about the epistemic shift in natural history—from "classical" to "modern"—and the sense in which "modern" natural history made the world "visible" is made in Foucault, *Order of Things,* and his *Archaeology of Knowledge.*

6. This point about the geographical nature of science helping to define Enlightenment natural history is made in Cooper, "From the Alps to Egypt (and Back Again)," and in Koerner, "Daedalus Hyperborus." For an earlier view along these lines, see Larson, "Not without a Plan." On natural history and visual description, see also Schiebinger, "Nature's Unruly Body."

7. On this term and on the shape-of-the-earth debate, see Greenberg, "Mathematical Physics in Eighteenth-Century France," and his *The Problem of the Earth's Shape from Newton to Clairault.* Good summaries are also provided by Chapin, "Shape of the Earth"; and Wilson, "Astronomy and Cosmology."

8. On Maupertuis's Lapland travels, see Terrall, "Representing the Earth's Shape." On La Condamine, see McConnell, "La Condamine's Scientific Journey down the River Amazon"; and Pratt, *Imperial Eyes,* 15–24. The point about "flattening the Earth and the Cassinis" I take from Terrall, *Man Who Flattened the Earth.*

9. Quoted in Livingstone, *Putting Science in Its Place,* 91–92.

10. Glick, *Comparative Reception of Darwinism;* Numbers and Stenhouse, *Disseminating Darwinism;* Livingstone, "Darwinism and Calvinism."

11. On this point, see, for example, Jacob, *Newtonians and the English Revolution;* Fissell and Cooter, "Exploring Natural Knowledge," 134–39; Schaffer, "Natural Philosophy and Public Spectacle in the Eighteenth Century"; Schaffer, "Newtonianism"; Schofield, "Evolutionary Taxonomy of Eighteenth-Century Newtonianisms"; Secord, "Newton in the Nursery"; and McNeil, "Newton as National Hero."

12. Gascoigne, "Ideas of Nature," 289.

13. This element of the story is clear in Iliffe, "Aplatisseur du Monde et de Cassini."

14. With the exception of Rennell, these examples are from Heilbron, "Measure of Enlightenment." Rennell's observation is from his *Memoir of a Map of Hindoostan,* 4.

15. Alder, "Revolution to Measure," and his *Measure of All Things.* On Enlightenment emphases on calculation as a basis to practical reason, see also Daston, *Classical Probability in the Enlightenment,* and her "Enlightenment Calculations." The contemporary quotation is from Boscovich and Maire, *Voyage astronomique et géographique dans l'Etat et d'Eglise,* 42.

16. My remarks on Linnaeus are based principally on the work of Lisbet Koerner: see "Carl Linnaeus in His Time and Place"; *Linnaeus: Nature and Nation;* and "Purposes of Linnaean Travel." I have also made use of Frängsmyr, *Linnaeus;* Larson, *Interpreting Nature;* and Spary, "Linnaeus." Linnaeus's use of the terms "map" and "Geographia Naturae" are cited in Koerner, "Carl Linnaeus in His Time and Place," 146.

17. As quoted in Koerner, "Carl Linnaeus in His Time and Place," 153.

18. As quoted in Koerner, *Linnaeus,* 94.

19. See, for example, Larson, "An Alternative Science."

20. On Buffon's work, see Lyon and Sloan, *From Natural History to the History of Nature;* Sloan, "Buffon"; Roger, *Buffon.* On critical reactions to Linnaeus's schemes, including from Buffon, see Spary, "Linnaeus" and her *Utopia's Gardens.*

21. On Soulavie, see Bourguet, "Landscape with Numbers"; and Rudwick, *Bursting the Limits of Time,* 214–18.

22. Nicolson, "Alexander von Humboldt, Humboldtian Science and the Origins of the Study of Vegetation"; Nicolson, "Alexander von Humboldt and the Geography of Vegetation"; Nicolson, "Humboldtian Plant Geography after Humboldt." On the general sense in which Humboldtian practices helped establish new conceptions of the physical world, see Dettelbach, "Humboldtian Science"; Dettelbach, "Global Physics and Aesthetic Empire"; and Godlewska, "From Enlightenment Vision to Modern Science?" On Humboldtian science, see Cannon, *Science in Culture.* On Humboldt and vitalism in Enlightenment science, see Reill, *Vitalizing Nature in the Enlightenment,* passim.

23. As quoted in Huxley, "Natural History Collectors and Their Collections," 90.

24. These connections made by Banks are stressed in Gascoigne, *Enlightenment and the Origins of Modern Australia.* On the HMS *Investigator* voyage and its role in plant collecting, see Edwards, "Robert Brown and the Natural History of Matthew Flinders' Voyage."

25. Drayton, *Nature's Government;* on forest clearance (in Enlightenment Europe as well as in Europe's colonies), see Williams, *Deforesting the Earth.*

26. On Zimmermann, see Bodenheimer, "Zimmermann's Specimen Zoologiae Geographiae Quadrupedum"; Browne, *Secular Ark,* 25–27; Barrett, "Finke's 1792 Map of Human Diseases," 917–19.

27. Gould, "Lisbon 1755"; the reported differences in The Hague between the general public's awareness and that of the business community is from D'Haen, "On How Not to Be Lisbon if You Want to Be Modern," 351–52.

28. Kendrick, *Lisbon Earthquake.* In addition to Gould, "Lisbon 1755," which refines and updates Kendrick's view, see also Maggs, "Eighteenth-Century Russian Reflections on the Lisbon Earthquake"; Carozzi, "Reaction of British Colonies in America to the Lisbon Earthquake"; Araujo, "European Public Opinion and the Lisbon Earthquake." Italian and Dutch reactions to the Lisbon earthquake are discussed in Adamo, "Constructing an Event, Contemplating Ruins, Theorizing Nature," and D'Haen, "On How Not to Be Lisbon if You Want to Be Modern."

29. Keller, "Sections and Views," is strong on the contrasting explanations for earthquakes in the Enlightenment. On Stukeley's views regarding earthquakes, see Hancock, *William Stukeley;* and on Benjamin Franklin's, see Dean, "Benjamin Franklin and Earthquakes."

30. Keller, "Sections and Views," 150. On the general development of visual languages in Enlightenment geology, see Rudwick, "Emergence of a Visual Language for Geological Science."

31. Rappaport, "Geology and Orthodoxy," and her "Earth Sciences."

32. Hamblyn, "Private Cabinets and Popular Geology."

33. Guettard and Monnet, *Atlas et description minéralogique de la France,* iii.

34. Ellenberger, "Recherches et réflexions sur la naissance de la cartographie géologique"; Rudwick, *Bursting the Limits of Time,* 71–84.

35. Hutton's work and importance have most recently been reviewed in Dean, *James Hutton and the History of Geology.* On Hutton and Werner, see Dean, "Age of the Earth Controversy." The working volcano on the Wörlitz estate is discussed in Umbach, "Visual Culture, Scientific Images and German Small-Scale Politics in the Enlightenment."

36. Rappaport, "Earth Sciences."

37. Kant's work in the earth sciences, often overlooked in Enlightenment studies, is the subject of articles by Reinhardt and Oldroyd: "Kant's Thoughts on the Ageing of the Earth"; "Kant's Theory of Earthquakes and Volcanic Action"; and "By Analogy with the Heavens." On Kant's physical geography lectures and Enlightenment natural philosophy, see Bowen, *Empiricism and Geographical Thought,* 206–9.

38. Dean, *James Hutton and the History of Geology.*

39. Rudwick makes just this point in his magnificent *Bursting the Limits of Time,* in mapping there "some of the famous localities [in Europe] that were visited, described and discussed in the later eighteenth century by savants with interests in the sciences of the earth" (fig. 2.4).

40. On the connections in the Enlightenment among terrestrial magnetism, hydrography, oceanography, and navigation, see Deacon, *Scientists and the Sea,* 154–219.

41. Cawood, "Terrestrial Magnetism and the Development of International Collaboration in the Early Nineteenth Century"; Yost, "Pondering the Imponderable."

42. Deacon, *Scientists and the Sea,* 175–98.

43. Le Guisguet, "Le Dépôt des Cartes, plans et journaux de la marine sous l'ancien régime"; Headrick, *When Information Came of Age,* 108–15. On the Gulf Stream in the Enlightenment, see Withers, "Science at Sea."

44. As quoted in Webster, "Cartographic Controversy," 38.

45. Pelizzari, "Enlightenment Intellectuals and Popular Mentality after the 1783 Calabrian Earthquake."

46. Jankovic, *Reading the Skies.* On the question of "rustic science" and its geographical implications, see Jankovic, "Place of Nature and the Nature of Place," and his "Arcadian Instincts."

47. Jankovic, *Reading the Skies,* 125–42; Jankovic, "Meteorology." On meteorological and other instruments, see G. L'E. Turner, "Eighteenth-Century Scientific Instruments and Their Makers," 522–30.

48. Zuidervaart, "An Eighteenth-Century Medical-Meteorological Society in the Netherlands."

49. Feldman, "Late Enlightenment Meteorology"; Feldman, "Ancient Climate in the Eighteenth and Early Nineteenth Century."

50. Grove, *Little Ice Age.*

51. Kington, *Weather of the 1780s over Europe,* 125.

52. Jordanova and Porter, *Images of the Earth;* Grove, *Green Imperialism,* passim; and Grove, *Ecology, Climate and Empire,* 124–46.

Chapter Seven

1. See Lamb, Smith, and Thomas, "Byron and the Patagonian Giants," in *Exploration and Exchange,* 46–56. Belief in "Patagonian giants" (Patagonia literally means "land of the big-footed") continued even after the publication of accounts from Bougainville's and Cook's voyages denied their existence.

2. Hume, *Treatise of Human Nature,* as quoted in Olson, "Human Sciences," 436.

3. On this point, see Olson, "Human Sciences"; Foucault, *Order of Things;* Kelley, "Problem of Knowledge and the Concept of Discipline"; and the essays in Graham, Lepenies, and Weingart, *Functions and Uses of Disciplinary Histories;* Kelley, *History and the Disciplines;* and Fox, Porter, and Wokler, *Inventing Human Science.*

4. Hume, *Treatise of Human Nature,* 46.

5. Olson, "Human Sciences"; Wood, "Science, Philosophy and the Mind," esp. 817–19.

6. Moravia, "Enlightenment and the Sciences of Man"; Wood, "Science of Man"; Wood, "Science, Philosophy and the Mind."

7. Christie, "Human Sciences."

8. The comment from Herder on Forster's Pacific work—Herder also termed his countryman the Ulysses of the Pacific—is cited in Iliffe, "Science and Voyages of Discovery," 630. The term "Reservoir of Human Experiments" I take from Olson, "Human Sciences," 442–44.

9. Olson, "Human Sciences," 431.

10. This point about Brosse is made in Ryan, "Président des Terres Australes." For remarks on Cook's use of Brosse, see Thomas, *Discoveries,* 16–19. It is likely that the maps in Brosse's work were of more interest to Cook than the text since he did not read French.

11. O'Brien, "Enlightenment History in Scotland," 468.

12. For a summary of Montesquieu's life, work, and influence, see Kingston, "Charles-Louis de Secondat Montesquieu."

13. This is less the case for Kingston's recent summary than it is of Carrithers's treatment of Montesquieu in "Enlightenment Science of Society." It is not true of the discussion of Montesquieu's attention to climate in Glacken, *Traces on the Rhodian Shore,* esp. 568–81.

14. Glacken, *Traces on the Rhodian Shore,* 566.

15. Montesquieu, *L'esprit des lois,* bk. 14, pp. 398; as quoted in Glacken, *Traces on the Rhodian Shore,* 570.

16. As quoted in Glacken, *Traces on the Rhodian Shore,* 573.

17. Feldman, "Ancient Climate in the Eighteenth and Early Nineteenth Century," 25–26; Miller, "'Airs, Waters, and Places' in History."

18. Kendrick, *Alejandro Malaspina,* 37.

19. Hume, "On National Characters," in *Essays, Moral, Political, and Literary,* pt. 1, 200, 204.

20. Ibid., 205.

21. Ibid., 213. For a summary of the place of Montesquieu and of Hume and of the persistence of such discourses into the nineteenth century, see Livingstone, "Race, Space and Moral Climatology."

22. Wertz, "Hume and the Historiography of Science." The claim by this author that "he [Hume] was apparently the first British historian to give geography any importance, either for its own sake or as a factor in history" (427) cannot be taken at face value.

23. Quoted in Wood, "Science, Philosophy and the Mind," 820.

24. Sloan, "Gaze of Natural History," 121.

25. Ibid., 122–26, esp. table 5.2, pp. 124–25, lists these classificatory changes in detail.

26. This quote, from Linnaeus's discussion of *Homo* in his *Systema Naturae,* is from Eze, *Race and the Enlightenment,* 13.

27. More detailed accounts on Linnaean ideas of "variety" and on the differences between Linnaeus and Buffon in this respect are provided in Farber, "Buffon and the Concept of Species"; Sloan, "Buffon-Linnaeus Controversy"; Larson, *Reason and Experience,* 94–121; Broberg, "*Homo Sapiens.*" The point about the geography of the opposing ideas over human difference—Linnaeus in Uppsala, Buffon in Paris—I owe to Sloan, "Gaze of Natural History," 139.

28. For a summary of Buffon's method and its implications, see Sloan, "Gaze of Natural History," 126–31.

29. These extracts, from Buffon's discussion of the geographical distribution of mankind in his *Histoire naturelle,* are quoted in Eze, *Race and the Enlightenment,* 22, 25. On Buffon, anthropology, and the "Natural History of Man," see Duchet, *Anthropologie et histoire au siècle des lumières;* and Blanckaert, "Buffon and the Natural History of Man." On the Buffonian idea of degeneracy, see Sloan, "Idea of Racial Degeneracy in Buffon's *Histoire Naturelle.*"

30. Hudson, "From 'Nation' to 'Race,'" 255.

31. Gascoigne, "Blumenbach, Banks, and the Beginnings of Anthropology at Göttingen." On the translation of Buffon for English audiences, see Loveland, "Georges-Louis Leclerc de Buffon's *Histoire Naturelle* in English."

32. Hudson, "From 'Nation' to 'Race.'" On the various definitions of race, see Sebastiani, "Race as a Construction of the Other."

33. These quotations are all taken from Eze, *Race and the Enlightenment.* On Kant's in relation to Forster's view on race, see Strack, "Philosophical Anthropology on the Eve of Biological Determinism."

34. The remark is from Popkin, "Philosophical Basis of Eighteenth-Century Racism."

35. The quotes from Forster and these interpretative remarks here and in the following paragraph I take from Ryan, "Président des Terres Australes," 182–83. The views of Georg Forster on Pacific natives are discussed in West, "Limits of Enlightenment Anthropology."

36. Quoted in Gascoigne, *Joseph Banks and the English Enlightenment,* 173. See also Douglas, "Seaborne Ethnography and the Natural History of Man"; Wilson, *Island Race,* 54–91.

37. This paragraph is based on Garrett, "Anthropology." The literature on these questions in the Scottish Enlightenment is extensive. For summaries, see Berry, *Social Theory of the Scottish Enlightenment,* and the essays in Wood, *Scottish Enlightenment.* Many original sources have been usefully compiled and discussed in Sher, *Conjectural History and Anthropology in the Scottish Enlightenment.* See also Höpfl, "From Savage to Scotsman."

38. The literature on the idea of "progress" in the Enlightenment, and on conjectural history and stadial theory, is enormous. For summaries, see Bury, *Idea of Progress;* Carrithers, "Enlightenment Science of Society"; Frankel, *Faith in Reason;* Meek, *Social Science and the Ignoble Savage;* Nisbet, *History of the Idea of Progress;* Olson, "Human Sciences"; Pollard, *Idea of Progress;* Spadafora, *Idea of Progress in Eighteenth-Century Britain;* Spadafora, "Progress"; Wokler, "Anthropology and Conjectural History in the Enlightenment."

39. Spadafora, "Progress," 367.

40. Dunbar, *Essays on the History of Mankind in Rude and Cultivated Ages,* 225. For Glacken, Dunbar's work stands as a synthesis of Montesquieu, Buffon, Hume, and Herder: Glacken, *Traces on the Rhodian Shore,* 596.

41. On the Gaelic world—and the different views held of "Celtic antiquity" and "Gaelic barbarism"—see Kidd, "Gaelic Antiquity and National Identity in Enlightenment Scotland and Ireland"; and Withers, *Gaelic Scotland.*

42. Heffernan, "On Geography and Progress," esp. 334–35.

43. Pagden, "Eighteenth-Century Anthropology and the 'History of Mankind,'" 226.

44. This suggestion is made by Emerson, "American Indians, Frenchmen, and Scots Philosophers," who notes of Lafitau how "his analyses of statuses, the roles of women, kinship patterns, Indian arts, and primitive medicine were to be carefully noted by the Scots, who read him with far more attention than did the French *philosophes*" (227). For support of this view, see Bosher, "Joseph-François Lafitau," who observes that "Montesquieu, Voltaire, Diderot, and Rousseau scarcely noticed Lafitau's work" (343).

45. Carey, "Reconsidering Rousseau," 26.

46. Gerbi, *Dispute of the New World,* 52–79.

47. On Robertson's historical writings in general, see O'Brien, "Between Enlightenment and Stadial History"; O'Brien, *Narratives of Enlightenment;* and the essays in Brown, *William Robertson and the Expansion of Empire.* The point about Robertson's French reception I take from Renwick, "Reception of William Robertson's Historical Writings in Eighteenth-Century France." Robertson's German reception is discussed in Kontler, "William Robertson and His German Audience on European and Non-European Civilisations."

48. For a discussion of Bartram and Jefferson in this context, see Kornfeld, "Encountering 'the Other.'"

49. These quotes and my remarks on the Enlightenment treatment of Africa are from Jacques, "From Savages and Barbarians to Primitives." The cartographic "emergence" of Africa in the Enlightenment—a history of continued activity by the French in particular, which somewhat qualifies Jacques's points here—is discussed in Stone, *Short History of the Cartography of Africa;* Relaño, *Shaping of Africa;* and Withers, "Mapping the Niger."

50. I take these remarks from Pratt, *Imperial Eyes,* 49–56; Merians, "What They Are, Who We Are"; and Van Wyk Smith, "Most Wretched of the Human Race."

51. Whelan, "Population and Ideology in the Enlightenment"; Glacken, *Traces on the Rhodian Shore,* 512–14 (on Süssmilch), and pt. 4 in general.

52. Flavell, "Mapping Faces."

53. I take these points from Wood, "Buffon's Reception in Scotland"; Wood, "Natural History of Man in the Scottish Enlightenment"; Wood, "Science of Man"; from Reill, "Buffon and Historical Thought in Germany and Great Britain"; and Zbinden, "Heterogeneity, Irony, Ambivalence."

Chapter Eight

1. Anonymous, review of *John Pinkerton's Modern Geography,* 67–81, quote at 67.

2. The works in support of this contention are cited in what follows. For illustrations of the neglect and misunderstanding of Enlightenment geography texts, see Bowen, *Empiricism and Geographical Thought;* and Downes, "Bibliographic Dinosaurs of Georgian Geography." In contrast, for a superb study of geography in France during the Enlightenment, see Broc, *Géographie des philosophes.* For more recent overviews, see Godlewska, *Geography Unbound;*

Livingstone, *Geographical Tradition,* 102–38; Livingstone and Withers, *Geography and Enlightenment,* 1–28; Mayhew, *Enlightenment Geography;* Withers, "Geography."

3. Sitwell, *Four Centuries of Special Geography,* quotes at 10, 11, and 266. The *Eighteenth-Century Short Title Catalogue* lists 657 works with the term "geography" in their title. Many are reprints, some not about geography despite the title, but it is still the case that each decade of the eighteenth century saw the publication of between two and fourteen special geographies: Sitwell, *Four Centuries of Special Geography,* 16–18; Mayhew, " Character of English Geography," 388.

4. May, "Observations on the Allegory."

5. On the *Encyclopédie* itself, see Lough, *Encyclopédie;* Darnton, *Business of the Enlightenment;* and Hahn, "Science and the Arts in France." On the *Encyclopédie* and its predecessors and successors, see Kafker, *Notable Encyclopedias of the Seventeenth and Eighteenth Centuries,* and his *Notable Encyclopedias of the Late Eighteenth Century.* On the *Encyclopédie* and the culture of encyclopedism, see Yeo, *Encyclopaedic Visions,* and his "Classifying the Sciences."

6. D'Alembert, "*Discours préliminaire,*" as quoted in Withers, "Geography in Its Time," 256.

7. Bates, "Cartographic Aberrations"; Withers, "Encyclopaedism, Modernism and the Classification of Geographical Knowledge."

8. On the Societé Académique des Beaux-Arts, see Hahn, "Science and the Arts in France." The remarks on the regional geography of the Encyclopedists are from Kafker, *Encyclopedists as a Group,* 19–35.

9. These points on the geography of reception are distilled from Darnton, *Business of the Enlightenment,* 278–323 and appendices B and C, 586–93 and 594–96, respectively; the quote is at 287.

10. Owen Gingerich traces the book geographies of Copernicus's *De Revolutionibus* in his *The Book Nobody Read* with particular reference to marginalia. James Secord discusses the private reception of Chamber's *Vestiges* in his *Victorian Sensation.* This point about the difficulties of discerning the geographies of reading is made in Darnton, *Business of the Enlightenment,* 319–23; and Roche, "Urban Reading Habits during the French Enlightenment."

11. Darnton, *Business of the Enlightenment,* 314; Donato, "Eighteenth-Century Encyclopedias and National Identity," 961–62; Kafker, *Encyclopedists as a Group,* 61. On Jaucourt and "Nation" in the *Encyclopédie,* see Rechniewski, "References to 'National Character' in the *Encyclopédie.*"

12. On atlas production and ancient geography, see Goffart, *Historical Atlases.*

13. These remarks are taken from Broc, *Géographie des philosophes,* 248–56; Darnton, *Great Cat Massacre,* 185–207; Dörflinger, *Die Geographie in der "Encyclopédie,"* 21–33; Kafker and Kafker, *Encyclopedists as Individuals,* 11–12, 103–7, 330–33; Moravia, "Philosophie et géographie à la fin du XVIIIe siècle"; Perla, "Géographie dans l'*Encyclopédie*"; Withers, "Geography in Its Time"; Withers, "Encyclopedism, Modernism and the Classification of Geographical Knowledge"; Laboulais, "Géographie dans les arbres encyclopédiques de la second moitiée du XVIIIe siècle."

14. Godlewska, *Geography Unbound,* 26–27, 34 (who does use this term).

15. These points on the connections between geography and history and their significance for an understanding of geography's history are from Mayhew: "Character of English Geography" (from whom the quote is taken, at 393), "Was William Shakespeare an Eighteenth-Century Geographer?" and "Geography, Print Culture and the Renaissance." On

Samuel Johnson and geography, see Mayhew, *Geography and Literature in Historical Context.* Gibbon's engagement with geography is discussed in Abbatista, "Establishing the 'Order of Time and Place.'"

16. Mayhew, "Geography Books and the Character of Georgian Politics."

17. Mayhew, Geography in Eighteenth-Century British Education," 733–42; Mayhew, "Character of English Geography," 395–97; Secord, "Newton in the Nursery."

18. On Bickham, see Ogborn, "Geographia's Pen." On writing and geography, see also Ogborn and Withers, "Knowing Other Places." The remarks about the nature of Salmon's work are from Mayhew, introduction to *A New Geographical and Historical Grammar,* xvii–xxxiv.

19. Mayhew, "Character of English Geography," 406.

20. Mayhew, introduction to *A New Geographical and Historical Grammar,* quotes at xxvii.

21. Konvitz, *Cartography in France;* Godlewska, *Geography Unbound,* 129–48.

22. This is the central argument of Godlewska, *Geography Unbound.* For a contra-view, see Mayhew, "Proleptic Locations."

23. Godlewska's work is important and substantial. In challenging its views over the "modernity" of French geography I do not mean to dismiss it tout court. For other works on this issue, see Broc, *Géographie des philosophes;* Daniel, "Géographie dans les collèges de Jesuites aux XVIIe et XVIIIe siècles"; Dainville, *Géographie des humanistes;* Moravia, "Philosophie et géographie à la fin du XVIIIe siècle"; Abbatista, "Establishing the 'Order of Time and Place.'" The example of Mentelle—who may indeed have been unfairly represented by Godlewska—is taken from Heffernan, "Edme Mentelle's Geographies and the French Revolution."

24. Bowen, *Empiricism and Geographical Thought,* 154–59, 206–9.

25. This discussion of Guthrie is taken from Mayhew, "William Guthrie's *Geographical Grammar,* the Scottish Enlightenment and the Politics of British Geography" (from which source I quote the extracts from Guthrie's work); and Mayhew, *Enlightenment Geography,* 168–80. Bowen, in her *Empiricism and Geographical Thought,* 164–68, ignores this element of Guthrie's work. For a contemporary review of Guthrie that endorses Mayhew's interpretation, see Gilbert Stuart's review in Zachs, *Edinburgh Magazine and Review,* 1:215–18.

26. On Coronelli, see Fuchs, "Nationality and Knowledge in Eighteenth-Century Italy"; on Gatterer, see Goffart, *Historical Atlases,* passim; on Moisiodax and others in Greece, see Kitromilides, *Enlightenment as Social Criticism,* 103–4, 108, 116, 149, 185; and his "Greece and Europe in the Modern Period," 3–4.

27. These figures and examples are taken from the *Pennsylvania Gazette,* 1728–1800. On geography in early America, see the essays in the first part of Blouet, *The Origins of Academic Geography in the United States.*

28. These remarks on Morse and in the paragraphs following are distilled from Brown, "American Geographies of Jedidiah Morse"; Brückner, "Lessons in Geography"; Brückner, "Geography, Reading, and the World of Novels in the Early Republic"; Livingstone, "Risen into Empire"; and Short, "New Mode of Thinking." In what follows, I here refer to Morse's 1789 work as [*The*] *American Geography.* Although Morse's definitive work is sometimes referred to as [*The*] *American Universal Geography,* that title was not properly employed until the 1793 edition: see Brown, "American Geographies of Jedidiah Morse," 214.

29. On these issues, see Lawson-Peebles, *Landscape and Written Expression in Revolutionary America;* Bowden, "Invention of American Tradition"; Johnson, "Towards a National Landscape"; Meinig, *Shaping of America,* vol. 1.

30. Brückner, "Lessons in Geography," 316–17. In more detail, see Brückner, *Geographic Revolution in Early America.*

31. Livingstone, "Risen into Empire."

32. Quotes from the second edition of Morse, *American Geography,* 214, 219, 389, 417, 445, 451. On critical reaction to it, see Brown, "American Geographies of Jedidiah Morse," 177–85.

33. Gilmore, *Reading Becomes a Necessity,* 297.

34. Morse, *Elements of Geography,* 121.

35. The phrase "voluntary self-enlightenment" is from Sitwell, *Four Centuries of Special Geography,* 9.

36. Gilmore, *Reading Becomes a Necessity,* 64–68.

37. Mayhew, "Mapping Science's Imagined Community."

Chapter Nine

1. Mayhew, "Proleptic Locations," 68, 72.

2. This phrase is part of the opening line of Fox, "Introduction: How to Prepare a Noble Savage," 1.

3. Forbes, "Mathematical Cosmography," 428. On the Cosmographical Society of Nuremberg, see also Edney, "Mathematical Cosmography and the Social Ideology of British Cartography," 101–3.

4. On Rennell's different geographies, see Mayhew, *Enlightenment Geography,* 193– 206; Edney, *Mapping an Empire,* passim, from which I take the view of Rennell as the father of Indian geography, 22–23, and Edney, "Bringing India to Hand." The view of Rennell as the father of English geography is established in Markham, *Major James Rennell and the Rise of Modern English Geography.* See also Baker, *History of Geography,* 130–57.

5. De Vorsey, "Maps in Colonial Promotion"; Gronim, "Geography and Persuasion." The point about the "deepening" of knowledge about America is from Colley, *Captives,* 175–98. Jefferys's words on Green/Mead are from a letter by Jefferys to the Earl of Morton, inserted in the copy of J. Green, *Remarks, in Support of the New Chart of North and South America* (1753), owned by Chicago's Newberry Library.

6. Godlewska, *Geography Unbound,* 156–64, quote at 158–59. See also Godlewska, "Napoleon's Geographers"; on military mapping in Enlightenment France, see Pansini, "Géographie appliquée à la guerre."

7. On this point, particularly for France, see Broc, *Géographie des philosophes;* Dainville, *Langage des géographes;* Godlewska, *Geography Unbound,* 129–48; Laboulais-Lesage, "Géographes française de la fin du XVIIIe siècle et le terrain"; and, more generally, Paxman, *Voyage into Language.*

8. Rusnock, "Biopolitics," 66.

9. These two quotes, one in 1774 by the English physician John Haygarth, the other by the French administrator Antione Auget in 1778, are taken from Rusnock, "Biopolitics," 49, 50, from whom I also quote Davenant. For political arithmetic and the related rise of numerical thinking in the Enlightenment, see Daston, *Classical Probability in the Enlightenment;* Poovey, *History of the Modern Fact;* and Rusnock, *Vital Accounts.* This point about the language of human classification I take from Corfield, "Class by Name and Number in Eighteenth-Century Britain."

10. Whelan, "Population and Ideology in the Enlightenment"; Buck, "People Who Counted."

11. Rusnock, "Biopolitics," 57; Ogborn discusses these "excise geographies" in his *Spaces of Modernity,* 158–200.

12. Mayhew, "Geography Books and Georgian Politics," 204–7.

13. This discussion of Sinclair's "Statistical Philosophy" and the *Statistical Account of Scotland* is from Withers, "How Scotland Came to Know Itself," 389–92 (emphasis in original sources). For the view that in its method, its attention to the people's happiness, and in future benefits Sinclair's project encapsulates the "true spirit of Enlightenment," see Withrington, "What Was Distinctive about the Scottish Enlightenment?" 16.

14. This vignette of the geography of Parisian *géographes* is taken from Pedley, "Map Trade in Paris."

15. For more on the often fractious social relationships within the Paris map world, see Pedley, "New Light on an Old Atlas," and her "Atlas Editing in Enlightenment France." On the social nature of the Enlightenment map world more generally, see Edney, "Mathematical Cosmography and the Social Ideology of British Cartography"; Withers, "Social Nature of Map Making in the Scottish Enlightenment"; Laboulais-Lesage, "Reading a Vision of Space."

16. Marshall, "Military Maps of the Eighteenth-Century and the Tower of London Drawing Room." On the social hierarchies of mapping see also Edney, "Mathematical Cosmography and the Social Ideology of British Cartography."

17. This paragraph is based on Harley, "Society of Arts and the Surveys of English Counties"; the quotes from Borlase and about Faden are at 43 and 273, respectively.

18. These points are distilled from Pedley, *Map Trade in the Late Eighteenth Century.*

19. On this example, see Harley, "Power and Legitimation in the English Geographical Atlases of the Eighteenth Century."

20. On correspondence networks, different forms of writing, and the making of geography in the Enlightenment, see Withers, "Writing in Geography's History"; and Ogborn and Withers, "Knowing Other Places." This idea of the "map world" is the subject of Pedley, *Commerce of Cartography.*

21. This list of topics informs the discussion of "Enlightenment" in Porter, *Greatest Benefit to Mankind,* 245–303.

22. Jordanova, "Earth Science and Environmental Medicine."

23. Ibid., 124–26; Rusnock, *Vital Accounts,* 109–36.

24. Goldschmid, "Nosologia Naturalis"; Stevenson, "Putting Disease on the Map."

25. Schaffer, "Measuring Virtue"; Jordanova, "Earth Science and Environmental Medicine." The term "biopolitics" is from Foucault, "Politics of Health in the Eighteenth Century"; see also Rusnock, "Biopolitics," and her *Vital Accounts,* 4, 109–36.

26. Volney, *View of the Climate and Soil of the United States of America,* 123, 234–35, 294–95.

27. Barrett terms Finke's work "one of the foundation pieces of modern medical geography," in his "Medical Geographical Anniversary," 701. See also Barrett, "Finke's 1792 Map of Human Diseases."

28. For recent summaries of this complexity, see Valenčius, "Histories of Medical Geography"; Barrett, *Disease and Geography.*

29. Jarcho, "Some Early Italian Epidemiological Maps."

30. Stevenson, "Putting Disease on the Map"; Rosen, "Noah Webster—Historical Epidemiologist"; Jarcho, "Yellow Fever, Cholera, and the Beginnings of Medical Cartography."

31. Barrett, "'Scurvy' Lind's Medical Geography," quotes at 348, 351. For others then working on the "geography of fevers," see Bynum, "Cullen and the Study of Fevers in Britain."

32. Rosen, "Leonhard Ludwig Finke and the First Medical Geography."

33. This summary discussion of Finke's work, with quotes from his 1792 book, is based on Barrett, "Medical Geographical Anniversary."

34. The Buache–Zimmermann–Finke connection (but not Gatterer's involvement) is discussed briefly in Barrett, "Finke's 1792 Map of Human Diseases," 917–19.

Chapter Ten

1. These examples are discussed in Reitinger, "Mapping Relationships" (from which I take the terms "geo-pornography" and "géographie galante"); and Harvey, "Spaces of Erotic Delight."

2. See Cohen, "Enlightenment and the Dirty Philosopher"; Hunt, *Invention of Pornography;* and Ormsby-Lennon and Pappa, "Pornography."

3. On geography and geographical knowledge within the economic societies of Enlightenment Germany, see Lowood, *Patriotism, Profit, and the Production of Science in the German Enlightenment.* The Maupertuis example is from Brown, "Maupertuis *Philosophe.*" The "general cartography" phrase is from Roche, *Siècle des lumières en province.*

4. Gascoigne, "Eighteenth-Century Scientific Community."

5. Philosophical Society of Edinburgh, *Essays and Observations, Physical and Literary,* 423–36.

6. On Paris, see McClellan, "Académie Royale des Sciences." The Edinburgh case is taken from Emerson, "Philosophical Society of Edinburgh, 1737–1747," "Philosophical Society of Edinburgh, 1748–1768," and "Philosophical Society of Edinburgh, 1768–1783." On geography in the Royal Society of London, see Sorrenson, "Towards a History of the Royal Society in the Eighteenth Century," quote at 39.

7. Shaw, "Geographical Practice and Its Significance in Peter the Great's Russia"; Schulze, "Russification of the St. Petersburg Academy of Sciences and Arts in the Eighteenth Century," 326; Goodman and Russell, *Rise of Scientific Europe,* 347–51; Frost, *Bering.*

8. Friis, "Role of Geographers and Geography in the Federal Government."

9. These details are taken from Withers, "Toward a Historical Geography of Enlightenment in Scotland"; Withers, "Situating Practical Reason"; and Withers and Mayhew, "Rethinking 'Disciplinary' History."

10. Moravia, "Philosophie et géographie à la fin du XVIIIe siècle," 949–58; Staum, Human Geography in the French Institute."

11. Bowen, *Empiricism and Geographical Thought,* 203–9; Büttner, "Kant and the Physico-Theological Consideration of the Geographical Facts."

12. Bowen, *Empiricism and Geographical Thought,* 154–59; Denecke, "Die Geschichte der Geographie in Göttingen"; Denecke, "Anwendungsorientierte Ansätze in der Frühzeit der Geographie in Göttingen."

13. The Saussure material is from Carozzi and Newman, *Lectures on Physical Geography Given in 1775 by Saussure.*

14. The examples of Priestley, Mair, and Hamilton, respectively, are from Mayhew, "Geography in Eighteenth-Century British Education," 747; Robinson, "Geography in the Dissenting Academies"; and Withers, *Geography, Science and National Identity,* 137. For other

work on school geography and the Enlightenment in Britain, see Adams, "Mass Distribution of Geographical Literature in Ulster"; Smith, "State Formation, Geography, and a Gentleman's Education."

15. On Dinwiddie, see Edney, *Mapping an Empire*, 303–4; and Withers, *Geography, Science and National Identity*, 128 (who also discusses MackGregory, at 119–21).

16. Withers, "Towards a History of Geography in the Public Sphere"; Withers, "Situating Practical Reason."

17. The facts and figures on Philadelphia are taken from analysis of the *Pennsylvania Gazette*, 1728–1800. On Virginia, see the *Virginia Gazette*, and for South Carolina, Dunbar, "Geographic Education in Early Charleston."

18. As is so obviously the case for Smith and Vining, *American Geographers*.

19. Ibid., 163.

20. Maxwell, *Pombal, Paradox of the Enlightenment*, 100; Carrato, "Enlightenment in Portugal and the Educational Reforms of the Marquis of Pombal."

21. In Germany, for example, Büsching's *Magazin für die Neue Historie und Geographie* was mainly concerned with the historical aspects of geography, whereas Friedrich Justin Bertuch's *Allgemeine Geographischen Ephemeriden* aimed at providing a unified geography but mainly included explorers' communications in letter form. In contrast, the *Gentleman's Magazine* offered maps and accompanying articles, which, between 1731 and 1754 at least, allowed the British public to understand and follow British military campaigns overseas: see Reitan, "Expanding Horizons."

22. Terrall, "Émilie du Châtelet and the Gendering of Science"; Terrall, "Salon, Academy, and Boudoir." On salon conversation, see also Chartier, "Man of Letters," 158–62.

23. Klein, "Liberty, Manners, and Politeness in Early Eighteenth-Century England"; Walters, "Conversation Pieces."

24. Secord, "Newton in the Nursery"; Heilbron, "Domesticating Science in the Eighteenth Century"; Schaffer, "Machine Philosophy," from whom I take the quote (181).

25. The Holmes example is cited in Mayhew, "Geography in Eighteenth-Century British Education," 743–44.

26. On Spilsbury and other geographical games, see Shefrin, *Neatly Dissected for the Instruction of Young Ladies and Gentlemen in the Knowledge of Geography*.

27. Benjamin Martin's work *Description and Use of Both the Globes* (1762), as quoted in Walters, "Conversation Pieces," 125.

28. Kitromilides, *Enlightenment as Social Criticism*, 149.

29. Cosgrove, *Apollo's Eye*, 176–77. More widely, see Shea, "Rhetoric of Experiments and Scientific Illustrations in the Enlightenment."

Chapter Eleven

1. On the Atlantic Enlightenment, see, for example, Dubois, "An Enslaved Enlightenment"; Withers, "Where Was the Atlantic Enlightenment?"; on the Baltic Enlightenment, see Koerner, "Daedalus Hyperborus."

2. I take this from Cañizares-Esguerra, *How to Write the History of the New World*, chapter 5 of which (266–345) is titled "Whose Enlightenment Was It Anyway?" The quotes are from 266 and 267.

3. These are issues explored by Glacken, *Traces on the Rhodian Shore*, 623–54.

4. Condorcet, *Outlines of an Historical View of the Progress of the Human Mind*, 317. As if to connect my claims across time, the copy in Edinburgh University Library I consulted in thinking about these matters was Dugald Stewart's.

5. Rose, "Utopianism."

6. As quoted in Withers, "Geography, Enlightenment, and the Paradise Question," 82.

7. Heffernan, "Historical Geographies of the Future."

8. Mercier, *L'an 2440*, quotes at 59, 172, and 200.

9. On this point about improvement, see Daston, "Afterword."

10. Outram, "Enlightenment Our Contemporary."

11. Whitaker, *Mapmaker's Wife*.

12. Dening, "Deep Times, Deep Spaces."

Bibliography

Abbattista, Guido. "Establishing the 'Order of Time and Place': 'Rational Geography,' French Erudition and the Emplacement of History in Gibbon's Mind." In *Edward Gibbon: Bicentenary Essays,* ed. David Womersley, 45–72. Oxford: Voltaire Foundation, 1997.

Adams, J. R. R. "The Mass Distribution of Geographical Literature in Ulster, 1750–1850." *Geographical Journal* 153 (1987): 383–87.

Adamo, Sergia. "Constructing an Event, Contemplating Ruins, Theorizing Nature: The Lisbon Earthquake and Some Italian Reactions." *European Review* 14 (2004): 339–50.

African Association. *Proceedings of the Association for Promoting the Discovery of the Interior Parts of Africa.* 2 vols. London: W. Bulmer, 1810.

Alder, Ken. *The Measure of All Things: The Seven-Year Odyssey That Transformed the World.* London: Little, Brown, 2002.

———. "A Revolution to Measure: The Political Economy of the Metric System in France." In *The Values of Precision,* ed. M. Norton Wise, 39–71. Princeton, NJ: Princeton University Press, 1995.

Alexander, Michael. *Omai "Noble Savage."* London: Collins & Harvill Press, 1977.

Allan, David. "The Scottish Enlightenment and the Politics of Provincial Culture: The Perth Literary and Antiquarian Society, ca. 1784–1790." *Eighteenth-Century Life* 27 (2003): 1–30.

Allen, David. "*Arcana ex Multitudine*: Prosopography as a Research Technique." *Archives of Natural History* 17 (1990): 349–59.

Allen, John L. "Thomas Jefferson and the Mountain of Salt: A Presidential Image of Louisiana Territory." *Historical Geography* 31 (2003): 7–20.

Anderson, Benedict. *Imagined Communities: Reflections on the Origins and Spread of Nationalism.* London and New York: Verso, 1991.

Andrewes, William J. H., ed. *The Quest for Longitude.* Cambridge, MA: Harvard University Press, 1996.

Anonymous. Review of *John Pinkerton's Modern Geography: A Description of the Empires, Kingdoms, States, and Colonies, with the Oceans, Seas, and Isles, in All Parts of the World: Including the Most Recent Discoveries, and Political Alterations. Edinburgh Review* 3 (1803): 67–81.

Araujo, Ana Christina. "European Public Opinion and the Lisbon Earthquake." *European Review* 14 (2006): 313–20.

Bajkó, Mátyás. "Ideas of Enlightenment in the Colleges of Hungary and of Transylvania." *Studies on Voltaire and the Eighteenth Century* 264 (1989): 742–44.

Baker, John N. L. *The History of Geography*. Oxford: Blackwell, 1963.

Baker, Keith Michael, and Peter Hanns Reill, eds. *What's Left of Enlightenment? A Postmodern Question*. Stanford, CA: Stanford University Press, 2000.

Ballantyne, Tony, ed. *Science, Empire and the European Exploration of the Pacific*. London: Ashgate, 2004.

Barley, Nigel. "Africa: In the Shadow of the Enlightenment." In *Enlightenment: Discovering the World in the Eighteenth Century*, ed. Kim Sloan, 270–75. London: British Museum Press, 2003.

Barrett, Frank. *Disease and Geography: The History of an Idea*. Geographical Monographs, no. 23. York University, 2000.

———"Finke's 1792 Map of Human Diseases: The First World Disease Map?" *Social Science and Medicine* 50 (2000): 915–21.

———. "A Medical Geographical Anniversary." *Social Science and Medicine* 37 (1993): 701–10.

———. "'Scurvy' Lind's Medical Geography." *Social Science and Medicine* 33 (1991): 347–53.

Bartlett, Roger, and Jane M. Hartley, eds. *Russia in the Age of the Enlightenment: Essays for Isabel de Madariaga*. London: Macmillan, 1990.

Barton, David, and Nigel Hall, eds. *Letter Writing as a Social Practice*. Amsterdam and Philadelphia: John Benjamins Publishing, 1999.

Bates, David. "Cartographic Aberrations: Epistemology and Order in the Encyclopedic Map." In *Using the* Encyclopédie*: Ways of Knowing, Ways of Reading*, ed. Daniel Brewer and Julie Candler Hayes, 1–20. Oxford: Voltaire Foundation, 2002.

Baudin, Nicolas. *The Journal of Post Captain Nicolas Baudin, Commander-in-Chief of the Corvettes Géographe and Naturaliste*. Adelaide: Libraries Board of South Australia, 1974.

Bazerman, Charles. "Letters and the Social Grounding of Differentiated Genres." In *Letter Writing as a Social Practice*, ed. David Barton and Nigel Hall, 15–29. Amsterdam and Philadelphia: John Benjamins Publishing, 1999.

Beaglehole, J. C., ed. *James Cook's Journals*. 3 vols. Cambridge: Cambridge University Press, 1965–68.

Behrens, Catherine B. A. *Society, Government and the Enlightenment: The Experiences of Eighteenth-Century France and Prussia*. London: Thames & Hudson, 1985.

Bellin, Jacques Nicolas. *Hydrographie française*. 2 vols. Paris: A. Baudet, 1792.

———. *Recueil des mémoires qui on été publies avec Les Cartes Hydrographique*. Paris: Librarie Imprimeur, 1751.

Belyea, Barbara. "Amerindian Maps: The Explorer as Translator." *Journal of Historical Geography* 18 (1992): 267–77.

———. "Inland Journeys, Native Maps." In *Cartographic Encounters: Perspectives on Native American Mapmaking and Map Use*, ed. G. Malcolm Lewis, 135–55. Chicago: University of Chicago Press, 1998.

Bender, John. "A New History of the Enlightenment?" *Eighteenth-Century Life* 16 (1992): 1–20.

Bennett, Jim. "The Travels and Trials of Mr. Harrison's Timekeeper." In *Instruments, Travel and Science: Itineraries of Precision from the Seventeenth to the Twentieth Century*, ed. Marie-Noëlle Bourguet, Christian Licoppe, and H. Otto Sibum, 75–95. London: Routledge, 2002.

Bereta, Marco, and Tore Frängsmyr, eds. *Sidereus Nuncius and Stella Polaris: The Scientific Relations between Italy and Sweden in Early Modern History.* Canton, MA: Science History Publications, 1997.

Berlin, Isaiah. *The Age of Enlightenment.* New York: New American Library, 1956.

Berry, Christopher. *Social Theory of the Scottish Enlightenment.* Edinburgh: Edinburgh University Press, 1997.

Berry, H. "Rethinking Politeness in Eighteenth-Century England: Moll King's Coffeehouse and the Significance of 'Flash Talk.'" *Transactions of the Royal Historical Society,* 5th ser., 6 (2001): 65–81.

Blanckaert, Claude. "Buffon and the Natural History of Man: Writing History and the 'Foundational Myth' of Anthropology." *History of the Human Sciences* 6 (1993): 13–50.

Blanning, T. C. W. "The Enlightenment in Catholic Germany." In *The Enlightenment in National Context,* ed. Roy Porter and Mikuláš Teich, 118–26. Cambridge: Cambridge University Press, 1981.

Blouet, Brian, ed. *The Origins of Academic Geography in the United States.* Hamden, CT: Archon Books.

Bodenheimer, F. S. "Zimmermann's Specimen Zoologiae Geographiae Quadrupedum: A Remarkable Zoogeographical Publication of the End of the Eighteenth Century." *Archives Internationales d'Histoire des Sciences* 34 (1955): 351–57.

Bodi, Leslie. "The Austrian Enlightenment: An Essay on Publications, 1975–1990." In *The Austrian Enlightenment and Its Aftermath,* ed. Ritchie Robertson and Edward Timms, 171–87. Edinburgh: Edinburgh University Press, 1991.

Boschung, Urs. "Göttingen, Hanover, and Europe: Haller's Correspondence." In *Göttingen and the Development of the Natural Sciences,* ed. Nicolaas Rupke, 33–49. Göttingen: Wallstein Verlag, 2002.

Boschung, Urs, Barbara Braun-Bucher, Stefan Hächler, Anne Kathrin Ott, Hubert Steinke, and Martin Stuber, eds. *Albrecht von Haller. Repertorium zu Albrecht von Hallers Korrespondenz, 1724–1777.* Basel: Schwabe, 2002.

Boscovich, R. J., and Christoph Maire. *Voyage astronomique et géographique dans l'Etat et d'Eglise: Enterpris par l'ordre at sous les auspices du Pape Benôit XIV.* Paris: Tilliard, 1770.

Bosher, J. F. "Joseph-François Lafitau." In *Encyclopedia of the Enlightenment,* ed. Alan C. Kors, 2:342–43. New York and Oxford: Oxford University Press, 2003.

Bourguet, Marie-Noëlle. "The Explorer." In *Enlightenment Portraits,* ed. Michel Vovelle and trans. Lydia Cochrane, 257–315. Chicago: University of Chicago Press, 1997.

———. "Landscape with Numbers: Natural History, Travel and Instruments in the Late Eighteenth and Early Nineteenth Centuries." In *Instruments, Travel and Science: Itineraries of Precision from the Seventeenth to the Twentieth Century,* ed. Marie-Noëlle Bourguet, Christian Licoppe, and H. Otto Sibum, 96–125. London: Routledge, 2002.

Bourguet, Marie-Noëlle, Christian Licoppe, and H. Otto Sibum, eds. *Instruments, Travel and Science: Itineraries of Precision from the Seventeenth to the Twentieth Century.* London: Routledge, 2002.

———. Introduction to *Instruments, Travel and Science: Itineraries of Precision from the Seventeenth to the Twentieth Century,* ed. Marie-Noëlle Bourguet, Christian Licoppe and H. Otto Sibum, 1–19. London: Routledge, 2002.

Bowen, Margarita. *Empiricism and Geographical Thought: From Francis Bacon to Alexander von Humboldt.* Cambridge: Cambridge University Press, 1981.

Bowden, Martin S. "The Invention of American Tradition." *Journal of Historical Geography* 18 (1992): 3–26.

Bravo, Michael. "The Anti-Anthropology of Highlanders and Islanders." *Studies in the History and Philosophy of Science* 29 (1998): 369–89.

———. "Ethnographic Navigation and the Geographical Gift." In *Geography and Enlightenment,* ed. David N. Livingstone and Charles W. J. Withers, 199–235. Chicago: University of Chicago Press, 1999.

———. "Precision and Curiosity in Scientific Travel: James Rennell and the Orientalist Geography of the New Imperial Age (1760–1830)." In *Voyages and Visions: Towards a Cultural History of Travel,* ed. Ja Elsner and Joan-Pau Rubiés, 162–83. London: Reaktion Books, 1999.

Brewer, Daniel. *Discourse of Enlightenment in Eighteenth-Century France: Diderot and the Art of Philosophizing.* Cambridge: Cambridge University Press, 1993.

———. "Lights in Space." *Eighteenth-Century Studies* 37 (2004): 171–86.

Briggs, Robin. "The Académie Royale des Sciences and the Pursuit of Utility." *Past and Present* 131 (1991): 38–88.

Broadie, Alexander. *The Cambridge Companion to the Scottish Enlightenment.* Cambridge: Cambridge University Press, 2003.

———. Introduction to *The Cambridge Companion to the Scottish Enlightenment,* ed. Alexander Broadie, 1–7. Cambridge: Cambridge University Press, 2003.

Broberg, Gunnar. "*Homo Sapiens*: Linnaeus's Classification of Man." In *Linnaeus: The Man and His Work,* ed. Tore Frängsmyr, 156–94. Berkeley, Los Angeles, and London: University of California Press, 1983.

Broc, Numa. *La géographie des philosophes, géographes et voyageurs français au XVIIIe siècle.* Paris: Edition Ophrys, 1974.

Brock, Helen. "North America, a Western Outpost of European Medicine." In *The Medical Enlightenment of the Eighteenth Century,* ed. Andrew Cunningham and Roger French, 194–216. Cambridge: Cambridge University Press, 1990.

Broman, Thomas. "The Habermasian Public Sphere and Science *in* the Enlightenment." *History of Science* 36 (1998): 123–49.

Brown, Harcourt. "Maupertuis *Philosophe*: Enlightenment and the Berlin Academy." *Studies on Voltaire and the Eighteenth Century* 24 (1963): 255–69.

Brown, Ralph H. "The American Geographies of Jedidiah Morse." *Annals of the Association of American Geographers* 31 (1941): 145–217.

Brown, Stewart J., ed. *William Robertson and the Expansion of Empire.* Cambridge: Cambridge University Press, 1997.

Browne, Janet. *The Secular Ark: Studies in the History of Biogeography.* New Haven, CT, and London: Yale University Press, 1983.

Brückner, Martin. *The Geographic Revolution in Early America: Maps, Literacy and National Identity.* Chapel Hill: University of North Carolina Press, 2006.

———. "Geography, Reading, and the World of Novels in the Early Republic." In *Early America Re-explored: New Readings in Colonial, Early National, and Antebellum Culture,* ed. Klaus H. Schmidt and Fritz Fleischmann, 385–410. New York: Lang, 2000.

———. "Lessons in Geography: Maps, Spellers, and Other Grammars of Nationalism in the Early Republic." *American Quarterly* 51 (1999): 311–43.

Bruyset, Jean-Marie. *Atlas des enfans, ou nouvelle méthode pour apprendre la géographie*. Lyon: Bruyset, 1790.

Buck, Peter. "People Who Counted: Political Arithmetic in the Eighteenth Century." *Isis* 73 (1982): 28–45.

Buffon, Georges-Louis LeClerc, comte de, Louis-Jean-Marie Daubenton, Philibert Guéneau de Montbeillard, and Gabrield-Léopold-Charles-Aimé Bexon. *Histoire naturelle, générale et particulière, avec la description du cabinet du roi*. 36 vols. Paris: Imprimerie Royales, 1749–89.

Buisman, J. W. "Some Considerations on the Social Diffusion of Enlightenment Ideas in the Netherlands." *Studies on Voltaire and the Eighteenth Century* 346 (1997): 354–57.

Burnett, D. Graham. *Masters of All They Surveyed: Exploration, Geography, and a British El Dorado*. Chicago: University of Chicago Press, 2000.

Bury, J. B. T. *The Idea of Progress: An Enquiry into Its Origins and Growth*. London: Macmillan, 1932.

Büttner, Manfred. "Kant and the Physico-Theological Consideration of the Geographical Facts." In *Science and Religion/ Wissenschaft und Religion*, ed. Anne Bäumer and Manfred Büttner, 82–92. Bochum: Brockmeyer, 1989.

Bynum, William F. "Cullen and the Study of Fevers in Britain, 1760–1820," *Medical History* (supplement no. 1) (1981): 135–47.

Calhoun, Craig, ed. *Habermas and the Public Sphere*. Cambridge: Cambridge University Press, 1992.

Camic, Charles. *Experience and Enlightenment: Socialization for Cultural Change in Eighteenth-Century Scotland*. Edinburgh: Edinburgh University Press, 1983.

Cañizares-Esguerra, Jorge. *How to Write the History of the New World: Histories, Epistemologies, and Identities in the Eighteenth-Century Atlantic World*. Stanford, CA: Stanford University Press, 2001.

Cannon, Susan Faye. *Science in Culture: The Early Victorian Period*. New York: Science History Publications, 1978.

Cardoso, José Luís. "Economic Thought in Late Eighteenth-Century Portugal." *History of Political Economy* 22 (1990): 429–41.

Carey, Daniel. "Reconsidering Rousseau: Sociability, Moral Sense and the American Indian from Hutcheson to Bartram." *British Journal for Eighteenth-Century Studies* 21 (1998): 25–38.

Carneiro, Ana, Ana Simões, and Maria Paula Diogo. "Enlightenment Science in Portugal: The *Estrangeirados* and Their Communication Networks." *Social Studies of Science* 30 (2000): 591–619.

Carozzi, Albert V., and John K. Newman. *Lectures on Physical Geography Given in 1775 by Horace-Bénédict de Saussure at the Academy of Geneva*. Geneva: Éditions Zoé, 2003.

Carozzi, Marguerite. "Reaction of British Colonies in America to the Lisbon Earthquake." *Earth Sciences History* 2 (1983): 17–27.

Carrato, José Ferreira. "The Enlightenment in Portugal and the Educational Reforms of the Marquis de Pombal." *Studies on Voltaire and the Eighteenth Century* 67 (1977): 359–93.

Carrithers, David. "The Enlightenment Science of Society." In *Inventing Human Science: Eighteenth-Century Domains*, ed. Christopher Fox, Roy Porter, and Robert Wokler, 232–70. Berkeley and London: University of California Press, 1996.

Cascardi, Anthony J. *Consequences of Enlightenment*. Cambridge: Cambridge University Press, 1999.

Cassirer, Ernst. *The Philosophy of the Enlightenment* (1932). Princeton, NJ: Princeton University Press, 1951.

Cavallo, Guglielmo, and Roger Chartier, eds. *A History of Reading in the West*. Trans. Lydia G. Cochrane. Cambridge, MA: Polity Press, 1999.

Cawood, John A. "Terrestrial Magnetism and the Development of International Collaboration in the Early Nineteenth Century." *Annals of Science* 34 (1977): 551–87.

Cerruti, Luigi. "Dante's Bones." In *The Sciences in the European Periphery during the Enlightenment*, ed. Kostas Gavroglu, 95–178. Dordrecht: Kluwer Academic Publishers, 1999.

Chadwick, Owen. "The Italian Enlightenment." In *The Enlightenment in National Context*, ed. Roy Porter and Mikuláš Teich, 90–105. Cambridge: Cambridge University Press, 1981.

Chapin, Seymour L. "The Shape of the Earth." In *Planetary Astronomy from the Renaissance to the Rise of Astrophysics, Part B: The Eighteenth and Nineteenth Centuries*, ed. René Taton and Curtis Wilson, 22–34. Cambridge: Cambridge University Press, 1995.

Chard, Chloe. *Pleasure and Guilt on the Grand Tour: Travel Writing and Imaginative Geography, 1600–1830*. Manchester: Manchester University Press, 1999.

Chard, Chloe, and Helen Langdon, eds. *Transports: Travel, Pleasure, and Imaginative Geography, 1600–1830*. New Haven, CT, and London: Yale University Press, 1996.

Chartier, Roger. "The Man of Letters." In *Enlightenment Portraits*, ed. Michel Vovelle and trans. Lydia G. Cochrane, 142–89. Chicago: University of Chicago Press, 1992.

Ching, Julia, and Willard G. Oxtoby, eds. *Discovering China: European Interpretations in the Enlightenment*. Rochester: University of Rochester Press, 1992.

Christensen, Thomas. "Music." In *Encyclopedia of the Enlightenment*, ed. Alan Kors, 3:104–10. New York and Oxford: Oxford University Press, 2003.

Christensson, Jakob. "Scandinavia." In *Encyclopedia of the Enlightenment*, ed. Alan Kors, 4:20–25. New York and Oxford: Oxford University Press, 2003.

Christian, Jessica. "Paul Sandby and the Military Survey of Scotland." In *Mapping the Landscape: Essays on Art and Cartography*, ed. Nicholas Alfrey and Stephen Daniels, 18–22. Nottingham, UK: Nottingham Art Gallery, 1990.

Christie, John. "The Human Sciences: Origins and Histories." *History of the Human Sciences* 6 (1993): 6–12.

Clark, Peter. *British Clubs and Societies, 1580–1800: The Origins of an Associational World*. Oxford: Clarendon Press, 2000.

Clark, William, "The Pursuit of the Prosopography of Science." In *The Cambridge History of Science*, vol. 4: *Eighteenth-Century Science*, ed. Roy Porter, 211–37. Cambridge: Cambridge University Press, 2003.

Clark, William, Jan Golinski, and Simon Schaffer. Introduction to *The Sciences in Enlightened Europe*, ed. William Clark, Jan Golinski, and Simon Schaffer, 3–31. Chicago: University of Chicago Press, 1999.

———, eds. *The Sciences in Enlightened Europe*. Chicago: University of Chicago Press, 1999.

Clarke, John J. *Oriental Enlightenment: The Encounter between Asian and Western Thought*. London: Routledge, 1997.

Clayton, Daniel W. *Islands of Truth: The Imperial Fashioning of Vancouver Island*. Vancouver: University of British Columbia Press, 2000.

Close, Charles F. *The Early Years of the Ordnance Survey*. With an introduction by J. B. Harley. Newton Abbot: David & Charles, 1969.

Cochrane, Eric W. *Tradition and Enlightenment in the Tuscan Academies, 1690–1800*. Chicago: University of Chicago Press, 1961; Rome: Edizioni di Storia e Letteratura, 1961.

Cohen, Emily Jane. "Enlightenment and the Dirty Philosopher." *Configurations* 5 (1997): 369–424.

Cole, Laurence. "Nation, Anti-Enlightenment, and Religious Revival in Austria: Tyrol in the 1790s." *Historical Journal* 43 (2000): 475–97.

Colley, Linda. *Captives: Britain, Empire and the World, 1600–1850*. Jonathan Cape: London, 2002.

Colten, Craig, and Lydia DeLyser. "The Louisiana Purchase Territory: Historical Geographers' Perspectives." *Historical Geography* 31 (2003): 3–6.

Commager, Henry S. *The Empire of Reason: How Europe Imagined and America Realized the Enlightenment*. Garden City, NY: Anchor Books/Doubleday, 1978.

Condorcet, Jean-Antoine-Nicolas de Caritat. *Outlines of an Historical View of the Progress of the Human Mind*. London: Printed for J. Johnston, 1795.

Cook, Elizabeth. *Epistolary Bodies: Gender and Genre in the Eighteenth-Century Republic of Letters*. Stanford, CA: Stanford University Press, 1996.

Cooper, Alix. "From the Alps to Egypt (and Back Again): Dolomieu, Scientific Voyaging, and the Construction of the Field in Eighteenth-Century Natural History." In *Making Space for Science: Territorial Themes in the Shaping of Knowledge*, ed. Crosbie Smith and Jon Agar, 39–63. London: Macmillan, 1998.

Cooter, Roger, and Stephen Pumfrey. "Separate Spheres and Public Places: Reflections on the History of Science Popularization and Science in Popular Culture." *History of Science* 32 (1994): 237–67.

Corfield, Penelope. "Class by Name and Number in Eighteenth-Century England." *History* 72 (1987): 38–61.

Cosgrove, Denis E. *Apollo's Eye: A Cartographic Genealogy of the Earth in the Western Imagination*. Baltimore and London: Johns Hopkins University Press, 2001.

———. "Global Illumination and Enlightenment in the Geographies of Vincenzo Coronelli and Athanasius Kircher." In *Geography and Enlightenment*, ed. David N. Livingstone and Charles W. J. Withers, 33–66. Chicago: University of Chicago Press, 1999.

Costa, Shelley. "Marketing Mathematics in Early Eighteenth-Century England: Henry Beighton, Certainty, and the Public Sphere." *History of Science* 40 (2002): 211–32.

Cowan, Barry. "What Was Masculine about the Public Sphere? Gender and the Coffee-house Milieu in Post-Restoration England." *History Workshop Journal* 51 (2001): 127–57.

Cranmer-Byng, J. L., and Trevor Levere. "A Case Study in Cultural Collision: Scientific Apparatus in the Macartney Embassy to China, 1793." *Annals of Science* 38 (1981): 503–26.

Cribb, Joe, Jessica Harrison-Hall, and Tim Clark. "Trade and Learning: The European 'Discovery' of the East." In *Enlightenment: Discovering the World in the Eighteenth Century*, ed. Kim Sloan, 258–69. London: British Museum Press, 2003.

Crocker, Lester G., ed. *The Age of Enlightenment*. London: J. M. Dent, 1969.

———. "The Enlightenment: What and Who?" *Studies in Eighteenth-Century Culture* 17 (1987): 335–47.

———. "Interpreting the Enlightenment: A Political Approach." *Journal of the History of Ideas* 46 (1985): 211–30.

———. Introduction to *The Blackwell Companion to the Enlightenment*, ed. John W. Yolton, 1–12. Oxford: Blackwell, 1991.

Cunningham, Andrew. "Medicine to Calm the Mind: Boerhaave's Medical System, and Why It Was Adopted in Edinburgh." In *The Medical Enlightenment of the Eighteenth Century*, ed. Andrew Cunningham and Roger French, 40–66. Cambridge: Cambridge University Press, 1990.

Cunningham, Andrew, and Roger French, eds. *The Medical Enlightenment of the Eighteenth Century*. Cambridge: Cambridge University Press, 1990.

Dainville, François de. *La géographie des humanistes*. Paris: Beauchesnes, 1940.

———. *Le langage des géographes: Terms, signes, couleurs des cartes anciennes, 1500–1800*. Paris: Harmattan, 1964.

Dampier, William. *A New Voyage round the World*. London: Printed for James Knapton, 1697.

Daniel, C. "La géographie dans les collèges de Jesuites aux XVIIe et XVIIIe siècles." *Etudes* 7 (1879): 801–24.

D'Anville, Jean Baptiste Bourguinon. *Nouvel atlas de la Chine, de la Tartarie Chinoise, et du Tibet*. Paris: Henri Scheurleer, 1737.

Darnton, Robert. *The Business of the Enlightenment: A Publishing History of the Encyclopédie, 1775–1800*. Cambridge, MA: Harvard University Press, 1979.

———. "An Early Information Society: News and the Media in Eighteenth-Century Paris." *American Historical Review* 105 (2000): 1–35.

———. "George Washington's False Teeth." *New York Review of Books*, 27 March 1997, 34–38.

———. *George Washington's False Teeth: An Unconventional Guide to the Eighteenth Century*. New York and London: Norton, 2003.

———. *The Great Cat Massacre and Other Episodes in French Cultural History*. Harmondsworth: Penguin, 1984.

———. "The History of Reading." In *New Perspectives in Historical Writing*, ed. Peter Burke, 140–67. Cambridge, MA: Polity Press, 1991.

———. "In Search of the Enlightenment: Recent Attempts to Create a Social History of Ideas." *Journal of Modern History* 63 (1971): 113–32.

———. "What Is the History of Books?" *Daedalus* 111 (1982): 65–83.

Daston, Lorraine. "Afterword: The Ethos of Enlightenment." In *The Sciences in Enlightened Europe*, ed. William Clark, Jan Golinski, and Simon Schaffer, 495–504. Chicago: University of Chicago Press, 1999.

———. *Classical Probability in the Enlightenment*. Princeton, NJ: Princeton University Press, 1988.

———. "Enlightenment Calculations." *Critical Inquiry* 21 (1994): 182–202.

———. "The Ideal and Reality of the Republic of Letters in the Enlightenment." *Science in Context* 4 (1991): 367–86.

David, Andrew, ed. *The Malaspina Expedition, 1789–1794: Journal of the Voyage by Alejandro Malaspina*. London: Hakluyt Society, 2001.

Davidson, Cathy. "Introduction: Towards a History of Books and Readers." *American Quarterly* 40 (1988): 7–17.

Davies, Alun C. "Testing a New Technology: Captain George Vancouver's Survey and Navigation in Alaskan Waters, 1794." In *Enlightenment and Exploration in the North Pacific, 1741–1805*, ed. Stephen Haycox, James K. Barnett, and Caedmon A. Liburd, 103–15. Seattle and London: University of Washington Press, 1997.

Davis, Natalie Zemon. *Culture and Society in Early Modern France*. Cambridge, MA: Polity Press in association with Blackwell, 1987.

Davis, Walter W. "China, the Confucian Ideal, and the European Age of Enlightenment." In *Discovering China: European Interpretations in the Enlightenment*, ed. Julia Ching and Willard G. Oxtoby, 1–26. Rochester, NY: University of Rochester Press, 1992.

Deacon, Margaret. *Scientists and the Sea, 1650–1900*. London: Academic Press, 1971.

Dean, Dennis. "The Age of the Earth Controversy." *Annals of Science* 38 (1981): 435–56.

———. "Benjamin Franklin and Earthquakes." *Annals of Science* 46 (1989): 481–95.

———. *James Hutton and the History of Geology*. Ithaca, NY, and London: Cornell University Press, 1992.

Delano-Smith, Catherine, and Roger J. P. Kain. *English Maps: A History*. Toronto and Buffalo: University of Toronto Press, 1999.

Delon, Michel, ed. *Encyclopedia of the Enlightenment*. 2 vols. Chicago: Fitzroy Daerborn, 2001.

De Madariaga, Isabel. "Russia." In *Encyclopedia of the Enlightenment*, ed. Alan Kors, 3:491–97. New York and Oxford: Oxford University Press, 2003.

Demel, Walter. "China's Changing Image during the Age of Enlightenment." *Studies on Voltaire and the Eighteenth Century* 348 (1997): 1231–35.

Denecke, Dietrich. "Anwendungsorientierte Ansätze in der Frühzeit der Geographie in Göttingen." *Göttinger Geographische Abhandlungen* 100 (1997): 111–26.

———. "Die Geschichte der Geographie in Göttingen." In *Die Geschichte der Verfassung und der Fachbereiche der Georg-August-Universitat zu Göttingen*, ed. Hans-Günther Schlotter, 198–204. Göttingen: Vandenhock & Ruprecht, 1998.

Dening, Greg. "Deep Times, Deep Spaces: Civilising the Sea." In *Sea Changes: Historicizing the Ocean*, ed. Bernhard Klein and Gesa Mackenthun, 13–36. New York: Routledge, 2004.

———. *Mr. Bligh's Bad Language: Passion, Power and Theatre on the Bounty*. Cambridge: Cambridge University Press, 1992.

———. *Presenting the Past; History's Theatre Performances*. Melbourne: University of Melbourne Press, 1996.

De Pauw, Abbé Cornelius. *Recherches philosophique sue les américains ou mémoire interassants pour server à l'histoire de l'espèce humaine*. Berlin: Schmidt, 1768.

Dettelbach, Michael. "Global Physics and Aesthetic Empire: Humboldt's Physical Portrait of the Tropics." In *Visions of Empire: Voyages, Botany, and Representations of Nature*, ed. David Philip Miller and Peter Hanns Reill, 258–92. Cambridge: Cambridge University Press, 1996.

———. "Humboldtian Science." In *Cultures of Natural History*, ed. Nicholas Jardine, James A. Secord, and Emma C. Spary, 287–304. Cambridge: Cambridge University Press, 1996.

De Vorsey, Louis. "Maps in Colonial Promotion: James Edward Oglethorpe's Use of Maps in 'Selling' the Georgia Scheme." *Imago Mundi* 38 (1986): 35–45.

D'Haen, Theo. "On How Not to Be Lisbon if You Want to Be Modern—Dutch Reactions to the Lisbon Earthquake." *European Review* 14 (2006): 351–58.

Dialetis, Dimitris, Kostas Gavroglu, and Manolis Patiniotis. "The Sciences in the Greek Speaking Regions during the 17th and 18th Centuries." In *The Sciences in the European Periphery during the Enlightenment*, ed. Kostas Gavroglu, 41–72. Dordrecht: Kluwer Academic Publishers, 1999.

Dias, Rosie. "'A World of Pictures': Pall Mall and the Topography of Display, 1780–99." In *Georgian Geographies: Essays on Space, Place and Landscape in the Eighteenth Century,* ed. Miles Ogborn and Charles W. J. Withers, 92–113. Manchester: Manchester University Press, 2004.

Dikötter, Frank. "China." In *The Cambridge History of Science,* vol. 4: *Eighteenth-Century Science,* ed. Roy Porter, 688–97. Cambridge: Cambridge University Press, 2003.

Dolan, Brian. *Exploring European Frontiers: British Travellers in the Age of Enlightenment.* Basingstoke: Macmillan, 2000.

Domingues, Francisco Contente. "Science and Nationalism: Portugal in the Late 18th Century." *History of European Ideas* 16 (1993): 91–96.

Donato, Clorinda. "Eighteenth-Century Encyclopedias and National Identity." *History of European Ideas* 16 (1993): 959–65.

Donovan, Arthur. "Scottish Responses to the New Chemistry of Lavoisier." *Studies in Eighteenth-Century Culture* 9 (1979): 237–49.

Dörflinger, J. *Die Geographie in der "Encyclopéedie" ein Wissenschaftsgeschichtlische studie.* Vienna, 1976.

Dorn, Harold. *The Geography of Science.* Baltimore: Johns Hopkins University Press, 1991.

Douglas, Bronwen. "Seaborne Ethnography and the Natural History of Man." *Journal of Pacific History* 38 (2003): 3–27.

Downes, Alan. "The Bibliographic Dinosaurs of Georgian Geography (1714–1830)." *Geographical Journal* 137 (1971): 379–87.

Drayton, Richard. "Knowledge and Empire." In *The Oxford History of the British Empire,* vol. 2: *The Eighteenth Century,* ed. P. J. Marshall, 231–52. Oxford: Oxford University Press, 1998.

———. *Nature's Government: Science, Imperial Britain and the 'Improvement' of the World.* New Haven: Yale University Press, 2000.

Dubois, Laurent. "An Enslaved Enlightenment: Rethinking the Intellectual History of the French Atlantic." *Social History* 31 (2006): 1–14.

Duchet, Michel. *Anthropologie et histoire au siècle des lumières.* Paris: Maspéro, 1971.

Dukes, Paul. "The Russian Enlightenment." In *The Enlightenment in National Context,* ed. Roy Porter and Mikuláš Teich, 176–91. Cambridge: Cambridge University Press, 1981.

Dunbar, Gary. "Geographic Education in Early Charleston." *Journal of Geography* 69 (1970): 348–50.

Dunbar, James. *Essays on the History of Mankind in Rude and Cultivated Ages.* London and Edinburgh: W. Strahan, T. Cadell, & J. Balfour, 1780.

Dunmore, John. *French Explorers in the Pacific.* 2 vols. Oxford: Clarendon Press, 1965–69.

Duţu, Alexandru. "National and European Consciousness in the Romanian Enlightenment." *Studies on Voltaire and the Eighteenth Century* 55 (1967): 463–79.

Earle, Rebecca, ed. *Epistolary Selves: Letters and Letter-Writers, 1600–1945.* Aldershot: Ashgate, 1999.

Edney, Matthew. "Bringing India to Hand: Mapping an Empire, Denying Space." In *The Global Eighteenth Century,* ed. Felicity A. Nussbaum, 65–78. Baltimore and London: Johns Hopkins University Press, 2003.

———. "Cartography: Disciplinary History." In *Sciences of the Earth: An Encyclopedia of Events, People, and Phenomena,* ed. Gregory A. Good, 1:81–85. New York: Garland, 1998.

———. *Mapping an Empire: The Geographical Construction of British India, 1765–1843.* Chicago: University of Chicago Press, 1997.

———. "Mathematical Cosmography and the Social Ideology of British Cartography, 1780–1820." *Imago Mundi* 46 (1994): 101–16.

Edwards, Phyllis. "Robert Brown (1773–1858) and the Natural History of Matthew Flinders' Voyage in H.M.S. *Investigator*." *Journal of the Society for the Bibliography of Natural History* 7 (1976): 385–407.

Eisenstein, Elizabeth. *The Printing Press as an Agent of Change: Communication and Cultural Transformation in Early Modern Europe.* 2 vols. Cambridge: Cambridge University Press, 1979.

Ellenberger, François. "Recherches et sur la naissance de la cartographie géologique, en Europe et plus particulièrement en France." *Histoire et Nature* 22–23 (1985): 3–54.

Elliott, Paul. "The Birth of Public Science in the English Provinces: Natural Philosophy in Derby, *c.* 1690–1760." *Annals of Science* 57 (2000): 61–100.

Ellis, Markman. *The Coffee House: A Cultural History.* London: Weidenfeld & Nicolson, 2004.

Emerson, Roger. "American Indians, Frenchmen, and Scots Philosophers." *Studies in Eighteenth-Century Culture* 9 (1979): 211–36.

———. "How Not to Deal with Enlightenments." *Historically Speaking* 3 (2002): 5–7.

———. "The Philosophical Society of Edinburgh, 1737–1747." *British Journal for the History of Science* 12 (1979): 154–91.

———. "The Philosophical Society of Edinburgh, 1748–1768." *British Journal for the History of Science* 14 (1981): 133–66.

———. "The Philosophical Society of Edinburgh, 1768–1783." *British Journal for the History of Science* 18 (1985): 255–303.

———. "The Scottish Enlightenment and the End of the Philosophical Society of Edinburgh." *British Journal for the History of Science* 21 (1988): 33–66.

Emerson, Roger, and Paul Wood. "Science and Enlightenment in Glasgow, 1680–1820." In *Science and the Medicine in the Scottish Enlightenment,* ed. Charles W. J. Withers and Paul Wood, 79–142. East Linton: Tuckwell Press, 2002.

Etlin, Richard A. *Symbolic Space: French Enlightenment Architecture and Its Legacy.* Chicago: University of Chicago Press, 1994.

Evans, Lewis. *Geographical, Historical, Political, Philosophical and Mechanical Essays.* Philadelphia: Printed for B. Franklin & D. Hall, 1755.

Eze, Emmanuel C., ed. *Race and the Enlightenment: A Reader.* Oxford: Blackwell, 1997.

Fara, Patricia. *Pandora's Breeches: Women, Science and Power in the Enlightenment.* London: Pimlico, 2004.

Farber, Paul L. "Buffon and the Concept of Species." *Journal of the History of Biology* 5 (1972): 259–84.

———. "Natural History." In *Encyclopedia of the Enlightenment,* ed. Alan Kors, 3:124–30. New York and Oxford: Oxford University Press, 2003.

Faull, Katherine M., ed. *Anthropology and the German Enlightenment: Perspectives on Humanity.* Lewisburg, PA: Bucknell University Press, 1995.

Febvre, Lucien, and Henri-John Martin. *The Coming of the Book: The Impact of Printing, 1450–1800.* Trans. David Gerrad. London: New Left Books, 1976.

Feldman, Theodore S. "The Ancient Climate in the Eighteenth and Early Nineteenth Century." In *Science and Nature: Essays in the History of the Environmental Sciences,* ed. Michael Shortland, 23–40. Oxford: British Society for the History of Science, 1993.

———. "Late Enlightenment Meteorology." In *The Quantifying Spirit in the Eighteenth Century*, ed. Tore Frängsmyr, J. L. Heilbron, and Robin Rider, 144–77. Berkeley and London: University of California Press, 1990.

Ferguson, Robert. *The American Enlightenment, 1750–1820*. Cambridge, MA: Harvard University Press, 1997.

Ferrone, Vincenzo. "The Accademia Reale delle Scienze: Cultural Sociability and Men of Letters in Turin of the Enlightenment under Vittorio Amedeo III." *Journal of Modern History* 70 (1998): 519–60.

———. *Intellectual Roots of the Italian Enlightenment: Newtonian Science, Religion and Politics in the Early Eighteenth Century*. Trans. Sue Brotherton. Atlantic Highlands, NJ; Humanities Press, 1999.

Findlen, Paula. "A Forgotten Newtonian: Women and Science in the Italian Provinces." In *The Sciences in Enlightened Europe*, ed. William Clark, Jan Golinski, and Simon Schaffer, 313–49. Chicago: University of Chicago Press, 1999.

———. "Science as a Career in Enlightenment Italy: The Strategies of Laura Bassi." *Isis* 84 (1993): 441–69.

Finkelstein, Gabriel, and Alistair McCleery, eds. *The Book History Reader*. London: Routledge, 2002.

Fisher, Robin, and Hugh Johnston, eds. *From Maps to Metaphors: The Pacific World of George Vancouver*. Vancouver: University of British Columbia Press, 1993.

Fissell, Mary, and Roger Cooter. "Exploring Natural Knowledge: Science and the Popular." In *The Cambridge History of Science*, vol. 4: *Eighteenth-Century Science*, ed. Roy Porter, 129–58. Cambridge: Cambridge University Press, 2003.

Flavell, K. "Mapping Faces: National Physiognomies as Cultural Prediction." *Eighteenth-Century Life* 18 (1994): 8–22.

Flavell, M. K. "The Enlightened Reader and the New Industrial Towns: A Study of the Liverpool Library, 1758–1790." *British Journal for Eighteenth-Century Studies* 8 (1985): 17–35.

Fleurieu, Charles Pierre Claret. *Discoveries of the French in 1768 and 1769, to the south-east of New Guinea, with the subsequent visit to the same lands by English Navigators, who gave them new Names*. London: Printed for J. Stockdale, 1791.

Forbes, Eric. "Mathematical Cosmography." In *The Ferment of Knowledge: Studies in the Historiography of Eighteenth-Century Science*, ed. George Rousseau and Roy Porter, 417–48. Cambridge: Cambridge University Press, 1980.

Forrest, Robert. "The Problem of Influence: The Case of French Impact on Eastern Europe in the Eighteenth Century." *Studies on Voltaire and the Eighteenth Century* 264 (1989): 996–1000.

Forster, Georg. *A Voyage round the World in His Britannic Majesty's Sloop, Resolution, Commanded by Capt. James Cook, during the Years 1772, 3, 4, and 5*. London: White, Robson, Elmsley & Robinson, 1777.

Forster, Johann Reinhold. *Observations Made during a Voyage round the World on Physical Geography, Natural History and Ethic Philosophy*. London: G. Robinson, 1778.

Foucault, Michel. *The Archaeology of Knowledge*. London: Routledge, 1972.

———. "Kant on Enlightenment and Revolution." *Economy and Society* 15 (1986): 88–95.

———. *The Order of Things: An Archaeology of the Human Sciences*. New York: Pantheon Books, 1970.

———. "The Politics of Health in the Eighteenth Century." In *Michel Foucault: Power/Knowledge,* ed. Colin Gordon, 166–82. New York and London: Harvester Wheatsheaf, 1980.

———. "What Is Enlightenment?" In *The Foucault Reader,* ed. Paul Rabinow, 45–56. London: Penguin, 1984.

Fox, Christopher. "Introduction: How to Prepare a Noble Savage." In *Inventing Human Science: Eighteenth-Century Domains,* ed. Christopher Fox, Roy Porter, and Robert Wokler, 1–30. Berkeley and London: University of California Press, 1996.

Fox, Christopher, Roy Porter, and Robert Wokler, eds. *Inventing Human Science: Eighteenth-Century Domains.* Berkeley and London: University of California Press, 1996.

France, Peter. "Lost for Words: Travel and Translation." *Studies on Voltaire and the Eighteenth Century* 10 (2001): 201–16.

Frängsmyr, Tore. "The Enlightenment in Sweden." In *The Enlightenment in National Context,* ed. Roy Porter and Mikuláš Teich, 164–75. Cambridge: Cambridge University Press, 1981.

———, ed. *Linnaeus: The Man and His Work.* Canton, MA: Science History Publications, 1994.

———. "Swedish Science in the Eighteenth Century." *History of Science* 12 (1974): 29–42.

Frängsmyr, Tore, John Heilbron, and Richard Rider, eds. *The Quantifying Spirit in the Eighteenth Century.* Berkeley and London: University of California Press, 1990.

Frankel, Charles. *The Faith in Reason: The Idea of Progress in the French Enlightenment.* New York: King's Crown Press, 1948.

Franklin, Benjamin. "A Letter from Dr. Benjamin Franklin, to Mr. Alphonsus le Roy, Member of Several Academies, at Paris. Containing Sundry Maritime Observations." *Transactions of the American Philosophical Society* 2 (1786): 294–324.

French, Roger. "Sickness and the Soul: Stahl, Hoffman and Sauvages on Pathology." In *The Medical Enlightenment of the Eighteenth Century,* ed. Andrew Cunningham and Roger French, 88–110. Cambridge: Cambridge University Press, 1990.

Friis, Herman R. "The Role of Geographers and Geography in the Federal Government 1771–1905." In *The Origins of Academic Geography in the United States,* ed. Brian W. Blouet, 37–56. New York: Archon Books, 1981.

Frost, Alan. "The Pacific Ocean: The Eighteenth Century's 'New World.'" *Studies on Voltaire and the Eighteenth-Century* 152 (1976): 779–822.

Frost, Orcutt. *Bering: The Russian Discovery of America.* New Haven, CT, and London: Yale University Press, 2003.

Fubini, Enrico. *Music and Culture in Eighteenth-Century Europe: A Source Book.* Chicago and London: University of Chicago Press, 1994.

Fuchs, James L. "Nationality and Knowledge in Eighteenth-Century Italy." *Studies in Eighteenth-Century Culture* 21 (1991): 207–18.

Fulford, Tim, with Carol Bolton, eds. *Travels, Explorations and Empires: Writings from the Era of Imperial Expansion, 1770–1835.* 4 vols. London: Pickering & Chatto, 2001.

Fulford, Tim, Debbie Lee, and Peter J. Kitson. *Literature, Science and Exploration in the Romantic Era.* Cambridge: Cambridge University Press, 2004.

Galison, Peter. *Einstein's Clocks, Poincaré's Maps.* London: Hodder & Stoughton, 2003.

Gargett, Graham. "Voltaire's Reception In Ireland." In *Ireland and the French Enlightenment, 1700–1800,* ed. Graham Gargett and Geraldine Sheridan, 67–89. Basingstoke: Macmillan, 1999.

———. "Voltaire's View of the Irish." In *Ireland and the French Enlightenment, 1700–1800,* ed. Graham Gargett and Geraldine Sheridan, 152–70. Basingstoke: Macmillan, 1999.

Gargett, Graham, and Geraldine Sheridan, eds. *Ireland and the French Enlightenment, 1700–1800.* Basingstoke: Macmillan, 1999.

Garrett, Aaron. "Anthropology: The 'Original' of Human Nature." In *The Cambridge Companion to the Scottish Enlightenment,* ed. Alexander Broadie, 79–93. Cambridge: Cambridge University Press, 2003.

Gascoigne, John. "Blumenbach, Banks, and the Beginnings of Anthropology at Göttingen." In *Göttingen and the Development of the Natural Sciences,* ed. Nicolaas Rupke, 86–98. Göttingen: Wallstein Verlag, 2002.

———. *Cambridge in the Age of the Enlightenment: Science, Religion and Politics from the Restoration to the French Revolution.* Cambridge: Cambridge University Press, 1989.

———. "The Eighteenth-Century Scientific Community: A Prosopographical Study." *Social Studies of Science* 25 (1995): 575–81.

———. *The Enlightenment and the Origins of European Australia.* Cambridge: Cambridge University Press, 2002.

———. "Ideas of Nature: Natural Philosophy." In *The Cambridge History of Science,* vol. 4: *Eighteenth-Century Science,* ed. Roy Porter, 285–304. Cambridge: Cambridge University Press, 2003.

———. *Joseph Banks and the English Enlightenment: Useful Knowledge and Polite Culture.* Cambridge: Cambridge University Press, 1994.

———. "Joseph Banks, Mapping and the Geographies of Natural Knowledge." In *Georgian Geographies: Essays on Space, Place and Landscape in the Eighteenth Century,* ed. Miles Ogborn and Charles W. J. Withers, 151–73. Manchester: Manchester University Press, 2004.

———. "The Royal Society and the Emergence of Science as an Instrument of State Policy." *British Journal for the History of Science* 32 (1999): 171–84.

———. *Science in the Service of Empire: Joseph Banks, the British State and the Uses of Science in the Age of Revolution.* Cambridge: Cambridge University Press, 1998.

Gavroglou, Kostas, ed. *The Sciences in the European Periphery during the Enlightenment.* Dordrecht: Kluwer Academic Publishers, 1999.

Gay, Peter. *The Enlightenment: An Interpretation.* 2 vols. New York: Knopf, 1966–69.

Gellner, Ernest. *Nations and Nationalism.* Oxford: Oxford University Press, 1983.

Geras, Norman, and Robert Wokler, eds. *Enlightenment and Modernity.* Basingstoke: Macmillan, 2000.

Gerbi, Antonello. *The Dispute of the New World: The History of a Polemic, 1750–1900.* Rev. and trans. J. Moyle. Pittsburgh: University of Pittsburgh Press, 1973.

Geyer-Kordesch, Joanna. "Georg Ernst Stahl's Radical Pietist Medicine and Its Influence on the German Enlightenment." In *The Medical Enlightenment of the Eighteenth Century,* ed. Andrew Cunningham and Roger French, 67–87. Cambridge: Cambridge University Press, 1990.

Gilmore, William J. *Reading Becomes a Necessity of Life: Material and Cultural Life in Rural New England, 1780–1835.* Knoxville: University of Tennessee Press, 1989.

Gilroy, Amanda, and W. M. Verhoeven, eds. *Epistolary Histories: Letters, Fiction, Culture.* Charlottesville and London: University Press of Virginia, 2000.

Gingerich, Owen. *The Book Nobody Read: In Pursuit of the Revolutions of Nicolaus Copernicus.* London: Heinemann, 2004.

Glacken, Clarence. *Traces on the Rhodian Shore: Nature and Culture in Western Shore from Ancient Times to the End of the Eighteenth Century*. Berkeley and London: University of California Press, 1967.

Glick, Thomas, ed. *The Comparative Reception of Darwinism*. Chicago: University of Chicago Press, 1974.

Godlewska, Anne Marie Claire. "From Enlightenment Vision to Modern Science? Humboldt's Visual Thinking." In *Geography and Enlightenment*, ed. David N. Livingstone and Charles W. J. Withers, 236–75. Chicago: University of Chicago Press, 1999.

———. *Geography Unbound: French Geographic Science from Cassini to Humboldt*. Chicago: University of Chicago Press, 1999.

———. "The Napoleonic Survey of Egypt; A Masterpiece of Cartographic Communication and Early Nineteenth Century Fieldwork." *Cartographica* 25 (1–2), monograph 38–39 (1988).

———. "Napoleon's Geographers: Imperialists and Soldiers of Modernity." In *Geography and Empire*, ed. Anne M. C. Godlewska and Neil Smith, 31–53. Oxford: Blackwell, 1994.

Goffart, Walter. *Historical Atlases: The First Three Hundred Years, 1570–1870*. Chicago: University of Chicago Press, 2003.

Goldenberg, L. A., and A. V. Postnikov. "Development of Mapping Methods in Russia in the Eighteenth Century." *Imago Mundi* 37 (1985): 63–80.

Goldgar, Anne. *Impolite Learning: Conduct and Community in the Republic of Letters, 1680–1850*. Yale: Yale University Press, 1995.

Goldschmid, Edgar. "Nosologia Naturalis." In *Science, Medicine and History: Essays in the Evolution of Scientific Thought and Medical Practice*, ed. E. Ashworth Underwood, 2:103–22. Oxford: Oxford University Press, 1953.

Golinski, Jan. "Conversations about Chemistry: Talk about Phlogiston in the Coffee House Society, 1780–1787." In *Discussing Chemistry and Steam: The Minutes of a Coffee House Philosophical Society, 1780–1787*, ed. Trevor Levere and G. Turner, 191–205. Oxford: Oxford University Press, 2002.

———. *Making Natural Knowledge: Constructivism and the History of Science*. Cambridge: Cambridge University Press, 1998.

———. *Science as Public Culture: Chemistry and Enlightenment in Britain*. Cambridge: Cambridge University Press, 1992.

———. "Science *in* the Enlightenment." *History of Science* 24 (1986): 411–24.

———. "Utility and Audience in Eighteenth-Century Chemistry: Case Studies of William Cullen and Joseph Priestley." *British Journal for the History of Science* 21 (1988): 1–31.

Gooday, Graeme. "The Premisses of Premises: Spatial Issues in the Historical Construction of Laboratory Credibility." In *Making Space for Science: Territorial Themes in the Shaping of Knowledge*, ed. Crosbie Smith and Jon Agar, 216–45. Basingstoke: Macmillan, 1998.

Goodman, David. "Science and the Clergy in the Spanish Enlightenment." *History of Science* 21 (1983): 111–40.

Goodman, David, and Colin A. Russell. *The Rise of Scientific Europe, 1500–1800*. London: Hodder & Stoughton, 1991.

Goodman, Dena. *Republic of Letters: A Cultural Study of the French Enlightenment*. Ithaca, NY, and London: Cornell University Press, 1994.

———. "Introduction: The Public and the Nation." *Eighteenth-Century Studies* 29 (1995): 1–4.

Gordin, Michael D. "The Importation of Being Earnest: The Early St. Petersburg Academy of Sciences." *Isis* 91 (2000): 1–31.

Gould, Peter. "Lisbon 1755: Enlightenment, Catastrophe, and Communication." In *Geography and Enlightenment,* ed. David N. Livingstone and Charles W. J. Withers, 399–413. Chicago: University of Chicago Press, 1999.

Graham, Loren, Wolf Lepenies, and Peter Weingart, eds. *The Functions and Uses of Disciplinary Histories.* Dordrecht: Kluwer Academic Publishers, 1983.

Gray, John. *Enlightenment's Wake: Politics and Culture at the Close of the Modern Age.* London: Routledge, 1995.

Greenberg, John L. "Mathematical Physics in Eighteenth-Century France." *Isis* 77 (1986): 59–78.

———. *The Problem of the Earth's Shape from Newton to Clairault: The Rise of Mathematical Science in Eighteenth-Century Paris and the Fall of "Normal" Science.* Cambridge: Cambridge University Press, 1995.

Griswold, Charles L. *Adam Smith and the Virtues of Enlightenment.* Cambridge: Cambridge University Press, 1951.

Gronim, Sara Stidstone. "Geography and Persuasion: Maps in British Colonial New York." *William and Mary Quarterly* 58 (201): 373–402.

Grove, Jean. *The Little Ice Age.* London: Methuen, 1988.

Grove, Richard H. *Ecology, Climate and Empire: The Indian Legacy in Global Environmental History, 1400–1940.* Delhi: Oxford University Press, 1998.

———. *Green Imperialism: Colonial Expansion, Tropical Island Edens and the Origins of Environmentalism, 1600–1860.* Cambridge: Cambridge University Press, 1995.

Guettard, Jean-Étienne, and Antoine Monnet. *Atlas et description minéralogique de la France.* Monnet: Paris, 1780.

Hahn, Roger. *The Anatomy of a Scientific Institution: The Paris Academy of Sciences, 1666–1803.* Berkeley: University of California Press, 1971.

———. "Science and the Arts in France: The Limitations of an Encyclopedic Ideology." *Studies in Eighteenth-Century Culture* 10 (1981): 77–93.

Haidt, Rebecca. *Embodying Enlightenment: Knowing the Body in Eighteenth-Century Spanish Literature and Culture.* Basingstoke: Macmillan, 1998.

Hall, Thadd. "The Development of Enlightened Interest in Eighteenth-Century Corsica." *Studies on Voltaire and the Eighteenth Century* 64 (1968): 165–85.

Hamblyn, Richard. "Private Cabinets and Popular Geology: The British Audiences for Volcanoes in the Eighteenth Century." In *Transports: Travel, Pleasure, and Imaginative Geography, 1600–1830,* ed. Chloe Chard and Helen Langdon, 179–206. New Haven, CT, and London: Yale University Press, 1996.

Hampson, Norman. *The Enlightenment.* London: Penguin, 1969.

———. "The Enlightenment in France." In *The Enlightenment in National Context,* ed. Roy Porter and Mikuláš Teich, 41–53. Cambridge: Cambridge University Press, 1981.

Hancock, David Boyd. *William Stukeley: Science, Religion and Archaeology in Eighteenth-Century England.* Woodbridge: Boydell Press, 2002.

Hankins, Thomas. *Science and the Enlightenment.* Cambridge: Cambridge University Press, 1985.

Harley, J. Brian. "Power and Legitimation in the English Geographical Atlases of the Eighteenth Century." In *Images of the World: The Atlas through History,* ed. John A. Wolter and Ronald E. Grim, 161–204. New York: McGraw-Hill, 1997.

———. "The Society of Arts and the Surveys of English Counties, 1759–1809." *Journal of the Royal Society of Arts* 112 (1963–64): 43–46 (pt. 1), 119–24 (pt. 2), 269–75 (pt. 3), 538–43 (pt. 4).

Harpham, Geoffrey G. "So . . . What *Is* Enlightenment? An Inquisition into Modernity." *Critical Inquiry* 20 (1994): 524–56.

Harris, Steven. "Long-Distance Corporations, Big Sciences, and the Geography of Knowledge." *Configurations* 6 (1998): 269–304.

Harvey, Karen. "Spaces of Erotic Delight." In *Georgian Geographies: Essays on Space, Place and Landscape in the Eighteenth Century,* ed. Miles Ogborn and Charles W. J. Withers, 131–50. Manchester: Manchester University Press, 2004.

Hayward, Rhodri. "Emmanuel Mendes da Costa (1717–1791): A Case Study in Scientific Reputation." In *Travels of Learning: A Geography of Science in Europe,* ed. Ana Simões, Ana Carneiro, and Maria Paula Diogo, 101–15. Dordrecht: Kluwer Academic Publishers, 2003.

Hazard, Paul. *The European Mind, 1680–1715.* Trans. James Lewis May. London: Hollis and Carter, 1953. First published as *La crise de la conscience européenne, 1680–1715* (1935).

Headrick, Daniel R. *When Information Came of Age: Technologies of Knowledge in the Age of Reason and Revolution, 1700–1850.* Oxford: Oxford University Press, 2000.

Heffernan, Michael. "The Changing Political Map: Geography, Geopolitics, and the Idea of Europe since 1500." In *An Historical Geography of Europe,* ed. Robin A. Butlin and Robert A. Dodgshon, 140–80. Oxford: Clarendon Press,1998.

———. "Edme Mentelle's Geographies and the French Revolution." In *Geography and Revolution,* ed. David N. Livingstone and Charles W. J. Withers, 273–303. Chicago: University of Chicago Press, 2005.

———. "Historical Geographies of the Future: Three Perspectives from France, 1750–1825." In *Geography and Enlightenment,* ed. David N. Livingstone and Charles W. J. Withers, 125–64. Chicago: University of Chicago Press, 1999.

———. *The Meaning of Europe: Geography and Geopolitics.* London: Edward Arnold, 1998.

———. "On Geography and Progress: Turgot's *Plan d'un Ouvrage sur la Géographie Politique* (1751) and the Origins of Modern Progressive Thought." *Political Geography* 13 (1994): 328–43.

Heilbron, J. L. "Domesticating Science in the Eighteenth Century." In *Science and the Visual Image in the Enlightenment,* ed. William R. Shea, 1–24. Canton, MA: Science History Publications, 2000.

———. "The Measure of Enlightenment." In *The Quantifying Spirit in the Eighteenth Century,* ed. Tore Frängsmyr, John Heilbron, and Robin Rider, 207–42. Berkeley and London: University of California Press, 1990.

———. "Natural Philosophy and Science." In *Encyclopedia of the Enlightenment,* ed. Alan Kors, 3:134–42. New York and Oxford: Oxford University Press, 2003.

Herman, Arthur. *The Scottish Enlightenment: The Scots' Invention of the Modern World.* London: Fourth Estate, 2002.

Hobsbawm, Eric J. *Nations and Nationalism since 1780.* Cambridge: Cambridge University Press, 1990.

Holub, Robert. *Reception Theory: A Critical Introduction.* Reprint. London: Routledge, 2003.

Höpfl, H. M. "From Savage to Scotsman: Conjectural History in the Scottish Enlightenment." *Journal of British Studies* 17 (1977–78): 19–40.

Horkheimer, Max, and Theodor Adorno. *Dialectic of Enlightenment.* New York: Herder & Herder, 1972. First published in 1947 (in German).

Hudson, Nicholas. "From 'Nation' to 'Race': The Origin of Racial Classification in Eighteenth-Century Thought." *Eighteenth-Century Studies* 29 (1996): 247–64.

Hufbauer, Karl. *The Formation of the German Chemical Community (1720–1795)*. Berkeley: University of California Press, 1982.

Hulme, Peter, and Ludmilla Jordanova, eds. *The Enlightenment and Its Shadows*. London: Routledge, 1990.

———. Introduction to *The Enlightenment and Its Shadows*, ed. Peter Hulme and Ludmilla Jordanova, 1–14. London: Routledge, 1990.

Hume, D. *An Abstract of a Treatise of Human Nature*. London, 1740.

———. *Essays Moral, Political, and Literary*. Edited with an introduction by Eugene Miller. Indianapolis: Liberty Classics, 1987.

Hunt, Lynn, ed. *The Invention of Pornography: Obscenity and the Origins of Modernity*. New York: Zone Books, 1996.

Hunt, Lynn, with Margaret Jacob. "Enlightenment Studies." In *Encyclopedia of the Enlightenment*, ed. Alan C. Kors, 1:418–30. New York and Oxford: Oxford University Press, 2003.

Hunt, Margaret. *Women and the Enlightenment*. New York: Institute for Research in History, 1984.

Huxley, Robert. "Natural History Collectors and Their Collections: 'Simpling Macaronis' and Instruments of Empire." In *Enlightenment: Discovering the World in the Eighteenth Century*, ed. Kim Sloan, 81–91. London: British Museum Press, 2003.

Hyland, Paul, with Olga Gomez and Francesca Greensides, eds. *The Enlightenment: A Sourcebook and Reader*. London: Routledge, 2003.

Iliffe, Rob. "'Aplatisseur du Monde et de Cassini': Maupertuis, Precision Measurement, and the Shape of the Earth in the 1730s." *History of Science* 31 (1993): 335–75.

———. "Science and the Voyages of Discovery." In *The Cambridge History of Science*, vol. 4: *Eighteenth-Century Science*, ed. Roy Porter, 618–45. Cambridge: Cambridge University Press, 2003.

Im Hof, Ulrich. *The Enlightenment*. Oxford: Blackwell, 1994.

Israel, Jonathan I. *Radical Enlightenment: Philosophy and the Making of Modernity, 1650–1750*. Oxford: Oxford University Press, 2001.

Jacob, Margaret. *The Enlightenment: A Brief History with Documents*. Boston: Bedford/St. Martin's, 2001.

———. *Living the Enlightenment: Freemasonry and Politics in Eighteenth-Century Europe*. New York: Oxford University Press, 1991.

———. "The Mental Landscape of the Public Sphere: A European Perspective." *Eighteenth-Century Studies* 28 (1994): 95–113.

———. *The Newtonians and the English Revolution, 1689–1720*. Hassocks, Sussex: Harvester Press, 1976.

Jacob, Margaret, and Wijnand W. Mijnhardt, eds. *The Dutch Republic in the Eighteenth Century: Decline, Enlightenment and Revolution*. Ithaca, NY, and London: Cornell University Press, 1992.

———. Introduction to *The Dutch Republic in the Eighteenth Century: Decline, Enlightenment and Revolution*, ed. Margaret Jacob and Wijnand Mijnhardt, 1–15. Ithaca, NY, and London: Cornell University Press, 1992.

Jacob, Margaret, and Dorothée Sturkenboom, "A Women's Scientific Society in the West: The Late Eighteenth-Century Assimiliation of Science." *Isis* 94 (2003): 217–52.

Jacques, T. Carlos. "From Savages and Barbarians to Primitives: Africa, Social Typologies, and History in Eighteenth-Century French Philosophy." *History and Theory* 36 (1997): 190–215.

Jaffee, David. "The Village Enlightenment in New England, 1760–1820." *William and Mary Quarterly* 47 (1990): 327–46.

Jankovic, Vladimir. "Arcadian Instincts: A Geography of Truth in Georgian England." In *Georgian Geographies: Essays on Space, Place and Landscape in the Eighteenth Century,* ed. Miles Ogborn and Charles W. J. Withers, 174–91. Manchester: Manchester University Press, 2004.

———. "Meteorology." In *Encyclopedia of the Enlightenment,* ed. Alan C. Kors, 3:171–73. New York and Oxford: Oxford University Press.

———. "The Place of Nature and the Nature of Place: The Chorographic Challenge to the History of British Provincial Science." *History of Science* 38 (2000): 79–113.

———. *Reading the Skies: A Cultural History of English Weather, 1650–1820.* Manchester: Manchester University Press, 2000.

Jarcho, Saul. "Some Early Italian Epidemiological Maps," *Imago Mundi* 35 (1983): 9–19.

———. "Yellow Fever, Cholera, and the Beginnings of Medical Cartography." *Journal of the History of Medicine and Allied Sciences* 25 (1970): 131–42.

Johns, Adrian. *The Nature of the Book: Print and Knowledge in the Making.* Chicago: University of Chicago Press, 1998.

Johnson, Hildegard B. "Towards a National Landscape." In *The Making of the American Landscape,* ed. Michael Conzen, 127–45. New York and London: Routledge, 1990.

Jones, Peter. "Living the Enlightenment and the French Revolution: James Watt, Matthew Boulton and Their Sons." *Historical Journal* 42 (1999): 157–82.

———. *The Peasantry in the French Revolution.* Cambridge: Cambridge University Press, 1998.

Joppien, Rüdiger. "Philippe Jacques de Loutherbourg's Pantomime *Omai, or A Trip Round the World* and the Artists of Captain Cook's Voyages." In *Captain Cook and the South Pacific,* ed. Timothy Mitchell, 81–136. London: British Museum, 1979.

Jordanova, Ludmilla. "Earth Science and Environmental Medicine: The Synthesis of the Late Enlightenment." In *Images of the Earth: Essays in the History of the Environmental Sciences,* ed. Ludmilla Jordanova and Roy Porter, 119–47. Chalfont St. Giles: British Society for the History of Science, 1979.

Jordanova, Ludmilla, and Roy Porter, eds. *Images of the Earth: Essays in the History of the Environmental Sciences.* Chalfont St. Giles: British Society for the History of Science, 1979.

Kafker, Frank A., ed. *The Encyclopedists as a Group: A Collective Biography of the Authors of the "Encyclopédie."* Oxford: Voltaire Foundation, 1996.

———, ed. *Notable Encyclopedias of the Late Eighteenth Century: Eleven Successors of the "Encyclopédie."* Oxford: Voltaire Foundation, 1994.

———, ed. *Notable Encyclopedias of the Seventeenth and Eighteenth Centuries: Nine Predecessors of the "Encyclopédie."* Oxford: Voltaire Foundation, 1981.

Kafker, Frank A., and Serena L. Kafker. *The Encyclopedists as Individuals: A Biographical Dictionary of the Authors of the "Encyclopédie."* Oxford: Voltaire Foundation, 1988.

Kaufmann, Paul. *Borrowings from the Bristol Library, 1773–1874.* Charlottesville: University of Virginia Press, 1960.

Kaye Lamb, W. *The Voyage of George Vancouver, 1791–1795.* 4 vols. London: Hakluyt Society, 1984.

Keller, Susanne B. "Sections and Views: Visual Representations in Eighteenth-Century Earthquake Studies." *British Journal for the History of Science* 31 (1998): 129–59.

Kelley, Donald, ed. *History and the Disciplines: The Reclassification of Knowledge in Early Modern Europe.* Rochester, NY: University of Rochester Press, 1997.

———. "The Problem of Knowledge and the Concept of Discipline." In *History and the Disciplines: The Reclassification of Knowledge in Early Modern Europe,* ed. Donald Kelley, 13–28. Rochester, NY: University of Rochester Press, 1997,.

Kelley, Donald R., and Richard H. Popkin, eds. *The Shapes of Knowledge from the Renaissance to the Enlightenment.* Dordrecht: Kluwer Academic Publishers, 1991.

Kendrick, John. *Alejandro Malaspina: Portrait of a Visionary.* Montreal and London: McGill–Queen's University Press, 1999.

Kendrick, Thomas D. *The Lisbon Earthquake.* London: Methuen, 1956.

Kennedy, Máire. "Readership in French: The Irish Experience." In *Ireland and the French Enlightenment, 1700–1800,* ed. Graham Gargett and Geraldine Sheridan, 3–20. Basingstoke: Macmillan, 1999.

Kidd, Colin. "Gaelic Antiquity and National Identity in Enlightenment Ireland and Scotland." *English Historical Review* 109 (1994): 1197–1214.

King, Pauline. "Some Thoughts on Native Hawaiian Attitudes towards Captain Cook." In *Captain Cook: Explorations and Reassessments,* ed. Glyndwr Williams, 94–109. Woodbridge and New York: Boydell Press, 2004.

Kingston, Rebecca. "Charles-Louis de Secondat Montesquieu." In *Encyclopedia of the Enlightenment,* ed. Alan Kors, 3:83–88. New York and Oxford: Oxford University Press, 2003.

Kington, John. *The Weather of the 1780s over Europe.* Cambridge: Cambridge University Press, 1987.

Kitromilides, Paschalis. *The Enlightenment as Social Criticism: Iosipos Moisiodax and Greek Culture in the Eighteenth Century.* Princeton, NJ: Princeton University Press, 1992.

———. "Europe and the Dilemmas of Greek Conscience." In *Greece and Europe in the Modern Period: Aspects of a Troubled Relationship,* ed. Philip Carabott, 1–15. London: King's College, 1995.

Klein, Lawrence. "Liberty, Manners, and Politeness in Early Eighteenth-Century England." *Historical Journal* 32 (1989): 583–605.

Klein, Lawrence, and Anthony J. La Vopa, eds. *Enthusiasm and the Enlightenment in Europe, 1650–1850.* San Marino, CA: Huntingdon Library, 1998.

Klemme, Heiner, ed. *Reception of the Scottish Enlightenment in Germany: Six Significant Translations, 1755–1782.* 7 vols. Bristol: Thoemmes Press, 2000.

Klonk, Charlotte. "Science, Art and the Representation of the Natural World." In *The Cambridge History of Science,* vol. 4: *Eighteenth-Century Science,* ed. Roy Porter, 584–617. Cambridge: Cambridge University Press, 2003.

Koch, Adrienne. "Aftermath of the American Enlightenment." *Studies on Voltaire and the Eighteenth Century* 56 (1967): 735–63.

———. "The Contest of Democracy and Aristocracy in the American Enlightenment." *Studies on Voltaire and the Eighteenth Century* 26 (1963): 999–1018.

Koerner, Lisbet. "Carl Linnaeus in His Time and Place." In *Cultures of Natural History,* ed. Nicholas Jardine, James A. Secord, and Emma C. Spary, 145–62. Cambridge: Cambridge University Press, 1996.

———. "Daedalus Hyperborus: Baltic Natural History and Mineralogy in the Enlightenment." In *The Sciences in Enlightened Europe,* ed. William Clark, Jan Golinski, and Simon Schaffer, 389–422. Chicago: University of Chicago Press, 1999.

———. *Linnaeus: Nature and Nation.* Cambridge, MA: Harvard University Press, 1999.

———. "Purposes of Linnaean Travel: A Preliminary Research Report." In *Visions of Empire: Voyages, Botany, and Representations of Nature,* ed. David P. Miller and Peter H. Reill, 117–52. Cambridge: Cambridge University Press, 1996.

———. "Women and Utility in Enlightenment Science." *Configurations* 2 (1995): 233–55.

Kontler, László. "Enlightenment, *Aufklärung* and Conservatism: Reflections on the Reception of the British Enlightenment and Early British Conservatism in Central Europe." *Studies on Voltaire and the Eighteenth Century* 303 (1992): 514–19.

———. "William Robertson and His German Audience on European and Non-European Civilisations." *Scottish Historical Review* 80 (2001): 63–89.

———. "William Robertson's History of Manners in German, 1770–95." *Journal of the History of Ideas* 58 (1997): 125–44.

Konvitz, Josef W. *Cartography in France, 1660–1848: Science, Engineering and Statecraft.* Chicago: University of Chicago Press, 1987.

———. "The Nation-State, Paris and Cartography in Eighteenth- and Nineteenth-Century France." *Journal of Historical Geography* 16 (1990): 3–16.

Kornfeld, Eve. "Encountering the 'Other': American Intellectuals and Indians in the 1790s." *William and Mary Quarterly* 52 (1995): 287–314.

Kors, Alan, ed. *Encyclopedia of the Enlightenment.* 4 vols. New York: Oxford University Press, 2003.

Kovács, Ilona. "Hungary." In *Encyclopedia of the Enlightenment,* ed. Alan Kors, 2:238–40. New York and Oxford: Oxford University Press, 2003.

Kumar, Deepak. "India." In *The Cambridge History of Science,* vol. 4: *Eighteenth-Century Science,* ed. Roy Porter, 669–87. Cambridge: Cambridge University Press, 2003.

Laboulais, Isabelle. "La géographie dans les arbres encyclopédique de la seconde moitié du XVIIIe siècle." In *Géographies plurielles: Les sciences géographiques au moment de l'emergence des sciences humaines (1750–1850),* ed. Hélène Blais and Isabelle Laboulais, 63–94. Paris: L'Harmattan, 2005.

Laboulais-Lesage, Isabelle. "Les géographes français de la fin du XVIIIe siècle et le terrain: Recherches sur une paradoxale absence." *L'Espace Geographique* 2 (2001): 97–110.

———. "Reading a Vision of Space: The Geographical Map Collection of Charles-Etienne Coquebert de Montbret (1755–1831)." *Imago Mundi* 56 (2004): 48–66.

Lamb, Jonathan, Vanessa Smith, and Nicholas Thomas, eds. *Exploration and Exchange: A South Seas Anthology, 1680–1900.* Chicago: University of Chicago Press, 2000.

Langer, Wolfhart. "German Translations of 18th Century Italian Geoscience Writing and Their Impact on Contemporaneous Science in Germany." In *Cosmographica et Geographica,* ed. Bernhard Fritscher and Gerhard Brey, 81–95. Munich: Institut für Geschichte der Naturwissenschaften, 1994.

Lanning, John Tate. "The Reception of the Enlightenment in Latin America." In *Latin America and the Enlightenment,* ed. Arthur Whitaker, 71–93. Ithaca, NY: Great Seal Books, 1961.

Larson, James. "An Alternative Science: Linnaean Natural History in Germany, 1770–1790." *Janus* 66 (1979): 267–83.

———. *Interpreting Nature: The Science of Living Form from Linnaeus to Kant.* Baltimore and London: Johns Hopkins University Press, 1994.

———. "Not without a Plan: Geography and Natural History in the Late Eighteenth Century." *Journal of the History of Biology* 19 (1986): 447–88.

———. *Reason and Experience: The Representation of Natural Order in the Work of Carl von Linné*. Berkeley and London: University of California Press, 1971.

Latour, Bruno. *Science in Action: How to Follow Scientists and Engineers through Society*. Cambridge, MA: Harvard University Press, 1987.

La Vopa, Anthony J. "Herder's *Publikum*: Language, Print, and Sociability in Eighteenth-Century Germany." *Eighteenth-Century Studies* 29 (1995): 5–24.

Lawson-Peebles, Robert. *Landscape and Written Expression in Revolutionary America: The World Turned upside Down*. Cambridge: Cambridge University Press, 1988.

Leask, Nigel. *Curiosity and the Aesthetics of Travel Writing, 1770–1840*. Oxford: Oxford University Press, 2002.

Le Guisguet, Bernard. "Le Dépôt des Cartes, Plans et Journaux de la Marine sous l'ancien régime." *Neptunia* 214 (1999): 3–21.

Levere, Trevor H. "Natural Philosophers in a Coffee House: Dissent, Radical Reform and Pneumatic Chemistry." In *Science and Dissent in England, 1688–1945*, ed. Paul Wood, 131–46. Aldershot: Ashgate, 2004.

Levere, Trevor H., and G. Turner, eds. *Discussing Chemistry and Steam: The Minutes of a Coffee House Philosophical Society, 1780–1787*. Oxford: Oxford University Press, 2002.

Licoppe, Christian. *La formation de la pratique scientifique: Le discours de l'expérience France et en Angleterre (1630–1820)*. Paris: Éditions de la Découverte, 1996.

———. "The Project for a Map of Languedoc in Eighteenth-Century France at the Contested Intersection between Astronomy and Geography: The Problem of Co-ordination between Philosophers, Instruments and Observations as a Keystone of Modernity." In *Instruments, Travel and Science: Itineraries of Precision from the Seventeenth to the Twentieth Century*, ed. Marie-Noëlle Bourguet, Christian Licoppe, and H. Otto Sibum, 51–74. London: Routledge, 2002.

Liedman, Sven-Eric, ed. *The Postmodern Critique of the Project of Enlightenment*. Amsterdam: Rodopi, 1997.

Livingstone, David N. "Darwinism and Calvinism: The Belfast–Princeton Connection." *Isis* 83 (1992): 408–28.

———. *The Geographical Tradition: Episodes in the History of a Contested Enterprise*. Oxford: Blackwell, 1992.

———. *Putting Science in Its Place: Geographies of Scientific Knowledge*. Chicago: University of Chicago Press, 2003.

———. "Race, Space and Moral Climatology: Notes toward a Genealogy." *Journal of Historical Geography* 28 (2002): 159–80.

———. "'Risen into Empire': Moral Geographies of the American Republic." In *Geography and Revolution*, ed. David N. Livingstone and Charles W. J. Withers, 304–35. Chicago: University of Chicago Press, 2005.

———. "Science, Religion and the Geography of Reading: Sir William Whitla and the Editorial Staging of Isaac Newton's Writings on Biblical Prophecy." *British Journal for the History of Science* 36 (2003): 27–42.

Livingstone, David N., and Charles W. J. Withers, eds. *Geography and Enlightenment*. Chicago: University of Chicago Press, 1999.

Logan, Gabriella Berti. "The Desire to Contribute: An Eighteenth-Century Italian Woman of Science." *American Historical Review* 99 (1994): 785–812.

Lough, John. *The Encyclopédie*. Geneva: Slatkine, 1989.

Loveland, Jeff. "Georges-Louis Leclerc de Buffon's *Histoire Naturelle* in English, 1775–1815." *Archives of Natural History* 31 (2004): 214–35.

Lowood, Henry. *Patriotism, Profit, and the Promotion of Science in the German Enlightenment: The Economic and Scientific Societies, 1760–1815*. New York and London: Garland Publishing, 1991.

Lux, David S., and Harold Cook. "Closed Circles or Open Networks? Communicating at a Distance during the Scientific Revolution." *History of Science* 36 (1998): 179–211.

Lyon, John, and Phillip R. Sloan, eds. *From Natural History to the History of Nature: Buffon and His Critics*. South Bend, IN: University of Notre Dame Press, 1981.

Mackay, David. "Agents of Empire: The Banksian Collectors and Evaluation of New Lands." In *Visions of Empire: Voyages, Botany, and Representations of Nature,* ed. David Philip Miller and Peter Hanns Reill, 38–57. Cambridge: Cambridge University Press, 1996.

———. *In the Wake of Cook: Exploration, Science and Empire, 1780–1801*. London: Croom Helm, 1985.

MacLeod, Roy, and Philip Rehbock, eds. *Nature in Its Greatest Extent: Western Science in the Pacific*. Honolulu: University of Hawaii Press, 1998.

MacPherson, Ryan Cameron. "Essay Review: When Evolution Becomes Conversation: *Vestiges of Creation,* Its Readers, and Its Respondents in Victorian Britain." *Journal of the History of Biology* 34 (2001), 565–79.

Maggs, Barbara W. "Asia." In *Encyclopedia of the Enlightenment,* ed. Alan Kors, 1:85–90. New York and Oxford: Oxford University Press, 2003.

———. "China." In *Encyclopedia of the Enlightenment,* ed. Alan Kors, 1:232–34. New York and Oxford: Oxford University Press, 2003.

———. "Eighteenth-Century Russian Reflections on the Lisbon Earthquake, Voltaire and Optimism." *Studies on Voltaire and the Eighteenth Century* 137 (1975): 7–29.

———. "Voltaire and the Balkans." *Studies on Voltaire and the Eighteenth Century* 189 (1980): 81–118.

Mah, Harold. "Phantasies of the Public Sphere: Rethinking the Habermas of Historians." *Journal of Modern History* 72 (2000): 153–81.

Marassinova, Elena. "The Russian Enlightenment and the Enlightened Nobility." *Studies on Voltaire and the Eighteenth Century* 347 (1997): 956–58.

Marchant, Alexander. "Aspects of the Enlightenment in Brazil." In *Latin America and the Enlightenment,* ed. Arthur Whitaker, 95–118. Ithaca, NY: Great Seal Books, 1961.

Markham, Clements R. *Major James Rennell and the Rise of Modern English Geography*. London: Allen & Unwin, 1895.

Marsak, Leonard M. *The Enlightenment*. Wiley: Chichester and New York: 1972.

Marshall, Douglas. "Military Maps of the Eighteenth-Century and the Tower of London Drawing Room." *Imago Mundi* 32 (1980): 21–44.

Marshall, Peter, and Glyndwyr Williams. *The Great Map of Mankind: British Perceptions of the World in the Age of Enlightenment*. London: J. M. Dent, 1982.

Martin, Julian. "Sauvages's Nosology: Medical Enlightenment in Montpellier." In *The Medical Enlightenment of the Eighteenth Century,* ed. Andrew Cunningham and Roger French, 111–37. Cambridge: Cambridge University Press, 1990.

Martins, Luciana. "The Art of Tropical Travel, 1768–1830." In *Georgian Geographies: Essays on Space, Place and Landscape in the Eighteenth Century,* ed. Miles Ogborn and Charles W. J. Withers, 72–91. Manchester: Manchester University Press, 2004.

Mason, Haydn, ed. *The Darnton Debate: Books and Revolution in the Eighteenth Century.* Oxford: Voltaire Foundation, 1998.

Mauviel, Maurice. "Volney, Constantin-François de Chasseboeuf." In *Encyclopedia of the Enlightenment,* ed. Alan Kors, 4:233–35. New York and Oxford: Oxford University Press, 2003.

Maxwell, Kenneth. *Pombal, Paradox of the Enlightenment.* Cambridge: Cambridge University Press, 1995.

May, Georges. "Observations on the Allegory: The Frontispiece of the *Encyclopédie.*" *Diderot Studies* 16 (1973): 159–74.

May, Henry F. *The Enlightenment in America.* New York and Oxford: Oxford University Press, 1976.

Mayhew, Robert J. "The Character of English Geography, *c.* 1660–1800: A Textual Approach." *Journal of Historical Geography* 24 (1998): 385–412.

———. *Enlightenment Geography: The Political Languages of British Geography, 1650–1850.* Basingstoke: Macmillan, 2000.

———. *Geography and Literature in Historical Context: Samuel Johnson and Eighteenth-Century English Conceptions of Geography.* University of Oxford School of Geography Research Papers No. 54. Oxford, 1997.

———. "Geography Books and the Character of Georgian Politics." In *Georgian Geographies: Essays on Space, Place and Landscape in the Eighteenth Century,* ed. Miles Ogborn and Charles W. J. Withers, 192–211. Manchester: Manchester University Press, 2004.

———. "Geography in Eighteenth-Century British Education." *Paedagogica Historica* 34 (1998): 731–69.

———. "Geography, Print Culture and the Renaissance: 'The Road Less Travelled By.'" *History of European Ideas* 27 (2001): 349–69.

———. Introduction to *A New Geographical and Historical Grammar,* by Thomas Salmon, ed. Robert Mayhew, v–xxxiv. Bristol: Thoemmes Press, 2003.

———. Mapping Science's Imagined Community: Geography as a Republic of Letters, 1600–1800." *British Journal for the History of Science* 38 (2005): 73–92.

———. "Proleptic Locations: Charting the Birth of Modern Geography." *History of European Ideas* 26 (2000): 67–73.

———. "Was William Shakespeare an Eighteenth-Century Geographer? Constructing Histories of Geographical Knowledge." *Transactions of the Institute of British Geographers* 23 (1998): 21–37.

———. "William Guthrie's *Geographical Grammar,* the Scottish Enlightenment and the Politics of British Geography." *Scottish Geographical Journal* 115 (1999): 19–34.

McClellan, James E. "The Académie Royale des Sciences, 1699–1793: A Statistical Portrait." *Isis* 72 (1981): 541–67.

———. "Learned Societies." In *Encyclopedia of the Enlightenment,* ed. Alan C. Kors, 2:371–77. New York and Oxford: Oxford University Press, 2003.

———. *Science Reorganized: Scientific Societies in the Eighteenth Century.* New York: Columbia University Press, 1985.

———. "Scientific Institutions and the Organisation of Science." In *The Cambridge History of Science,* vol. 4: *Eighteenth-Century Science,* ed. Roy Porter, 87–106. Cambridge: Cambridge University Press, 2003.

McConnell, Anita. "La Condamine's Scientific Journey down the River Amazon, 1743–1744." *Annals of Science* 48 (1991): 1–19.

McCormick, E. H. *Omai Pacific Envoy*. Auckland: Auckland University Press, 1977.

McDermott, John Francis. "The Enlightenment on the Mississippi Frontier, 1763–1804." *Studies on Voltaire and the Eighteenth Century* 26 (1963): 1129–42.

McDougall, Warren. "Charles Elliot's Medical Publications and the International Book Trade." In *Science and Medicine in the Scottish Enlightenment*, ed. Charles W. J. Withers and Paul Wood, 215–54. East Linton: Tuckwell Press, 2002.

McMaster, Robert B., and Eric Sheppard. "Introduction: Scale and Geographic Enquiry." In *Scale and Geographic Inquiry: Nature, Society and Method*, ed. Eric Sheppard and Robert B. McMaster, 1–22. Oxford: Blackwell, 2004.

McNeil, Marion. "Newton as National Hero." In *Let Newton Be! A New Perspective on his Life and Works*, ed. J. Fauvel, R. Flood, M. Shortland, and R. Wilson, 213–40. Oxford: Oxford University Press, 1989.

Meek, Ronald. *Social Science and the Ignoble Savage*. Cambridge: Cambridge University Press, 1976.

Meinig, Donald W. *The Shaping of America: A Geographical Perspective on 500 Years of History*, vol. 1: *Atlantic America, 1492–1800*. New Haven, CT, and London: Yale University Press, 1986.

Melton, James Van Horn. *The Rise of the Public in Enlightenment Europe*. Cambridge: Cambridge University Press, 2001.

Mercier, Louis-Sébastian. *L'an 2440: Rêve s'il en fut jamais*. Amsterdam: Van Harrevelt, 1770.

Merians, Linda. 1994 "What They Are, Who We Are: Representations of the 'Hottentot' in Eighteenth-Century Britain." *Eighteenth-Century Life* 17 (1994): 14–39.

Mijnand, Wijnand. "The Dutch Enlightenment: Humanism, Nationalism and Decline." In *The Dutch Republic in the Eighteenth Century: Decline, Enlightenment, and Revolution*, ed. Margaret Jacob and Wijnand Mijnhardt, 197–223. Ithaca, NY, and London: Cornell University Press, 1992.

Mikeshin, Mikhail. "Enlightened Russians in the Eighteenth Century and Today." *Studies on Voltaire and the Eighteenth Century* 348 (1997): 1574–77.

Millburn, John R. "James Ferguson's Lecture Tour of the English Midlands in 1771." *Annals of Science* 42 (1985): 397–415.

———. "The London Evening Courses of Benjamin Martin and James Ferguson, Eighteenth-Century Lecturers on Experimental Philosophy." *Annals of Science* 40 (1983): 437–55.

Miller, David Philip. "'Into the Valley of Darkness': Reflections on the Royal Society in the Eighteenth Century." *History of Science* 27 (1989): 155–66.

———. "Joseph Banks, Empire, and 'Centers of Calculation' in Late Hanoverian London." In *Visions of Empire: Voyages, Botany, and Representations of Nature*, ed. David Philip Miller and Peter Hanns Reill, 21–37. Cambridge: Cambridge University Press, 1996.

———. "The Usefulness of Natural Philosophy: The Royal Society and the Culture of Practical Utility in the Later Eighteenth Century." *British Journal for the History of Science* 32 (1999): 185–201.

Miller, David Philip, and Peter Hanns Reill, eds. *Visions of Empire: Voyages, Botany, and Representations of Nature*. Cambridge: Cambridge University Press, 1996.

Miller, Genevieve. "'Airs, Waters, and Places' in History," *Journal of the History of Medicine* 17 (January 1962): 129–40.

Money, John. "From Leviathan's Air Pump to Britannia's Voltaic Pile: Science, Public Life, and the Forging of Britain, 1660–1820." *Canadian Journal of History/Annales Canadiennes d'Histoire* 28 (1993): 521–44.

———. "Science, Technology and Dissent in English Provincial Culture: From Newtonian Transformation to Agnostic Incarnation." In *Science and Dissent in England, 1688–1945*, ed. Paul Wood, 67–112. Aldershot: Ashgate, 2004.

Montgomery, Scott L. *Science in Translation: Movements of Knowledge through Cultures and Time*. Chicago: University of Chicago Press, 2000.

Moracci, Giovanna. "The Influence on the Russian Enlightenment of the Cultural Policy of Catherine II." *Studies on Voltaire and the Eighteenth Century* 346 (1997): 268–70.

Moravia, Sergio. "The Enlightenment and the Sciences of Man." *History of Science* 17 (1980): 247–68.

———. "Philosophie et géographie à la fin du XVIIIe siècle." *Studies in Voltaire and the Eighteenth Century* 57 (1967): 937–1011.

Morse, Jedidiah. *The American Geography, or, A View of the Present Situation of the United States of America*. 2nd ed. London: Printed for John Stockdale, 1792.

———. *Elements of Geography, containing a Concise and Comprehensive View of That Useful Science*. Boston: J. Thomas & E. T. Andrews, 1795.

Morton, A. Q. "Lectures on Natural Philosophy in London: 1750–1765: S. C. T. Demainbray (1710–1782) and the 'Inattention' of his Countrymen." *British Journal for the History of Science* 23 (1990): 411–34.

Murray, Stuart. "'Notwithstanding Our Signs to the Contrary': Textuality and Authority at the Endeavour River, June to August 1771." In *Captain Cook: Explorations and Reassessments*, ed. Glyndwr Williams, 59–76. Woodbridge: Boydell Press, 2004.

Muthu, Sankar. *Enlightenment against Empire*. Princeton, NJ: Princeton University Press, 2003.

Nakayama, Shigeru. "Japan." In *The Cambridge History of Science*, vol. 4: *Eighteenth-Century Science*, ed. Roy Porter, 698–717. Cambridge: Cambridge University Press, 2003.

Naylor, Simon. "Introduction: Historical Geographies of Science—Places, Contexts, Cartographies." *British Journal for the History of Science* 38 (2005): 1–12.

Nicolson, Malcolm. "Alexander von Humboldt and the Geography of Vegetation." In *Romanticism and the Sciences*, ed. Nicholas Jardine and Andrew Cunningham, 169–88. Cambridge: Cambridge University Press, 1990.

———. "Alexander von Humboldt, Humboldtian Science and the Origins of the Study of Vegetation." *History of Science* 25 (1987): 167–94.

———. "Humboldtian Plant Geography after Humboldt: The Link to Ecology." *British Journal for the History of Science* 29 (1996): 289–310.

Nisbet, Robert. *History of the Idea of Progress*. London: Heinemann, 1980.

Nisbett, Richard E. *The Geography of Thought: How Asians and Westerners Think Differently—and Why*. London: Nicholas Brealey, 2003.

Nordman, Daniel. *Frontières de France: De l'espace au territoire, XVIe–XIXe siècle*. Paris: Gallimard, 1998.

Numbers, Ronald L., and John Stenhouse, eds. *Disseminating Darwinism: The Role of Place, Race, Religion and Gender*. New York and Cambridge: Cambridge University Press, 1999.

Nybakken, Elizabeth. "The Enlightenment and Calvinism: Mutual Support Systems for the Eighteenth-Century American Wilderness." *Studies on Voltaire and the Eighteenth Century* 192 (1980): 1126–35.

Obeyesekere, Gananath. *The Apotheosis of Captain Cook*. Princeton, NJ: University of Princeton Press, 1992.

O'Brien, Karen. "Between Enlightenment and Stadial History: William Robertson on the History of Europe." *British Journal for Eighteenth-Century Studies* 16 (1993): 53–63.

———. "Enlightenment History in Scotland: The Case of William Robertson." *Studies on Voltaire and the Eighteenth-Century* 303 (1992): 467–70.

———. *Narratives of Enlightenment: Cosmopolitan History from Voltaire to Gibbon*. Cambridge: Cambridge University Press, 1997.

Ó Ciosáin, Éamon. "Attitudes towards Ireland and the Irish in Enlightenment France." In *Ireland and the French Enlightenment, 1700–1800,* ed. Graham Gargett and Geraldine Sheridan, 129–51. Basingstoke, Macmillan, 1999.

Ogborn, Miles. "Geographia's Pen: Writing and the Arts of Commerce, 1660–1760." *Journal of Historical Geography* 30 (2004): 294–315.

———. *Spaces of Modernity: London's Geographies, 1680–1780*. New York and London: Guilford Press, 1998.

Ogborn, Miles, and Charles W. J. Withers, eds. *Georgian Geographies: Essays on Space, Place and Landscape in the Eighteenth Century*. Manchester: Manchester University Press, 2004.

———. "Knowing Other Places: Travel, Trade and Empire, 1660–1800." In *Concise Companion to the Restoration and Eighteenth Century,* ed. Cynthia Wall, 14–36. Oxford: Blackwell, 2005.

Okamoto, Masao. "The Enlightenment in Japan." *Studies on Voltaire and the Eighteenth Century* 26 (1963): 1263–73.

Oldroyd, David R. "Geology." In *Encyclopedia of the Enlightenment,* ed. Alan C. Kors, 2:117–19. New York and Oxford: Oxford University Press, 2003.

Olson, R. "The Human Sciences." In *The Cambridge History of Science,* vol. 4: *Eighteenth-Century Science,* 436–62. ed. Roy Porter, Cambridge: Cambridge University Press, 2003.

Ormsby-Lennon, H. J., and Joseph Pappa, "Pornography." In *Encyclopedia of the Enlightenment,* ed. Alan Kors, 3:334–37. New York and Oxford: Oxford University Press, 2003.

Osborne, Thomas. *Aspects of Enlightenment: Social Theory and the Ethics of Truth*. London: UCL Press, 1998.

Outram, Dorinda. *The Enlightenment*. Cambridge: Cambridge University Press, 1995.

———. "The Enlightenment Our Contemporary." In *The Sciences in Enlightened Europe,* ed. William Clark, Jan Golinski, and Simon Schaffer, 32–42. Chicago: University of Chicago Press, 1999.

———. "On Being Perseus: New Knowledge, Dislocation, and Enlightenment Exploration." In *Geography and Enlightenment,* ed. David N. Livingstone and Charles W. J. Withers, 281–294. Chicago: University of Chicago Press, 1999.

Ozouf, Mona. *Festivals and the French Revolution*. Cambridge, MA, and London: Harvard University Press, 1988.

Oz-Salzberger, Fania. *Translating the Enlightenment: Scottish Civic Discourse in Eighteenth-Century Germany*. Oxford: Clarendon Press, 1995.

———. "Translation." In *Encyclopedia of the Enlightenment,* ed. Alan C. Kors, 4:181–88. New York and Oxford: Oxford University Press, 2003.

Pagden, Anthony. "Eighteenth-Century Anthropology and the 'History of Mankind.'" In *History and the Disciplines: The Reclassification of Knowledge in Early Modern Europe,* ed. Donald Kelley, 223–35. Rochester, NY: University of Rochester Press, 1997.

———. *European Encounters with the New World: From the Renaissance to Romanticism*. New Haven, CT, and London: Yale University Press, 1993.

———. “The Reception of the ‘New Philosophy’ in Eighteenth-Century Spain.” *Journal of the Warburg and Courtauld Institutes* 51 (1988): 126–40.

Pajkossy, Gábor. “Western European Models and National Traditions: Reform Projects in Late Eighteenth-Century Hungary.” *Studies on Voltaire and the Eighteenth Century* 263 (1989): 349–52.

Pancaldi, Guilano. “Appropriating Invention: The Reception of the Voltaic Battery in Europe.” In *Instruments, Travel and Science: Itineraries of Precision from the Seventeenth to the Twentieth Century*, ed. Marie-Noëlle Bourguet, Christian Licoppe, and H. Otto Sibum, 126–55. London: Routledge, 2002.

———. *Volta: Science and Culture in the Age of Enlightenment*. Princeton, NJ: Princeton University Press, 2003.

Pansini, Valeria. “La géographie appliquée à la guerre: Le travail des topographes militaires, 1760–1820.” In *Géographie plurielles: Les sciences géographiques au moment de l'emergence des sciences humaines (1750–1850)*, ed. Hélène Blais and Isabelle Laboulais, 167–84. Paris: L'Harmattan, 2005.

Paul, Ina. “‘Europe the Way It Was and Is’—The Change in the Conception of Europe in Encyclopedias from the Eighteenth Century to the Nineteenth.” *Studies on Voltaire and the Eighteenth Century* 346 (1997): 192–93.

Paxman, David B. *Voyage into Language: Space and the Linguistic Encounter, 1500–1800*. London: Ashgate, 2003.

Pearson, W. H. “Hawkesworth's Alterations.” *Pacific History* 7 (1972): 45–71.

Pedley, Mary. “Atlas Editing in Enlightenment France.” *Journal of Scholarly Publishing* 27 (1996): 100–117.

———. *The Commerce of Cartography: Making and Marketing Maps in Eighteenth-Century France and England*. Chicago and London: University of Chicago Press, 2005.

———. “Mapping Brittany: An Eighteenth Century Mémoire.” *Map Collector* 48 (1989): 14–18.

———, ed. *The Map Trade in the Late Eighteenth Century: Letters to the London Map Sellers Jefferys and Faden*. Oxford: Voltaire Foundation, 2000.

———. “The Map Trade in Paris, 1650–1825.” *Imago Mundi* 33 (181): 33–45.

———. “New Light on an Old Atlas: Documents concerning the Publication of the *Atlas Universel* (1757).” *Imago Mundi* 36 (1984): 48–63.

Pelizzari, Maria Rosaria. “Enlightenment Intellectuals and Popular Mentality after the 1783 Calabrian Earthquake.” *Studies in Voltaire and the Eighteenth Century* 303 (1992): 523–27.

Pelletier, Monique. *La carte de Cassini: L'extraordinaire aventure de la carte de France*. Paris: École Nationale des Ponts et Chaussées, 1990.

Perdiguero, Enrique. “The Popularization of Medicine during the Spanish Enlightenment.” In *The Popularization of Medicine, 1650–1850*, ed. Roy Porter, 160–93. London: Routledge, 1992.

Pereira, José Esteves. “Portuguese Enlightenment.” *Studies on Voltaire and the Eighteenth Century* 264 (1989): 184–86.

Perla, Georges A. “La géographie dans l'*Encyclopédie*.” *Revue de Synthese* 115 (1984): 299–311.

Phillipson, Nicholas. “The Scottish Enlightenment.” In *The Enlightenment in National Context*, ed. Roy Porter and Mikuláš Teich, 19–40. Cambridge: Cambridge University Press, 1981.

Philosophical Society of Edinburgh. *Essays and Observations, Physical and Literary.* Edinburgh: Printed for John Balfour, 1771.

Pimentel, Juan. "The Iberian Vision: Science and Empire in the Framework of a Universal Monarchy, 1500–1800." *Osiris* 15 (2000): 17–30.

Pinkerton, John. *Modern Geography.* 2 vols. London: Strahan, Cadell, Davies, Longman, & Rees, 1802.

Pocock, John G. A. *Barbarism and Religion.* 2 vols. Cambridge: Cambridge University Press, 1999.

———. "Enlightenment and Revolution: The Case of English-Speaking North America." *Studies on Voltaire and the Eighteenth Century* 263 (1989): 249–61.

———. *Virtue, Commerce, and History: Essays on Political Thought and History Chiefly in the Eighteenth Century.* Cambridge: Cambridge University Press, 1985.

Pole, J. R. "Enlightenment and the Politics of American Nature." In *The Enlightenment in National Context,* ed. Roy Porter and Mikuláš Teich, 192–214. Cambridge: Cambridge University Press, 1981.

Pollard, Sidney. *The Idea of Progress.* Harmondsworth: Penguin, 1968.

Poovey, Mary. *A History of the Modern Fact: Problems of Knowledge in the Sciences of Wealth and Society.* Chicago: University of Chicago Press, 1998.

Popkin, Jeremy. "Periodical Publication and the Nature of Knowledge in Eighteenth-Century Europe." In *The Shapes of Knowledge from the Renaissance to the Enlightenment,* ed. Donald R. Kelley and Richard H. Popkin, 203–14. Dordrecht: Kluwer Academic Publishers, 1991.

———. "Robert Darnton's Alternative (to the) Enlightenment." In *The Darnton Debate: Books and Revolution in the Eighteenth Century,* ed. Haydn T. Mason, 105–28. Oxford: Voltaire Foundation, 1998.

Popkin, Richard. "The Philosophical Basis of Eighteenth-Century Racism." *Studies in Eighteenth-Century Culture* 3 (1973): 245–62.

Porter, Roy, ed. *The Cambridge History of Science,* vol. 4: *The Eighteenth Century.* Cambridge: Cambridge University Press, 2003.

———. "England." In *Encyclopedia of the Enlightenment,* ed. Alan Kors, 1:409–15. New York and Oxford: Oxford University Press, 2003.

———. *The Enlightenment.* London: Macmillan, 1990.

———. *Enlightenment: Britain and the Creation of the Modern World.* London: Penguin, 2000.

———. "The Enlightenment in England." In *The Enlightenment in National Context,* ed. Roy Porter and Mikuláš Teich, 1–18. Cambridge: Cambridge University Press, 1981.

———. *Flesh in the Age of Reason.* London: Penguin, 2003.

———. *The Greatest Benefit to Mankind: A Medical History of Humanity from Antiquity to the Present.* London: Harper Collins, 1997.

———. Introduction to *The Cambridge History of Science,* vol. 4: *Eighteenth-Century Science,* 1–20. ed. Roy Porter, Cambridge: Cambridge University Press, 2003.

———. "Medical Lecturing in Georgian London." *British Journal for the History of Science* 28 (1995): 91–99.

———. "Medical Science and Human Science in the Enlightenment." In *Inventing Human Science: Eighteenth-Century Domains,* ed. Christopher Fox, Roy Porter, and Robert Wokler, 53–87. Berkeley, University of California Press, 1995.

———, ed. *Medicine in the Enlightenment*. Amsterdam: Rodopi, 1995.

———, ed. *The Popularization of Medicine, 1650–1850*. Routledge: London: 1992.

"Science, Provincial Culture and Public Opinion in Enlightenment England." *British Journal for Eighteenth-Century Studies* 3 (1980): 20–46.

———. "Spreading Medical Enlightenment: The Popularization of Medicine in Georgian England." In *The Popularization of Medicine 1650–1850*, ed. Roy Porter, 215–32. London: Routledge, 1992.

———. "The Terraqueous Globe." In *The Ferment of Knowledge: Studies in the Historiography of Eighteenth-Century Science*, ed. G. S. Rousseau and R. Porter, 285–326. Cambridge: Cambridge University Press, 1980.

Porter, Roy, and Mikuláš Teich, eds. *The Enlightenment in National Context*. Cambridge: Cambridge University Press, 1981.

Pratt, Mary Louise. *Imperial Eyes: Travel Writing and Transculturation*. London: Routledge, 1992.

Preston, Diana, and Michael Preston. *A Pirate of Exquisite Mind: The Life of William Dampier, Explorer, Naturalist and Buccaneer*. London and New York: Doubleday, 2004.

Pyenson, Lewis. "An End to National Science: The Meaning and the Extension of Local Knowledge," *History of Science* 40 (2002): 251–90.

———. "Over the Bounding Main." *Historical Studies in the Physical and Biological Sciences* 20 (1990): 407–22.

Quill, Humphrey. *John Harrison: The Man Who Found Longitude*. London: Baker, 1966.

Racevskis, Karlis. *Postmodernism and the Search for Enlightenment*. Charlottesville and London: University of Virginia Press, 1993.

Raeff, Marc. "The Enlightenment in Russia and Russian Thought in the Enlightenment." In *The Eighteenth Century in Russia*, ed. J. G. Garrard, 25–47. Oxford: Clarendon Press, 1973.

Raj, Kapil. "Bounded Landscapes." *History Workshop Journal* 54 (2002): 242–45.

———. "Colonial Encounters and the Forging of New Knowledge and National Identities: Great Britain and India, 1760–1850." *Osiris* 15 (2000): 119–34.

———. "Refashioning Civilities, Engineering Trust: William Jones, Indian Intermediaries and the Production of Reliable Legal Knowledge in Late Eighteenth-Century Bengal." *Studies in History* 17 (2001): 175–209.

Rappaport, Rhoda. "The Earth Sciences." In *The Cambridge History of Science*, vol. 4: *Eighteenth-Century Science*, ed. Roy Porter, 419–35. Cambridge: Cambridge University Press, 2003.

———. "Geology and Orthodoxy: The Case of Noah's Flood in Eighteenth-Century Thought." *British Journal for the History of Science* 11 (1978): 1–19.

Ravel, Jeffrey. "France." In *Encyclopedia of the Enlightenment*, ed. Alan Kors, 2:60–65. New York and Oxford: Oxford University Press, 2003.

Ray, John. *The Wisdom of God Manifested in the Works of the Creation. In Two Parts*. London: Rivington, Ward & Richardson, 1681.

Rechniewski, Elizabeth. "References to 'National Character' in the *Encyclopédie*: The Western Nations." *Studies on Voltaire and the Eighteenth Century* 12 (2003): 221–38.

Reill, Peter Hanns. "Buffon and Historical Thought in Germany and Great Britain." In *Buffon 88*, ed. J. Gayon, 667–79. Paris: Vrin, 1992.

———. *Encyclopedia of the Enlightenment*. New York: Facts on File, 1996.

———. "The Legacy of the 'Scientific Revolution': Science and the Enlightenment." In *The Cambridge History of Science*, vol. 4: *Eighteenth-Century Science*, ed. Roy Porter, 23–43. Cambridge: Cambridge University Press, 2003.

———. *Vitalizing Nature in the Enlightenment*. Berkeley: University of California Press, 2005.

Reinhardt, O., and David R. Oldroyd. "By Analogy with the Heavens: Kant's Theory of the Earth." *Annals of Science* 41 (1984): 203–21.

———. "Kant's Theory of Earthquakes and Volcanic Action." *Annals of Science* 40 (1983): 247–72.

———. "Kant's Thoughts on the Ageing of the Earth." *Annals of Science* 39 (1982): 349–69.

Reitan, E. A. "Expanding Horizons: Maps in the *Gentleman's Magazine*, 1731–1754." *Imago Mundi* 37 (1985): 54–62.

Reitinger, Franz. "Mapping Relationships: Allegory, Gender and the Cartographical Image in Eighteenth-Century France and England." *Imago Mundi* 51 (1999): 106–30.

Relaño, Francesco. *The Shaping of Africa: Cosmographic Discourse and Cartographic Science in Late Medieval and Early Modern Europe*. Aldershot: Ashgate, 2002.

Rennell, James. *Memoir of a Map of Hindoostan*. London: G. Nicol & W. Richardson, 1788.

———. "On the Rate of Travelling, as Performed by Camels; and Its Application, as a Scale, to the Purpose of Geography." *Philosophical Transactions* 81 (1791): 129–45.

Renwick, John, ed. *L'invitation au Voyage: Studies in Honour of Peter France*. Oxford: Voltaire Foundation, 2000.

———. "The Reception of William Robertson's Historical Writings in Eighteenth-Century France." In *William Robertson and the Expansion of Empire*, ed. Stewart J. Brown, 145–63. Cambridge: Cambridge University Press, 1997.

Ricuperati, Giuseppe. "Enlightenment in the Romance Countries." *Studies on Voltaire and the Eighteenth Century* 346 (1997): 232–35.

Robert de Vaugondy, Didier. *Essai sur l'histoire de la géographie, ou sur son origine, ses progrès et son état actuel*. Paris: Antoine Boudet, 1755.

Roberts, Lissa. "Going Dutch: Situating Science in the Dutch Enlightenment." In *The Sciences in Enlightened Europe*, ed. William Clark, Jan Golinski, and Simon Schaffer, 350–88. Chicago: University of Chicago Press, 1999.

Robertson, John. *The Case for the Enlightenment: Scotland and Naples, 1680–1760*. Cambridge: Cambridge University Press, 2005.

———. "The Enlightenment above National Context: Political Economy in Eighteenth-Century Scotland and Naples." *Historical Journal* 40 (1997): 667–97.

———. "Franco Venturi's Enlightenment." *Past and Present* 137 (1992): 183–206.

Robertson, Richie, and Edward Timms, eds. *Austrian Enlightenment and Its Aftermath*. Edinburgh: Edinburgh University Press, 1981.

Robinson, Edward Forbes. *The Early History of Coffee Houses in London*. London: Kegan Paul, 1893.

Robinson, Harry. "Geography in the Dissenting Academies." *Geography* 36 (1951): 179–86.

Roche, Daniel. *France in the Enlightenment*. Trans. Arthur Goldhammer. Cambridge, MA, and London: Harvard University Press, 1998.

———. *Le siècle des lumières en province: Académies et académiciens provinciaux, 1680–1789*. 2 vols. Paris: Mouton, 1978.

———. "Urban Reading Habits during the French Enlightenment." *British Journal for Eighteenth-Century Studies* 2 (1979): 138–49 and 220–31.

Roger, Jacques. *Buffon: A Life in Natural History.* Trans. Sarah Bonnefoi. Ithaca, NY: Cornell University Press, 1997.

Rose, Jonathan. "The History of Books: Revised and Enlarged." In *The Darnton Debate: Books and Revolution in the Eighteenth Century,* ed. Haydn Mason, 83–104. Oxford: Voltaire Foundation

Rose, R. B. "Utopianism." In *Encyclopedia of the Enlightenment,* ed. Alan Kors, 4:215–18. New York and Oxford: Oxford University Press, 2003.

Rosen, George. "Leonhard Ludwig Finke and the First Medical Geography." In *Science, Medicine and History: Essays on the Evolution of Scientific Thought and Medical Practice,* ed. E. Ashworth Underwood, 2:186–93. Oxford: Oxford University Press, 1953.

———. "Noah Webster—Historical Epidemiologist." *Journal of the History of Medicine and Allied Sciences* 20 (1965): 97–114.

Rosner, Lisa. *Medical Education in the Age of Improvement: Edinburgh Students and Apprentices, 1760–1826.* Edinburgh: Edinburgh University Press, 1991.

Rousseau, George. *Perilous Enlightenment: Pre- and Post-Modern Discourses—Sexual, Historical.* Manchester: Manchester University Press, 1991.

———. "Whose Enlightenment? Not Man's: The Case of Michel Foucault." *Eighteenth-Century Studies* 6 (1972–73): 238–56.

Rousseau, George, and Roy Porter, eds. *Exoticism and the Enlightenment.* Manchester: Manchester University Press, 1990.

———, eds. *The Ferment of Knowledge: Studies in the Historiography of Eighteenth-Century Science.* Cambridge: Cambridge University Press, 1980.

———, eds. *Sexual Underworlds of the Enlightenment.* Manchester: Manchester University Press, 1987.

Roy, William. *An Account of the Mode Proposed to Be Followed in the Trigonometrical Operation for determining the Relative Situation of the Royal Observatories of Greenwich and Paris, with Observations on the Magnitude and Figure of the Earth.* London: John Nicols, 1787.

Rudwick, Martin J. S. *Bursting the Limits of Time: The Reconstruction of Geohistory in the Age of Revolution.* Chicago: University of Chicago Press, 2005.

———. "The Emergence of a Visual Language for Geological Science, 1760–1840." *History of Science* 14 (1976): 149–95.

Rupke, Nicolaas. *Alexander von Humboldt: A Metabiography.* Göttingen: Peter Lang, 2005.

———. "A Geography of Enlightenment: The Critical Reception of Alexander von Humboldt's Mexico Work." In *Geography and Enlightenment,* ed. David N. Livingstone and Charles W. J. Withers, 319–39. Chicago: University of Chicago Press, 1999.

Rupp, Jan. "The New Science and the Public Sphere in the Premodern Era." *Science in Context* 8 (1995): 487–507.

Rusnock, Andrea A. "Biopolitics: Political Arithmetic in the Enlightenment." In *The Sciences in Enlightened Europe,* ed. William Clark, Jan Golinski, and Simon Schaffer, 49–68. Chicago: University of Chicago Press, 1999.

———. *Vital Accounts: Quantifying Health and Population in Eighteenth-Century England and France.* Cambridge: Cambridge University Press, 2002.

Ryan, Tom. "'Le président des Terres Australes': Charles de Brosses and the French Enlightenment Beginnings of Oceanic Anthropology." *Journal of Pacific History* 37 (2002): 157–86.

Sahlins, Marshall. *How "Natives" Think: About Captain Cook, for Example*. Chicago: University of Chicago Press, 1996.

Sahlins, Peter. "Natural Frontiers Revisited: France's Boundaries since the Seventeenth Century." *American Historical Review* 95 (1990): 1423–51.

Said, Edward. *Culture and Imperialism*. London: Verso, 1991.

———. *Orientalism*. London: Penguin Books, 1978.

———. *The World, The Text and the Critic*. London: Vintage, 1991.

Saiedi, Nader. *The Birth of Social Theory: Social Thought in the Enlightenment and Romanticism*. Lanham, MD, and London: University Press of America, 1993.

Salmond, Anne. *Between Worlds: Early Exchanges between Maori and Europeans, 1773–1815*. Honolulu: University of Hawaii Press, 1997.

———. *Two Worlds: First Meetings between Maori and Europeans, 1642–1772*. Auckland: Viking, 1991.

Sambrook, James, ed. *The Task and Selected Other Poems*. London: Longman, 1994.

Savage-Smith, Emilie. "Islam." In *The Cambridge History of Science*, vol. 4: *Eighteenth-Century Science*, ed. Roy Porter, 649–68. Cambridge: Cambridge University Press, 2003.

Savours, Ann. "'A Very Interesting Point in Geography': The 1773 Phipps Expedition towards the North Pole." *Arctic* 37 (1984): 402–28.

Schaffer, Simon. "The Enlightenment Brought down to Earth." *History of Science* 41 (2003): 257–68.

———. "Machine Philosophy: Demonstration Devices in Georgian Mechanics." *Osiris* 9 (1994): 157–82.

———. "Measuring Virtue: Eudiometry, Enlightenment and Pneumatic Medicine." In *The Medical Enlightenment of the Eighteenth Century*, ed. Andrew Cunningham and Roger French, 281–318. Cambridge: Cambridge University Press, 1990.

———. "Natural Philosophy and Public Spectacle in the Eighteenth Century." *History of Science* 21 (1983):1–43.

———. "Newtonianism." In *Companion to the History of Modern Science*, ed. R. C. Olby, Geoffrey N. Cantor, John R. R. Christie, and Martin J. S. Hodge, 610–26. London: Routledge, 1990.

Schama, Simon. "The Enlightenment in the Netherlands." In *The Enlightenment in National Context*, ed. Roy Porter and Mikuláš Teich, 54–71. Cambridge: Cambridge University Press, 1981.

Schiebinger, Londa. "European Women in Science." *Science in Context* 15 (2002): 473–81.

———. *The Mind Has No Sex? Women in the Origin of Modern Science*. Cambridge, MA: Harvard University Press, 1989.

———. "Nature's Unruly Body: The Limits of Scientific Description." In *Regimes of Description: In the Archive of the Eighteenth Century*, ed. John Bender and Michael Marrinan, 25–43. Stanford, CA: Stanford University Press, 2005.

———. "The Philosopher's Beard: Women and Gender in Science." In *The Cambridge History of Science*, vol. 4: *Eighteenth-Century Science*, ed. Roy Porter, 185–210. Cambridge: Cambridge University Press, 2003.

Schlafly, Daniel. "Russian Travellers Discover Western Europe in the Age of the Enlightenment." *Studies on Voltaire and the Eighteenth Century* 305 (1992): 1657–60.

Schmidt, James. "Introduction: What Is Enlightenment? A Question, Its Context, and Some Consequences." In *What Is Enlightenment? Eighteenth-Century Answers and*

Twentieth-Century Questions, ed. James Schmidt, 1–44. Berkeley: University of California Press, 1996.

———, ed. *What Is Enlightenment? Eighteenth-Century Answers and Twentieth-Century Questions*. Berkeley: University of California Press, 1996.

Schmidt, Leigh Eric. *Hearing Things: Religion, Illusion and the American Enlightenment*. Cambridge, MA: Harvard University Press, 2000.

Schofield, Robert. "An Evolutionary Taxonomy of Eighteenth-Century Newtonianisms." *Studies in Eighteenth-Century Culture* 7 (1978): 175–92.

Schulze, Ludmilla. "The Russification of the St. Petersburg Academy of Sciences and Arts in the Eighteenth Century." *British Journal for the History of Science* 18 (1985): 305–35.

Sebastiani, Silvia. "Race as a Construction of the Other: 'Native Americans' and 'Negroes' in the Eighteenth-Century Editions of the *Encyclopedia Britannica*." *In Europe and the Other and Europe as the Other*, ed. Bo Stråth, 195–228. Brussels: Peter Lang, 2000.

Secord, James. "Knowledge in Transit." *Isis* 95 (2004): 654–72.

———. "Newton in the Nursery: Tom Telescope and the Philosophy of Tops and Balls." *History of Science* 23 (1985): 127–51.

———. *Victorian Sensation: The Extraordinary Publication, Reception, and Secret Authorship of "Vestiges of the Natural History of Creation."* Chicago: University of Chicago Press, 2000.

Serman, Il'ia. "Russian National Consciousness and Its Development in the Eighteenth Century." In *Russia in the Age of the Enlightenment*, ed. Roger Bartlett and Janet M. Hartley, 40–56. London: Macmillan, 1990.

Shafer, Robert J. *The Economic Societies in the Spanish World (1763–1821)*. Syracuse, NY: Syracuse University Press, 1958.

Shapin, Steven. "The Image of the Man of Science." In *The Cambridge History of Science*, vol. 4: *Eighteenth-Century Science*, ed. Roy Porter, 159–83. Cambridge: Cambridge University Press, 2003.

———. "Placing the View from Nowhere: Historical and Sociological Problems in the Location of Science." *Transactions of the Institute of British Geographers* 23 (1998): 5–12.

———. "Rarely Pure and Never Simple: Talking about Truth." *Configurations* 7 (1999): 1–14.

———. *The Scientific Revolution*. Chicago: University of Chicago Press, 1996.

———. *A Social History of Truth: Civility and Science in Seventeenth-Century England*. Chicago: University of Chicago Press, 1994.

———. "Social Uses of Science." In *The Ferment of Knowledge: Studies in the Historiography of Eighteenth-Century Science*, ed. G. S. Rousseau and Roy Porter, 93–139. Cambridge: Cambridge University Press, 1980.

Shaw, Denis J. B. "Geographical Practice and Its Significance in Peter the Great's Russia." *Journal of Historical Geography* 22 (1996): 160–76.

Shea, William R. "The Rhetoric of Experiments and Scientific Illustrations in the Enlightenment." In *Science and the Visual Image in the Enlightenment*, ed. William R. Shea, 39–56. Canton, MA: Science History Publications, 2000.

———, ed. *Science and the Visual Image in the Enlightenment*. Canton, MA: Science History Publications, 2000.

Shea, William, and Peter Huff. Introduction to *Knowledge and Belief in America: Enlightenment Traditions and Modern Religious Thought*, ed. William Shea and Peter Huff, 1–14. New York: Cambridge University Press, 1995.

———. *Knowledge and Belief in America: Enlightenment Traditions and Modern Religious Thought*. New York: Cambridge University Press, 1995.

Shefrin, Jill. *Neatly Dissected for the Instruction of Young Ladies and Gentlemen in the Knowledge of Geography: John Spilsbury and Early Dissected Puzzles*. Cotsen Library Occasional Publication. Los Angeles, 1999.

Sher, Richard., ed. *Conjectural History and Anthropology in the Scottish Enlightenment*. 7 vols. Bristol: Thoemmes Press, 1995.

———. "Science and Medicine in the Scottish Enlightenment: The Lessons of Book History." In *The Scottish Enlightenment: Essays in Reinterpretation*, 99–156. Rochester NY: University of Rochester Press, 2000.

Sheridan, Geraldine. "Irish Literary Review Magazines and Enlightenment France: 1730–1790." In *Ireland and the French Enlightenment, 1700–1800*, ed. Graham Gargett and Geraldine Sheridan, 21–46. Basingstoke, Macmillan, 1999.

Sherratt, Yvonne. "Adorno and Horkheimer's Concept of 'Enlightenment.'" *British Journal for the History of Philosophy* 8 (2000): 521–44.

Shevchenko, Nataliya. "The Contribution of Ukrainian Thinkers to the Political Culture of the Enlightenment," *Studies on Voltaire and the Eighteenth Century* 346 (1997): 286–89.

Short, John Rennie. "A New Mode of Thinking: Creating a National Geography in the Early Republic." In *Surveying the Record: North American Scientific Exploration to 1930*, ed. Edward C. Carter, 1–50. Philadelphia: American Philosophical Society, 1999.

Simões, Ana, Ana Carneiro, and Maria Paula Diogo. "Constructing Knowledge: Eighteenth-Century Portugal and the New Sciences." In *The Sciences in the European Periphery during the Enlightenment*, ed. Kostas Gavroglu, 1–40. Dordrecht: Kluwer Academic Publishers, 1999.

———, eds. *Travels of Learning: A Geography of Science in Europe*. Dordrecht: Kluwer Academic Publishers, 2003.

Sitwell, O. Francis G. *Four Centuries of Special Geography*. Vancouver: University of British Columbia Press, 1993.

Sloan, Phillip. "Buffon." In *Encyclopedia of the Enlightenment*, ed. Alan Kors, 1:176–80. New York and Oxford: Oxford University Press, 2003.

———. "The Buffon-Linnaeus Controversy." *Isis* 67 (1976): 356–73.

———. "The Gaze of Natural History." In *Inventing Human Science: Eighteenth-Century Domains*, ed. Christopher Fox, Roy Porter, and Robert Wokler, 112–51. Berkeley and London: University of California Press, 1996.

———. "The Idea of Racial Degeneracy in Buffon's *Histoire Naturelle*." In *Racism in the Eighteenth Century*, ed. Harold Pagliaro, 293–321. Cleveland: Press of Case Western Reserve University, 1973.

Smith, Anthony. *Theories of Nationalism*. London: Duckworth, 1971.

Smith, Ben, and James W. Vining. *American Geographers, 1784–1812: A Bio-Bibliographical Study*. Westport, CT, and London: Praeger Books, 2003.

Smith, Bernard. *European Vision and the South Pacific, 1768–1850: A Study in the History of Art and Ideas*. 2nd ed. Sydney: Harper & Row, 1985.

———. *Imagining the Pacific: In the Wake of the Cook Voyages*. New Haven: Yale University Press, 1992.

Smith, Crosbie, and Jon Agar, eds. *Making Space for Science: Territorial Themes in the Shaping of Knowledge*. Basingstoke: Macmillan, 1998.

Smith, Jonathan. "State Formation, Geography, and a Gentleman's Education." *Geographical Review* 86 (1996): 91–100.

Smith, Preserved. *The Enlightenment, 1687–1776*. New York: Knopf, 1962.

Smoliy, V. "The Enlightenment and State Formation in Eighteenth-Century Ukraine." *Studies on Voltaire and the Eighteenth Century* 346 (1997): 284–86.

Snelders, H. A. M. "Professors, Amateurs, and Learned Societies: The Organisation of the Natural Sciences." In *The Dutch Republic in the Eighteenth Century: Decline, Enlightenment, and Revolution*, ed. Margaret Jacob and Wijnand Mijnhardt, 308–23. Ithaca, NY, and London: Cornell University Press, 1992.

Sobel, Dava. *Longitude: The True Story of a Lone Genius Who Solved the Greatest Scientific Problem of His Time*. London: Fourth Estate, 1995.

Sobel, Dava, and William J. H. Andrewes. *Illustrated Longitude*. London: Fourth Estate, 1998.

Solkin, David H. *Painting for Money: The Visual Arts and the Public Sphere in Eighteenth-Century England*. New Haven: Yale University Press, 1993.

———, ed. *Art on the Line: The Royal Academy Exhibitions at Somerset House, 1780–1836*. New Haven, CT, and London: Yale University Press, 2001.

Sörlin, Sverker. "National and International Aspects of Cross-Boundary Science: Scientific Travel in the 18th Century." In *Denationalizing Science: The Contexts of International Scientific Practice*, ed. Elisabeth Crawford, Terry Shinn, and Sverker Sörlin, 43–72. Dordrecht: Kluwer Academic Publishers, 1993.

Sorrenson, Richard. "The Ship as a Scientific Instrument in the Eighteenth Century." *Osiris* 11 (1996): 221–36.

———. "Towards a History of the Royal Society in the Eighteenth Century." *Notes and Records of the Royal Society of London* 50 (1996): 29–46.

Spadafora, David. *The Idea of Progress in Eighteenth-Century Britain*. New Haven: Yale University Press, 1990.

———. "Progress." In *Encyclopedia of the Enlightenment*, ed. Alan C. Kors, 3:367–72. New York and Oxford: Oxford University Press, 2003.

Spary, Emma C. "Botany." In *Encyclopedia of the Enlightenment*, ed. Alan C. Kors, 1:167–68. New York and Oxford: Oxford University Press, 2003.

———. "Linnaeus." In *Encyclopedia of the Enlightenment*, ed. Alan C. Kors, 2:410–13. New York and Oxford: Oxford University Press, 2003.

———. *Utopia's Gardens: French Natural History from Old Regime to Revolution*. Chicago: University of Chicago Press, 2000.

Spencer, Lloyd, and Andrzej Krauze, eds. *The Enlightenment for Beginners*. Cambridge: Icon Books, 1997.

Stafford, Barbara Maria. *Body Criticism: Imaging the Unseen in Enlightenment Art and Medicine*. Cambridge, MA, and London: MIT Press, 1993.

———. *Voyage into Substance: Art, Science, Nature, and the Illustrated Travel Account, 1760–1840*. Cambridge, MA: MIT Press, 1984.

Staum, Martin S. "Human Geography in the French Institute: New Discipline or Missed Opportunity?" *Journal of the History of the Behavioral Sciences* 23 (1987): 332–40.

Stevenson, Lloyd G. "Putting Disease on the Map: The Early Use of Spot Maps in the Study of Yellow Fever." *Journal of the History of Medicine and Allied Sciences* 20 (1965): 226–61.

Stewart, Larry. "Global Pillage: Science, Commerce, and Empire." In *The Cambridge History of Science*, vol. 4: *Eighteenth-Century Science*, ed. Roy Porter, 825–44. Cambridge: Cambridge University Press, 2003.

———. "A Meaning for Machines: Modernity, Utility, and the Eighteenth-Century British Public." *Journal of Modern History* 70 (1998): 259–94.

———. "Other Centres of Calculation, or, Where the Royal Society Didn't Count: Commerce, Coffee-Houses and Natural Philosophy in Early Modern London." *British Journal for the History of Science* 32 (1999): 133–53.

———. "The Public Culture of Radical Philosophers in Eighteenth-Century London." In *Science and Dissent in England, 1688–1945,* ed. Paul Wood, 113–30. Aldershot: Ashgate, 2004.

———. "Public Lectures and Private Patronage in Newtonian England." *Isis* 77 (1986): 47–58.

———. *The Rise of Public Science: Rhetoric, Technology, and Natural Philosophy in Newtonian Britain, 1660–1750.* Cambridge: Cambridge University Press, 1992.

Stewart, Larry, and Paul Weindling. "Philosophical Threads: Natural Philosophy and Public Experiment among the Weavers of Spitalfields." *British Journal for the History of Science* 28 (1995): 37–62.

Stoddart, David R. "Captain Cook and How We Understand Him." *Geographical Review* 87 (1997): 537–41.

———. *On Geography.* Oxford: Blackwell, 1986.

Stone, Jeffrey. *A Short History of the Cartography of Africa.* Lewiston, NY: Edwin Mellen Press, 1995.

Storrs, Christopher. "Portugal." In *Encyclopedia of the Enlightenment,* ed. Alan Kors, 3:338–40. New York and Oxford: Oxford University Press, 2003.

———. "Spain." In *Encyclopedia of the Enlightenment,* ed. Alan Kors, 4:109–12. New York and Oxford: Oxford University Press, 2003.

Strack, Thomas. "Philosophical Anthropology on the Eve of Biological Determinism: Immanuel Kant and Georg Forster on the Moral Qualities and Biological Characteristics of the Human Race." *Central European History* 29 (1996): 285–308.

Sturdy, David J. *Science and Social Status: The Members of the Académie des Sciences, 1666–1750.* Woodbridge: Boydell Press, 1995.

Sutton, Geoffrey V. *Science for a Polite Society: Gender, Culture, and the Demonstration of Enlightenment.* Boulder, CO: Westview Press, 1995.

Szlatky, Maria, "Tissot as Part of the Medical Enlightenment in Hungary." In *The Popularization of Medicine, 1650–1850,* ed. Roy Porter, 194–214. London: Routledge, 1992.

Tabaki, Anna. "Greece." In *Encyclopedia of the Enlightenment,* ed. Alan Kors, 2:157–60. New York and Oxford: Oxford University Press, 2003.

Taylor, Samuel S. B. "The Enlightenment in Switzerland." In *The Enlightenment in National Context,* ed. Roy Porter and Mikuláš Teich, 72–89. Cambridge: Cambridge University Press, 1981.

Teich, Mikuláš. "Bohemia: From Darkness into Light." In *The Enlightenment in National Context,* ed. Roy Porter and Mikuláš Teich, 141–63. Cambridge: Cambridge University Press, 1981.

Terrall, Mary. "The Culture of Science in Frederick the Great's Berlin." *History of Science* 28 (1990): 333–64.

———. "Émilie du Châtelet and the Gendering of Science." *History of Science* 33 (1995): 283–310.

———. "Experimenting with Enlightenment." *Studies in the History and Philosophy of Science* 29 (1998): 319–25.

———. "Gendered Spaces, Gendered Audiences: Inside and Outside the Paris Academy of Sciences." *Configurations* 2 (1995): 207–32.

———. "Heroic Narratives of Quest and Discovery." *Configurations* 6 (1998): 223–42.

———. *The Man Who Flattened the Earth: Maupertuis and the Sciences in the Enlightenment.* Chicago: University of Chicago Press, 2003.

———. "Representing the Earth's Shape: The Polemics Surrounding Maupertuis's Expedition to Lapland." *Isis* 83 (1992): 218–37.

———. "Salon, Academy, and Boudoir: Generation and Desire in Maupertuis's Science of Life." *Isis* 87 (1996): 217–29.

Thomas, Nicholas. *Discoveries: The Voyages of Captain Cook.* London: Allen Lane, 2003.

Thomas, R. George. "The Enlightenment and Wales in the 18th Century." *Studies on Voltaire and the Eighteenth Century* 27 (1963): 1575–91.

Thompson, Martyn. "Reception Theory and the Interpretation of Historical Meaning." *History and Theory* 32 (1993): 248–72.

Thomson, Ann. "North Africa and the Levant." In *Encyclopedia of the Enlightenment,* ed. Alan Kors, 3:190–92. New York and Oxford: Oxford University Press, 2003.

———. "Ottoman Empire." In *Encyclopedia of the Enlightenment,* ed. Alan Kors, 3:220–22. New York and Oxford: Oxford University Press, 2003.

Till, Nicholas. *Mozart and the Enlightenment: Truth, Virtue and Beauty in Mozart's Operas.* London: Faber & Faber, 1992.

Tilly, Charles, ed. *The Formation of National States in Western Europe.* Princeton, NJ: Princeton University Press, 1975.

Todorov, Tzvetan. *The Deflection of the Enlightenment.* Stanford, CA: Stanford Humanities Center, 1989.

Tomaselli, Sylvana. "The Enlightenment Debate on Women." *History Workshop Journal* 20 (1985): 101–24.

Tribe, Keith. "Economic Thought." *Encyclopedia of the Enlightenment,* ed. Alan Kors, 1:371–79. New York and Oxford: Oxford University Press, 2003.

———. *Governing Economy: The Reformation of German Economic Discourse, 1750–1840.* Cambridge: Cambridge University Press.

Tucker, Bruce. "Beyond Reason and Revelation: Perspectives on the Puritan Enlightenment." *Studies in Eighteenth-Century Culture* 10 (1981): 165–79.

Turner, G. L'E. "Eighteenth-Century Scientific Instruments and Their Makers." In *The Cambridge History of Science,* vol. 4: *Eighteenth-Century Science,* ed. Roy Porter, 511–35. Cambridge: Cambridge University Press, 2003.

Uglow, Jennifer S. *The Lunar Men: The Friends Who Made the Future.* London: Faber, 2002.

Umbach, Maiken. *Federalism and Enlightenment in Germany, 1740–1806.* London and Rio Grande: Hambledon Press, 2000.

———. "Visual Culture, Scientific Images and German Small-Scale Politics in the Enlightenment." *Past and Present* 158 (1998): 110–45.

Valenčius, Conevery Bolton. "Histories of Medical Geography," in "Medical Geography in Historic al Perspective," ed. Nicolaas A. Rupke, supplement no. 20. *Medical History* 2000: 3–30.

Van Damme, Stephane. "Reason and Sentiment: The Enlightenment, Golden Age of the Translation of the Sciences?" *Studies in the History and Philosophy of Science* 35 (2004): 383–89.

Van Wyk Smith, M. "'The Most Wretched of the Human Race': The Iconography of the Khoikoin (Hottentots), 1500–1800." *History and Anthropology* 5 (1992): 285–330.

Vartanian, Aram. "The Annales School and the Enlightenment." *Studies in Eighteenth-Century Culture* 13 (1984): 233–47.

Venturi, Franco. *Italy and the Enlightenment.* Ed. and with an introduction by Stuart Woolf and trans. Susan Corsi. London: Longman, 1972.

———. *Utopia and Reform in the Enlightenment.* Cambridge: Cambridge University Press, 1971.

Verba, Cynthia. *Music and the French Enlightenment: Reconstructions of a Dialogue, 1750–1764.* Oxford: Clarendon Press,1993.

Vila, Anne. *Enlightenment and Pathology: Sensibility in the Literature and Medicine of Eighteenth-Century France.* Baltimore and London: Johns Hopkins University Press, 1998.

———. *Enlightenment and Religion: Rational Dissent in Eighteenth-Century Britain.* Cambridge: Cambridge University Press, 1996.

Vlahakis, George. "The Greek Enlightenment in Science: *Hermes the Scholar* and Its Contribution to Science in Early Nineteenth-Century Greece." *History of Science* 37 (1999): 319–45.

———. "A Note on the Penetration of Newtonian Physics in Greece." *Nuncius* 8 (1993): 645–56.

Vogel, Ursula. "The Sceptical Enlightenment: Philosopher Travellers Look Back at Europe." In *The Enlightenment and Modernity,* ed. Norman Geras and Robert Wokler, 3–24. London: Macmillan, 2000.

Voitle, Robert. "The Reason of the English Enlightenment." *Studies on Voltaire and the Eighteenth Century* 27 (1963): 1735–74.

Volney, Constantin-François de Chasseboeuf, comte de. *View of the Climate and Soil of the United States of America.* Reprint. London: J. Johnson, 1804.

Vovelle, Michel, ed. *Enlightenment Portraits.* Chicago: University of Chicago Press, 1992.

Walters, Alice. "Conversation Pieces: Science and Politeness in Eighteenth-Century England." *History of Science* 35 (1997): 121–54.

Wangermann, Ernst. "The Conditions of National Consciousness in the Epoch of Enlightenment." *Studies on Voltaire and the Eighteenth Century* 265 (1989): 1523–33.

———. "Reform Catholicism and Political Radicalism in the Austrian Enlightenment." In *The Enlightenment in National Context,* ed. Roy Porter and Mikuláš Teich, 127–40. Cambridge: Cambridge University Press, 1981.

Webster, Diana C. F. "A Cartographic Controversy: In Defence of Murdo Mackenzie." In *Togail Tir: The Map of the Western Isles,* ed. Finlay MacLeod, 33–42. Stornoway, Isle of Lewis: Acair Ltd., 1989.

Weinberg, Gregorio, "The Enlightenment and Some Aspects of Culture and Higher Education in Spanish America." *Studies on Voltaire and the Eighteenth Century* 167 (1977): 491–522.

Wertz, S. K. "Hume and the Historiography of Science." *Journal of the History of Ideas* 54 (1993): 411–36.

West, Hugh. "The Limits of Enlightenment Anthropology: Georg Forster and the Tahitians." *History of European Ideas* 10 (1989): 147–60.

Whaley, Joachim. "The Protestant Enlightenment in Germany." In *The Enlightenment in National Context,* ed. Roy Porter and Mikuláš Teich, 106–17. Cambridge: Cambridge University Press, 1981.

Whelan, Frederick. "Population and Ideology in the Enlightenment." *History of Political Thought* 12 (1991): 35–72.

Whelan, Ruth. "Republic of Letters." In *Encyclopedia of the Enlightenment,* ed. Alan C. Kors, 3:436–40. New York and Oxford: Oxford University Press, 2003.

Whitaker, Arthur, ed. *Latin America and the Enlightenment.* Ithaca, NY: Great Seal Books, 1961.

Whitaker, Robert. *The Mapmaker's Wife: A True Tale of Love, Murder and Survival in the Amazon.* New York: Doubleday, 2004.

Widmalm, Sven. "Accuracy, Rhetoric and Triangulation: The Paris–Greenwich Triangulation, 1784–88." In *The Quantifying Spirit in the Eighteenth Century,* ed. Töre Frangsmyr, John Heilbron, and Robin Rider, 179–206. Berkeley and Oxford: University of California Press, 1990.

———. "Professor Celsius and Don Andrea: North–South Dynamics in the Early Enlightenment." In *Sidereus Nuncius and Stella Polaris: The Scientific Relations between Italy and Sweden in Early Modern History,* ed. Marco Beretta and Tore Frängsmyr, 121–34. Canton, MA: Science History Publications, 1997.

Williams, David, ed. *The Enlightenment.* Cambridge: University of Cambridge Press, 1999.

Williams, Glyndwr. "The Pacific: Exploration and Exploitation." In *The Oxford History of the British Empire,* vol. 2: *The Eighteenth Century,* ed. P. J. Marshall, 552–75. Oxford: Oxford University Press, 1998.

———. *Voyages of Delusion: The Search for the Northwest Passage in the Age of Reason.* London: Harper Collins, 2003.

Williams, John. *The Rise, Progress, and Present State of the Northern Governments, viz., The United Provinces, Denmark, Sweden, Russia, and Poland.* London: White, 1777.

Williams, Michael. *Deforesting the Earth: From Prehistory to Global Crisis.* Chicago: University of Chicago Press, 2003.

Williams, Roger. *French Botany in the Enlightenment: The Ill-Fated Voyages of* La Pérouse *and His Rescuers.* Dordrecht: Kluwer Academic Publishers, 2003.

Wilson, Curtis. "Astronomy and Cosmology." In *The Cambridge History of Science,* vol. 4: *Eighteenth-Century Science,* ed. Roy Porter, 328–53. Cambridge: Cambridge University Press, 2003.

Wilson, Kathleen. *The Island Race: Englishness, Empire and Gender in the Eighteenth Century.* London: Routledge, 2003.

Wilson, Philip. "Medicine." In *Encyclopedia of the Enlightenment,* ed. Alan C. Kors, 3:42–53. New York and Oxford: Oxford University Press.

Withers, Charles W. J. "Encyclopaedism, Modernism and the Classification of Geographical Knowledge." *Transactions of the Institute of British Geographers* 21 (1996): 275–98.

———. *Gaelic Scotland: The Transformation of a Culture Region.* London: Routledge, 1988.

———. "Geography." In *Encyclopedia of the Enlightenment,* ed. Alan Kors, 2:114–17. New York and Oxford: Oxford University Press, 2003.

———. "Geography, Enlightenment, and the Paradise Question." In *Geography and Enlightenment,* ed. David N. Livingstone and Charles W. J. Withers, 67–92. Chicago: University of Chicago Press, 1999.

———. "Geography in Its Time: Geography and Historical Geography in Diderot and d'Alembert's *Encyclopédie.*" *Journal of Historical Geography* 19 (1993): 255–64.

———. "Geography, Natural History and the Eighteenth-Century Enlightenment: Putting the World in Place." *History Workshop Journal* 39 (1995): 136–63.

———. *Geography, Science and National Identity: Scotland since 1520.* Cambridge: Cambridge University Press, 2001.

———. "How Scotland Came to Know Itself: Geography, National Identity and the Making of a Nation, 1680–1790." *Journal of Historical Geography* 21 (1995): 371–97.

———. "Mapping." In *Encyclopedia of the Enlightenment,* ed. Alan Kors, 3:18–20. New York and Oxford: Oxford University Press, 2003.

———. "Mapping the Niger, 1798–1832: Truth, Testimony and 'Ocular Demonstration' in the Enlightenment." *Imago Mundi* 56 (2004): 170–93.

———. "The Rev. Dr. John Walker and the Practice of Natural History in Late Eighteenth-Century Scotland." *Archives of Natural History* 28 (1991): 201–20.

———. "Science at Sea: Charting the Gulf Stream in the Late Enlightenment." *Interdisciplinary Science Reviews* 31 (2006): 58–76.

———. "Situating Practical Reason: Geography, Geometry and Mapping in the Scottish Enlightenment." In *Science and Medicine in the Scottish Enlightenment,* ed. Charles W. J. Withers and Paul Wood, 54–78. East Linton: Tuckwell Press, 2002.

———. "The Social Nature of Map Making in the Scottish Enlightenment, *c.* 1682–*c.* 1832." *Imago Mundi* 54 (2002): 46–66.

———. "Toward a Historical Geography of Enlightenment in Scotland." In *The Scottish Enlightenment: Essays in Reinterpretation,* ed. Paul Wood, 63–98. Rochester, NY: University of Rochester Press, 2000.

———. "Towards a Geography of Trust." *Géographie et Culture* 33 (2000): 3–17.

———. "Towards a History of Geography in the Public Sphere." *History of Science* 37 (1999): 45–77.

———. "Where Was the Atlantic Enlightenment?—Questions of Geography." In *Placing the Atlantic Enlightenment,* ed. Frank Cogliano and Susan Manning. Aldershot: Ashgate, 2007, forthcoming.

———. "Writing in Geography's History: *Caledonia,* Networks of Correspondence and Geographical Knowledge in the Late Enlightenment." *Scottish Geographical Journal* 120 (2004): 33–45.

Withers, Charles W. J., and David N. Livingstone. "Introduction: On Geography and the Enlightenment." In *Geography and Enlightenment,* ed. David N. Livingstone and Charles W. J. Withers, 1–28. Chicago: University of Chicago Press, 1999.

Withers, Charles W. J., and Robert Mayhew. "Rethinking 'Disciplinary' History: Geography in British Universities, *c.* 1580–1887." *Transactions of the Institute of British Geographers* 27 (2002): 11–29.

Withers, Charles W. J., and Paul Wood, eds. *Science and Medicine in the Scottish Enlightenment.* East Linton: Tuckwell Press, 2002.

Withrington, Donald. "What Was Distinctive about the Scottish Enlightenment?" In *Aberdeen and the Enlightenment,* ed. Jennifer J. Carter and Joan H. Pittock, 9–19. Aberdeen: Aberdeen University Press, 1987.

Wokler, Robert. "Anthropology and Conjectural History in the Enlightenment." In *Inventing Human Science: Eighteenth-Century Domains,* ed. Christopher Fox, Roy Porter, and Robert Wokler, 31–52. Berkeley and London: University of California Press, 1996.

Wolff, Larry. *Inventing Eastern Europe: The Map of Civilisation on the Mind of the Enlightenment.* Stanford, CA: Stanford University Press, 1994.

———. "Travel Literature." In *Encyclopedia of the Enlightenment,* ed. Alan Kors, 4:188–94. New York and Oxford: Oxford University Press, 2003.

Wood, Paul. "Buffon's Reception in Scotland: The Aberdeen Connection." *Annals of Science* 44 (1987): 169–90.

———. "Introduction: Dugald Stewart and the Invention of the 'Scottish Enlightenment.'" In *The Scottish Enlightenment: Essays in Reinterpretation,* ed. Paul Wood, 1–35. Rochester, NY: Rochester University Press, 2000.

———. "The Natural History of Man in the Scottish Enlightenment." *History of Science* 28 (1990): 89–123.

———. "The Science of Man." In *Cultures of Natural History,* ed. Nicholas Jardine, James A. Secord, and Emma C. Spary, 197–210. Cambridge: Cambridge University Press, 1996.

———. "Science, Philosophy and the Mind." In *The Cambridge History of Science,* vol. 4: *Eighteenth-Century Science,* ed. Roy Porter, 800–24. Cambridge: Cambridge University Press, 2003.

———. "Science, the Universities, and the Public Sphere in Eighteenth-Century Scotland." *History of Universities* 13 (1994): 99–135.

———. *The Scottish Enlightenment: Essays in Reinterpretation.* Rochester, NY: University of Rochester Press, 2000.

Wuthnow, Robert. *Communities of Discourse: Ideology and Social Structure in the Reformation, the Enlightenment, and European Socialism.* Cambridge, MA: Harvard University Press, 1989.

Yeo, Richard. "Classifying the Sciences." In *The Cambridge History of Science,* vol. 4: *The Eighteenth Century,* ed. Roy Porter, 241–66. Cambridge: Cambridge University Press, 2003.

———. *Encyclopedic Visions: Scientific Dictionaries and Enlightenment Culture.* Cambridge: Cambridge University Press, 2001.

Yost, Robinson M. "Pondering the Imponderable: John Robison and Magnetic Theory in Britain (*c.* 1775–1805)." *Annals of Science* 56 (1999): 143–74.

Young, B. W. *Religion and Enlightenment in Eighteenth-Century England: Theological Debate from Locke to Burke.* Oxford: Clarendon Press, 1998.

Zachs, William, ed. *The Edinburgh Magazine and Review.* 5 vols. Bristol: Thoemmes Press, 2002.

Zbinden, Jürg. "Heterogeneity, Irony, Ambivalence: The Idea of Progress in the German Universal Histories and the Histories of Mankind in the German Enlightenment." *Storia della Storiografia* 29 (1996): 21–45.

Zuidervaart, Huib J. "An Eighteenth-Century Medical-Meteorological Society in the Netherlands: An Investigation of Early Organization, Instrumentation and Quantification. Part I." *British Journal for the History of Science* 38 (2005): 379–410.

———. "An Eighteenth-Century Medical-Meteorological Society in the Netherlands: An Investigation of Early Organization, Instrumentation and Quantification. Part II." *British Journal for the History of Science* 39 (2006): 49–66.

Index